The Diversity of Wood and Non-Wood Forest Products: Anatomical, Physical and Chemical Properties, and Potential Applications

The Diversity of Wood and Non-Wood Forest Products: Anatomical, Physical and Chemical Properties, and Potential Applications

Editors

Vicelina Sousa
Helena Pereira
Teresa Quilhó
Isabel Miranda

Basel • Beijing • Wuhan • Barcelona • Belgrade • Novi Sad • Cluj • Manchester

Editors
Vicelina Sousa
Forest Research Center
School of Agriculture
University of Lisbon
Lisbon
Portugal

Helena Pereira
Forest Research Center
School of Agriculture
University of Lisbon
Lisbon
Portugal

Teresa Quilhó
Forest Research Center
School of Agriculture
University of Lisbon
Lisbon
Portugal

Isabel Miranda
Forest Research Center
School of Agriculture
University of Lisbon
Lisbon
Portugal

Editorial Office
MDPI
St. Alban-Anlage 66
4052 Basel, Switzerland

This is a reprint of articles from the Special Issue published online in the open access journal *Forests* (ISSN 1999-4907) (available at: https://www.mdpi.com/journal/forests/special_issues/wood_application).

For citation purposes, cite each article independently as indicated on the article page online and as indicated below:

Lastname, A.A.; Lastname, B.B. Article Title. *Journal Name* **Year**, *Volume Number*, Page Range.

ISBN 978-3-0365-9258-9 (Hbk)
ISBN 978-3-0365-9259-6 (PDF)
doi.org/10.3390/books978-3-0365-9259-6

Cover image courtesy of Vicelina Sousa

Contents

About the Editors

Vicelina Sousa

Vicelina Sousa has a PhD in Forestry Engineering and Natural Resources from the School of Agriculture, University of Lisbon. She is a researcher at the Forest Research Centre and Associate Laboratory TERRA – Laboratory for Sustainability of Land Use and Ecosystem Services, School of Agriculture at the University of Lisbon. Her area of expertise is wood science and technology, namely wood properties and variability, wood formation and structure, wood anatomy and wood identification. Her other research interests are related to bark anatomy and properties and tree and plant responses to different environmental conditions.

Helena Pereira

Helena Pereira is Emeritus Professor at the University of Lisbon, School of Agriculture, with a degree in chemical engineering (Instituto Superior Técnico); a doctorate (Dr. rcr. nat.) from the Biology Faculty, University of Hamburg; and habilitation from the School of Agriculture, Technical University of Lisbon. Her academic and scientific path included teaching and research, as well as several academic management positions in different institutions, namely as Rector and vice-Rector of the Technical University of Lisbon and President of the Nation Science Funding Agency. Her research interests cross several disciplines, from chemistry, biology and forestry to materials science and engineering, focusing on forests and forest products, biomass, bioenergy and biorefineries, with leading international expertise in barks, especially cork.

Teresa Quilhó

Teresa Quilhó has a PhD in Forestry Engineering and Natural Resources from the School of Agriculture, University of Lisbon. She is a researcher at the Forest Research Centre and Associate Laboratory TERRA – Laboratory for Sustainability of Land Use and Ecosystem Services, School of Agriculture at the University of Lisbon. Her area of scientific activity is wood science, and her domains of specialization include wood and bark anatomy, wood identification, fiber identification and tropical wood. Her present research interests include the anatomy and variability of bark and other tree components, as well as non-wood fiber raw materials, including their integration into eco-physiological studies.

Isabel Miranda

Isabel Miranda has a PhD in Forestry Engineering and Natural Resources from the School of Agriculture, University of Lisbon. She is a researcher at the Forest Research Centre and Associate Laboratory TERRA – Laboratory for Sustainability of Land Use and Ecosystem Services, School of Agriculture at the University of Lisbon. Her research interests are focused on the chemical and structural characterization of lignocellulosic materials, including wood, barks, cork and unconventional biomass wastes and residues of cork within a biorefinery approach.

Preface

A present and urgent challenge is to follow the major guidelines set out for forest resource sustainability management in the context of both ecosystem-related global warming threats and green economy needs. For instance, we may need to pay special attention to threatened species and lesser-known endemic species, as well as to invasive species that may be profitable when managed, so as to avoid the overexploitation of current forest species. Additionally, the full resource use approach requires consideration of the residual materials from forest management and related industries, taking into account different tree components and non-wood products that present enormous diversity.

Specific knowledge on these diverse materials regarding their structure, anatomy and properties is, therefore, an essential tool in assessing the potential and suitability of these different forest resources. For the most profitable forest species, research over the years has produced a number of studies related to general wood growth tendencies and characteristics. However, different species might need to be involved in the framework of new ecosystem-oriented approaches for sustainable resource management. The high diversity within and between species also requires careful study of their variability, which may contribute toward a better understanding their properties and potential applications, thereby supporting improved tree selection and breeding programs.

This Special Issue aims to gather research-based data on wood and on non-wood forest products, with a special emphasis on bark—specifically, its anatomical, physical and chemical properties and their relationships—to provide tools and background information to assess the forest resource potential of different species. Our aim is twofold, as we look to contribute to efforts to increase forest sustainability and diversity through preserving species with potential for high-quality end use and to diversify wood supply using knowledge related to wood and non-wood characteristics and their diversity.

Vicelina Sousa, Helena Pereira, Teresa Quilhó, and Isabel Miranda
Editors

Editorial

The Diversity of Wood and Non-Wood Forest Products: Anatomical, Physical, and Chemical Properties, and Potential Applications

Vicelina Sousa *, Isabel Miranda, Teresa Quilhó and Helena Pereira

Centro de Estudos Florestais, Laboratório Associado TERRA, Instituto Superior de Agronomia, Universidade de Lisboa, 1349-017 Lisboa, Portugal; imiranda@isa.ulisboa.pt (I.M.); terisantos@isa.ulisboa.pt (T.Q.); hpereira@isa.ulisboa.pt (H.P.)
* Correspondence: vsousa@isa.ulisboa.pt

Citation: Sousa, V.; Miranda, I.; Quilhó, T.; Pereira, H. The Diversity of Wood and Non-Wood Forest Products: Anatomical, Physical, and Chemical Properties, and Potential Applications. *Forests* **2023**, *14*, 1988. https://doi.org/10.3390/f14101988

Received: 28 September 2023
Accepted: 29 September 2023
Published: 3 October 2023

Forests are continuously changing, as is the related gap in our understanding The resources diversity of forests is enormous because of the existence of temperate, boreal, and tropical forests. Before the implementation of any forest or agroforestry practices, the resources were mainly used based only on resource proximity and availability. Systematic forestry knowledge allows us to increase wood production and ensure species conservation. The present and urgent challenge is to follow the major guidelines set out for forest resources sustainability management in the context of both ecosystem-related global warming threats and green-economy needs. Yet, there are several forest resources that are not well known, and specific knowledge regarding the structure, anatomy, and properties is therefore an essential tool in assessing the potential and suitability of these different forest resources. Understanding and predicting effects on the growth and quality of each forest resource due to abiotic and biotic factors and conditions is complex and of crucial importance for sustainable forest resources management.

This Special Issue, "The Diversity of Wood and Non-Wood Forest Products: Anatomical, Physical, and Chemical Properties, and Potential Applications", features a collection of articles on diverse species ranging from temperate to tropical regions (endemic species, lesser-known timber species, invasive species, and non-woody species), highlighting their potential as wood and non-wood forest products. The research findings contained herein fulfill the Special Issue's proposed aim, i.e., to contribute to diversifying the lignocellulosic supply and to preserve species with the potential for high-quality end use, thereby increasing forest sustainability and diversity.

Rattan are monocotyledonous plants, also known as palms, mainly distributed in tropical forests in Southeast Asia and in West and Central Africa, with a high economic value due to their use in furniture production [1]. Rattan species are considered alternatives to timber species. However, studies on rattan species belonging to the *Calamus* genus, the most rich and traded genus, are still quite scarce [1]. Addressing this research gap, Ahmed et al. [1] studied the within-stem variability of two rattan species, *Calamus zollingeri* Beccari and *Calamus ornatus* Blume, growing in Indonesia, to analyse the potential of lignocellulosic material based on its biological durability and anatomical/physical properties.

Mediterranean oaks, and in particular the cork oak, *Quercus suber* L., even if well-adapted to the adverse conditions of the Mediterranean climate, are considered to be under threat due to climate change and pests [2,3]. Simões et al. [2] studied the seasonal variation of the chemical composition of cuticular waxes in cork oak leaves due to their ability to control water loss during abiotic stresses such as drought and high temperatures in an effort to contribute to climate change adaptation measures. Moreover, the major biotic threats are related to the cork borer, *Coraebus undatus* Fabricius (Coleoptera, Buprestidae), named "cobrilha da cortiça" in Portugal and "culebrilla" in Spain, the impact of which on cork quality was also studied by Simões et al. [3], analysing the cork chemical profile of secondary metabolites.

Another species considered a source of cork is the Chinese cork oak, *Quercus variabilis* Blume, also distributed in Korea, even if there is no reproduction cork from this species yet available for industrial use [4]. With the aim of exploring the further utilization of domestic cork resources in Korean cork industries, Prasetia et al. [4] presented and discussed the quantitative anatomical characteristics of virgin cork from *Quercus variabilis* growing in Korea and compared it to the reproductive cork from *Quercus suber* from Portugal. The authors suggest processing treatments and applications for the virgin cork based on its structural characteristics.

In the tropical forests of Gabon, a new morphospecies within the *Dialium* genus was introduced for the first time and studied by Doucet et al. [5], focusing on its discriminant leaf and wood composition traits. Since a limited number of timber species are presently used, the newly presented species, as well as other species such as *Dialium pachyphyllum* Harms and *Dialium lopense* Breteler, are important alternative species [5]. However, the authors mention the need to implement strategies for the sustainable wood production and conservation of these species.

The variability in the wood properties of several tropical species from India, East-Timor, and Mozambique, including important commercial species, such as *Tectona grandis* L. as well as lesser-known species have been studied by Bessa et al. [6], aiming to contribute to design targeted production and increase wood tropical resources diversity. Species with different geographical origins were grouped based on anatomical and physical similarities, describing the main wood characteristics and properties responsible for the wood variability, and referring to lesser-known species as alternatives for CITES-listed species such as *Cedrela odorata* and *Dalbergia melanoxylon*.

The eco-valorization of the renewable resources based on invasive species such as the Scotch broom, *Cytisus scoparius* (L.) Link, was addressed by Cruz-Lopes et al. [7], who studied the chemical composition of the branches from shrubs growing in Portugal to evaluate the potential for liquid mixture conversion and improvement regarding the further manufacture of valuable products, namely, as suggested by the authors, alternatives for petroleum-derived fuels. According to the authors, the results were promising for industry processing; they exclude the time-consuming and complex transformation of the shrubs to dust, in contrast to what is typically expected of these heterogenous lignocellulosic materials.

The chemical and anatomical characterization of stems from Asparagaceae species, comprising *Agave* spp., *Beaucarnea gracilis* Lem., *Furcraea longaeva* Karw. and Zucc., *Nolina excelsa* García-Mend. and E. Solano, and *Yucca* spp., widely distributed in Mexico, have not yet been studied, as indicated by Maceda et al. [8], despite the potential of these succulent species suggested by their resistance to drought and extreme temperatures. Structural heterogeneity was found between the different species but above all, the findings related to tissue distribution, fiber characteristics, and crystallinity indices suggest a potential for biofuel and lignocellulosic materials production [8].

Temperate species such as *Tilia amurensis* Rupr., *Tilia mandshurica* Rupr., and Maxim. have garnered significant commercial interest in China owing to their stemwood, while less interest has been given to their branchwood's potential, which could play a crucial role as sustainable forest resources in light of the current wood shortage in the country [9]. The authors found different degrees of suitability for each species, noting the high potential of branchwood from *Tilia amurensis* in mechanical pulp yield with specific requirements.

Four tropical timber wood species (*Daniellia oliveri, Isoberlinia doka, Khaya senegalensis,* and *Pterocarpus erinaceus*) from Mali, Africa, were studied by Traoré and Cortizas [10] in an effort to further our understanding of the effects of environmental factors on wood characteristics and chemical properties, including wood colour, which is less studied in comparison with tree growth markers. The results showed that tree species were not equally affected by the different local environmental conditions, and extractive compounds potential as markers by which wood from different provenance areas may be differentiated [10].

The Chinese weeping cypress (*Cupressus funebris* Endl.) is one of the representative evergreen coniferous species in China's subtropics under wood production management [11]. The effects of different type of knots, live and dead, on the physical properties and colour of this wood were discussed by Lyu et al. [11], and different commercial applications were suggested for clear wood and for wood with different knot presences. Moreover, the use of colorimetry methods allowed for the differentiation of different knot presences in wood [11].

Our hope is that this Special Issue will be of interest not only to researchers but also to students, forest managers, industrial practitioners, and society in general.

The editors are grateful to the authors and reviewers for their patience, especially during our illness periods, as well as for their confidence and assistance in the significant improvements made to the manuscripts contained herein.

Author Contributions: Conceptualization, V.S.; writing—original draft preparation, V.S.; writing—review and editing, V.S., I.M., T.Q. and H.P.; funding acquisition, V.S. All authors have read and agreed to the published version of the manuscript.

Funding: This work was supported by funding to Centro de Estudos Florestais (UIDB/00239/2020) from Fundação para a Ciência e a Tecnologia (FCT), Portugal. The first author also acknowledges a research contract (DL57/2016/CP1382/CT0004).

Conflicts of Interest: The authors declare no conflict of interest.

References

1. Ahmed, S.A.; Hosseinpourpia, R.; Brischke, C.; Adamopoulos, S. Anatomical, Physical, Chemical, and Biological Durability Properties of Two *Rattan* Species of Different Diameter Classes. *Forests* **2022**, *13*, 132. [CrossRef]
2. Simões, R.; Miranda, I.; Pereira, H. Effect of Seasonal Variation on Leaf Cuticular Waxes' Composition in the Mediterranean Cork Oak (*Quercus Suber* L.). *Forests* **2022**, *13*, 1236. [CrossRef]
3. Simões, R.; Branco, M.; Nogueira, C.; Carvalho, C.; Santos-Silva, C.; Ferreira-Dias, S.; Miranda, I.; Pereira, H. Phytochemical Composition of Extractives in the Inner Cork Layer of Cork Oaks with Low and Moderate *Coraebus undatus* Attack. *Forests* **2022**, *13*, 1517. [CrossRef]
4. Prasetia, D.; Purusatama, B.D.; Kim, J.-H.; Yang, G.-U.; Jang, J.-H.; Park, S.-Y.; Lee, S.-H.; Kim, N.-H. Quantitative Anatomical Characteristics of Virgin Cork in *Quercus variabilis* Grown in Korea. *Forests* **2022**, *13*, 1711. [CrossRef]
5. Doucet, R.; Bibang Bengono, G.; Ruwet, M.; Van De Vreken, I.; Lecart, B.; Doucet, J.-L.; Fernandez Pierna, J.A.; Lejeune, P.; Jourez, B.; Souza, A.; et al. Highlighting a New Morphospecies within the *Dialium* Genus Using Leaves and Wood Traits. *Forests* **2022**, *13*, 1339. [CrossRef]
6. Bessa, F.; Sousa, V.; Quilhó, T.; Pereira, H. An Integrated Similarity Analysis of Anatomical and Physical Wood Properties of Tropical Species from India, Mozambique, and East Timor. *Forests* **2022**, *13*, 1675. [CrossRef]
7. Cruz-Lopes, L.; Almeida, D.; Dulyanska, Y.; Domingos, I.; Ferreira, J.; Fragata, A.; Esteves, B. Chemical Composition and Optimization of Liquefaction Parameters of *Cytisus scoparius* (Broom). *Forests* **2022**, *13*, 1772. [CrossRef]
8. Maceda, A.; Soto-Hernández, M.; Terrazas, T. Chemical-Anatomical Characterization of Stems of *Asparagaceae* Species with Potential Use for Lignocellulosic Fibers and Biofuels. *Forests* **2022**, *13*, 1853. [CrossRef]
9. Guo, P.; Zhao, X.; Feng, Q.; Yang, Y. Branchwood Properties of Two *Tilia* Species Collected from Natural Secondary Forests in Northeastern China. *Forests* **2023**, *14*, 760. [CrossRef]
10. Traoré, M.; Martínez Cortizas, A. Color and Chemical Composition of Timber Woods (*Daniellia oliveri, Isoberlinia doka, Khaya senegalensis*, and *Pterocarpus erinaceus*) from Different Locations in Southern Mali. *Forests* **2023**, *14*, 767. [CrossRef]
11. Lyu, J.; Qu, H.; Chen, M. Influence of Wood Knots of Chinese Weeping Cypress on Selected Physical Properties. *Forests* **2023**, *14*, 1148. [CrossRef]

Article

Anatomical, Physical, Chemical, and Biological Durability Properties of Two Rattan Species of Different Diameter Classes

Sheikh Ali Ahmed [1], Reza Hosseinpourpia [1], Christian Brischke [2] and Stergios Adamopoulos [3,*

[1] Department of Forestry and Wood Technology, Faculty of Technology, Linnaeus University, Georg Lückligs Plats 1, 351 95 Växjö, Sweden; sheikh.ahmed@lnu.se (S.A.A.); reza.hosseinpourpia@lnu.se (R.H.)

[2] Wood Biology and Wood Products, University of Goettingen, Buesgenweg 4, 37077 Goettingen, Germany; christian.brischke@uni-goettingen.de

[3] Department of Forest Biomaterials and Technology, Division of Wood Science and Technology, Swedish University of Agricultural Sciences, Vallvägen 9C, 750 07 Uppsala, Sweden

* Correspondence: stergios.adamopoulos@slu.se; Tel.: +46-018-67-2474

Abstract: Rattan cane is an important forest product with economic value. Its anatomical, physical, and biological properties vary with the cane height. This makes it difficult to select the appropriate cane diameter for harvesting. Understanding the material properties of rattan cane with different diameter sizes is important to enhance its utilization and performance for different end uses. Thus, the present study was performed on two rattan species, *Calamus zollingeri* and *Calamus ornatus,* at two different cane heights (bottom/mature and top/juvenile). *Calamus zollingeri* was studied at diameter classes of 20 mm and 30 mm, while *Calamus ornatus* was analyzed at a diameter class of 15 mm. The anatomical properties, basic density, volumetric swelling, dynamic moisture sorption, and biological durability of rattan samples were studied. The results showed that *C. zollingeri* with a 20 mm diameter exhibited the highest basic density, hydrophobicity, dimensional stability, and durability against mold and white-rot (*Trametes versicolor*) fungi. As confirmed by anatomical studies, this could be due to the higher vascular bundle frequency and longer thick-walled fibers that led to a denser structure than in the other categories. In addition, the lignin content might have a positive effect on the mass loss of different rattan canes caused by white-rot decay.

Keywords: *Calamus zollingeri*; *Calamus ornatus*; dynamic vapor sorption; basic density; volumetric swelling; white rot; mold

Citation: Ahmed, S.A.; Hosseinpourpia, R.; Brischke, C.; Adamopoulos, S. Anatomical, Physical, Chemical, and Biological Durability Properties of Two Rattan Species of Different Diameter Classes. *Forests* 2022, 13, 132. https://doi.org/10.3390/f13010132

Academic Editors: Vicelina Sousa, Helena Pereira and Teresa Quilhó

Received: 16 November 2021
Accepted: 14 January 2022
Published: 17 January 2022

Publisher's Note: MDPI stays neutral with regard to jurisdictional claims in published maps and institutional affiliations.

1. Introduction

Rattan are spiny and climbing palms mainly distributed in Southeast Asia and in West and Central Africa [1]. With diverse uses, rattan have a high economic value and the most popular use is in furniture production. Depending on the stem diameter, rattan canes are generally classified into three categories: i.e., large (>18 mm), medium (10–18 mm), and small (<10 mm) diameter rattan. The medium and large diameter canes are used for furniture frames, walking sticks, sporting goods, and rural bridges, while small diameter canes are used either in round or split form, in basket making, chair seats, lampshades, and a variety of handicraft items [2]. Rattan canes are also widely used as excellent natural materials for ropes, decorative items, housing, craft products, umbrella handles, sporting goods, hats, cordage, birdcages, matting, baskets, hoops, and as implant materials [3,4]. In addition, other parts of the rattan, such as roots, fruits, and leaves, are used in folk medicines, while the shoots are edible because of the presence of high amounts of proteins, carbohydrates, vitamins, and other nutrients [5]. There are 12 different genera of rattan, which include more than 550 species [6], and only four of the genera are traded commercially, i.e., *Calamus, Daemonorops, Korthalsia,* and *Plectocomia* [7]. Among them, *Calamus* is the most species-rich genus of all palms with an estimated 374 species. *Calamus* grows over a large area and has excellent properties and extensive commercial uses. However, studies on this genus are still quite scarce. Commercially harvested rattan undergoes physico-chemical treatment and mechanical processing. Physico-chemical treatment involves seasoning or curing, which is boiling in oil, followed by drying, fumigation, bleaching, and deglazing. Deglazing or polishing is the mechanical removal of the highly silicified epidermis, enabling an easy bending of the cane to desired shapes [1]. Furthermore, scraping and polishing are used to smoothen the surfaces. Semi- or full polishing is done at various stages of the workflow and can be done after cutting, bending, molding, or assembling, and during the finishing process [8].

Rattan are monocotyledonous plants, and their stems are composed of vascular bundles embedded in the parenchymatous tissue. Rattan lacks the necessary lateral meristems to undergo secondary growth and thus only limited diameter growth is possible through ground parenchyma cell enlargement. Properties of the vascular bundle, such as the cell wall thickness of fibers, fiber ratio, and diameter of vessel elements in metaxylem, affect the moisture properties and volumetric shrinkage of rattan [9]. Vascular bundles in rattan canes provide the main reinforcing elements, together with their surrounding fiber sheaths. Remarkably, these reinforcing elements are densely distributed in the outer region and sparsely in the inner region [10]. Like other lignocellulosic materials, the physico-mechanical properties of rattan are the most important factors for determination of cane processing and utilization. Those properties are influenced by the species, stem position, diameter, internode length, and density [5]. Different properties of rattan have previously been investigated, including moisture content (MC), density, bending modulus of elasticity, modulus of rupture, axial compression strength, and tensile strength [11]. Also, the factors influencing mechanical properties, such as species [12], microfibril angle [13], anatomical characteristics, temperature, MC, and test methods [14,15], have been discussed. However, it is hardly possible to make a comprehensive conclusion since the properties of rattan canes vary considerably with plant height [11,16]. Trees undergo secondary growth and age-related variability in wood properties is evident, but such variability is not apparent in rattan canes because of the absence of secondary growth [17].

Rattan canes have long been used in different applications; however, they exhibit the typical drawbacks of lignocellulosic materials, i.e., dimensional instability, low resistance against mold and rot fungi decay, and strength loss with increasing moisture content [18,19]. It is therefore required to complete our knowledge of those fundamental properties as they vary with cane diameter and vertical position. Thus, the aim of this study is to analyze the anatomical variations, basic density, volumetric swelling, chemical composition, vapor sorption, and resistance against mold and fungal decay of two rattan species, *Calamus zollingeri* and *Calamus ornatus*, in two different heights (bottom/mature and top/juvenile), and three diameter classes. The results of this study will provide the scientific basis for a wiser utilization of rattan canes, such as allocating the right parts and dimensions for specific final products.

2. Materials and Methods

2.1. Source of Material

Rattan canes of *Calamus zollingeri* Beccari and *Calamus ornatus* Blume grown in Sulawesi, Indonesia were collected from local suppliers for this study. Those canes were commercially polished using a custom-made sanding machine upon which sandpaper was used with a grit size of 220. The canes were 3000 mm long and belonged to three diameter classes, viz. 30, 20, and 15 mm. Both semi- and fully polished samples were used for *C. zollingeri* while only fully polished samples were used for *C. ornatus*. Samples of about 400 mm long were cut from the bottom and top ends of rattan canes to separate the mature and juvenile portions, respectively. Working samples for physical and anatomical studies were converted from these 400 mm internode samples (Figure 1). Only fully polished samples were used for testing anatomical characteristics and vapor sorption as no differences were expected between semi- and fully polished samples, except from removal of the epidermis and some part of the cortex tissues, as seen in Figure 1.

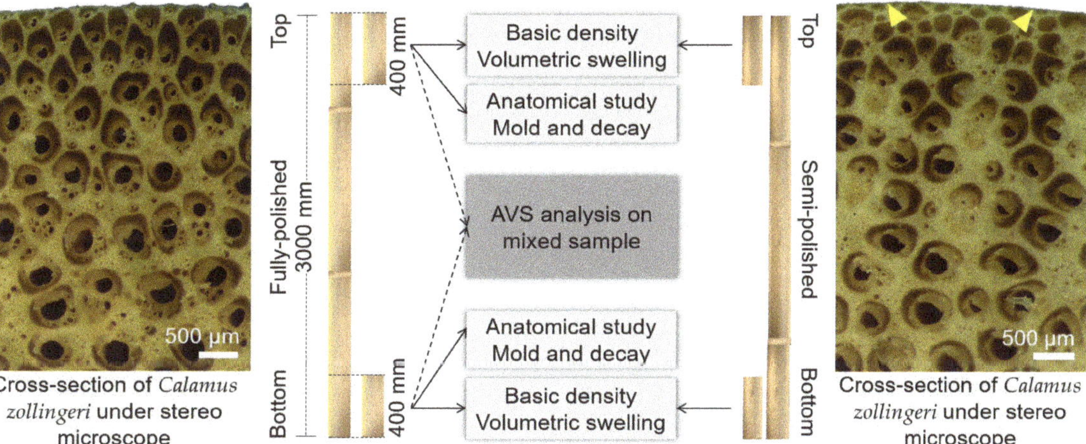

Figure 1. Schematic diagram of sample preparation and analyses. AVS: automated vapor sorption. Arrowheads in semi-polished sample show some cortex tissues. Cane length and diameter are not true to scale.

The number of samples used in this experiment per rattan species and diameter classes are shown in Table 1.

Table 1. Number of samples used per rattan species and diameter classes. Values in parentheses are standard deviations.

Sample	Diameter Class	Mean Diameter mm	Semi-Polished		Fully-Polished	
			Bottom	Top	Bottom	Top
C. zollingeri	30	30.33 (±1.41)	8	8	8	8
C. zollingeri	20	18.62 (±1.29)	8	8	8	8
C. ornatus	15	14.76 (±0.29)	-	-	8	8

2.2. Frequency of Vascular Bundles and Fiber Length

For analyzing the frequency of vascular bundles, 20 mm thick discs were cut from the cane and then blocks about 10 mm wide along the diameter were prepared. Sample blocks were then soaked in water for 8 h for softening. Cross-sections of 15–20 µm thick strips from periphery to periphery were prepared from the bottom and top samples of *C. zollingeri* and *C. ornatus* using a sledge microtome (WSL, Birmensdorf, Switzerland). One (1) sample per rattan species and portion (bottom and top) was used, i.e., a total of six randomly selected samples. The sections were stained with 1:1 solution of 1% Safranin O (Merck KGaA, Darmstadt, Germany) and Astra blue (Carl Roth GmbH, Karlsruhe, Germany) solution, washed with a series of alcohol solutions, and finally fixed on glass slides with Entellan™ mounting agent (Merck KGaA, Darmstadt, Germany). The glass slides were observed under a motorized Olympus BX63F light microscope (Tokyo, Japan) equipped with a DP73 color CCD cooled camera (max. 17.28 megapixel) and software program OLYMPUS cellSens Dimension, version 1.18 (Olympus, Tokyo, Japan). For every sample, the numbers of vascular bundles/mm^2 area were counted from periphery to periphery. The vascular bundle consists of the conducting tissue, phloem, and xylem with fibers that provide structural support. Vascular bundles were counted in one mm^2 rectangle area.

Those vascular bundles were also counted as one when at least 50% of the area fell in one mm^2 rectangle.

For analyzing the rattan fiber length, matchstick-sized samples were cut with a razor blade and put in glass vials containing 50:50 solutions by volume of hydrogen peroxide 20% and glacial acetic acid. The vials were placed in an oven at 60 °C for 2 days to macerate the material. After that period, the macerations were washed with de-ionized water and were stored in small glass containers with ethanol absolute. One droplet of the suspension containing rattan cells was placed on a slide and was observed under the microscope. Fifty intact fibers per sample (duplicate) were randomly selected to measure the length.

2.3. Density and Volumetric Swelling

Basic density values were measured for 80 rattan samples (400 mm long with dimeters of 15, 20, and 30 mm). Samples corresponded to the bottom (mature) and top (juvenile) parts of the canes.

Discs about 20 mm thick were cut from the middle of the original samples to obtain representative values of these cane portions (bottom and top). Samples were weighed on a balance with 0.001 g accuracy, put in an oven at 103 ± 2 °C until constant masses were attained and then the oven-dry masses were recorded. After that, all samples were submerged in a water bath for 14 days for complete saturation of the cell walls [20]. The samples' green volumes were recorded by the water-displacement method. Basic density (R) and volumetric swelling (S) determinations were performed using the following equations:

$$R \ (\text{g/cm}^3) = \frac{M_o}{V_2} \tag{1}$$

$$S \ (\%) = \frac{V_2 - V_1}{V_1} \times 100 \tag{2}$$

where M_o is the oven-dry mass of each sample in g, V_2 is the fully saturated (maximum), and V_1 is the oven-dried volume in cm^3.

2.4. Chemical Composition Analysis

Fully polished samples of 30 (length) × 10 (width) × 5 (thickness) mm^3 each from the bottom and top of rattan canes were used. The chemical composition of the rattan samples was analyzed according to Ghavidel et al. 2021 [21] by grinding them in a Willey mill through a No. 40 mesh sieve (0.425 mm). Ground powder from the bottom and top samples were mixed homogenously and then analyzed for chemical compositions. The extractive content was determined on 2–3 g of oven-dried samples (at 103 °C for 24 h) using cyclohexane-ethanol (50:50 v/v) solution for 4 h. Then, the samples were oven-dried again and used for the assessment of holocellulose content by transferring to an Erlenmeyer flask, adding 80 mL of hot water (70 °C), 1 g of sodium chlorite, and 0.5 mL of acetic acid, keeping it for 6 h in a water bath at 70 °C. Ash content was determined according to the standard EN 15403 [22]. The lignin amount was assessed by subtracting the sum of all other components, i.e., holocellulose, ash, and extractive contents, from 100 wt%. Samples were measured in duplicate to calculate average values.

2.5. Vapor Sorption Analysis

The water vapor sorption behaviors of the bottom and top rattan samples were determined using an automated vapor sorption (AVS) apparatus (Q5000 SA, TA Instruments, New Castle, USA) following the procedure described in Hosseinpourpia et al. [23]. Approximately 8 mg of rattan powder (as made for chemical composition analysis) were used for each measurement. The samples were exposed to relative humidity (RH) that was increased from 0% to 90% in stepped sequences of 15% and of 5% from 90% to 95% RH, decreasing to 0% RH in a reverse order, at a constant temperature of 25 °C. The instrument maintained a constant target RH until the mass change in the sample (dm/dt) was less than 0.01% per min over a 10 min period. The equilibrium moisture content (EMC) of the rattan samples was calculated based on their equilibrium weight at each given RH step throughout the sorption run measured by a micro balance.

2.6. Durability Tests

2.6.1. Mold

Fully polished samples of 30 (length) × 10 (width) × 5 (thickness) mm^3 were used for both mold and decay tests. All samples were conditioned at 20 °C and 65% RH in a climate chamber for three weeks before testing. An accelerated laboratory mold test was performed in a Memmert HCP 246 humidity chamber (Memmert GmbH, Germany), according to a previous study [24]. Five replications from each group giving a total of 30 samples were used for the accelerated mold test. Samples were suspended from the top of the chamber's support bars with the flat surface set vertical and parallel to the other sample surfaces with a 10 mm gap between in a randomly ordered fashion. The temperature and RH were then set to be 27 °C and 95%, respectively. Three Scots pine sapwood (*Pinus sylvestris* L.) samples infested mainly with *Aspergillus*, *Rhizopus*, and *Penicillium* genera were placed on the bottom of the climate chamber to be the sources of mold inocula. After 10 days of incubation, the experiment was stopped because of abundant mold growth on some of the sample surfaces. Both flat surfaces of the samples were evaluated visually and graded on a scale of 0 (no infestation) to 6 (extremely heavy infestation). More about the grading and method can be found in Sehlstedt-Persson et al. [25].

2.6.2. Basidiomycete Decay Test

The resistance of the two rattan species against white-rot fungi decay was evaluated with a malt agar incubation test according to EN 113-2 [26]. Deviating from the standard, mini-block specimens of 30 (length) × 10 (width) × 5 (thickness) mm^3 [27] were used and incubated for 8 weeks. Fifteen to 25 replicates per group were oven-dried at 103 ± 2 °C till constant mass and weighed to the nearest 0.001 g to determine the oven-dry mass.

After steam sterilization in an autoclave at 120 °C for 30 min, sets of two specimens of the same group were placed on fungal mycelium in Kolle flasks (100 ml malt extract agar, 4%). The white-rot fungus *Trametes versicolor* (L.) Lloyd. F strain was used for the tests. Furthermore, 10 replicates made from Scots pine sapwood and beech (*Fagus sylvatica* L.) were used as virulence controls. All specimens were incubated for 8 weeks at 22 °C and 70% RH.

After incubation, specimens were cleansed of adhering fungal mycelium, weighed to the nearest to 0.001 g, oven-dried at 103 ± 2 °C, and weighed again to the nearest 0.001 g to determine mass loss through wood-destroying basidiomycetes as follows:

$$ML_F = \frac{M_0 - M_{0,\,inc}}{M_0} \times 100 \tag{3}$$

where ML_F is the mass loss by fungal decay (%), $M_{0,inc}$ is the oven-dry mass after incubation (g), and M_0 is the oven-dry mass before incubation (g).

2.7. Statistical Analysis

To determine any statistically significant differences between the bottom and top samples, mean values were compared by a two-tailed group t-test at a 0.05 significance level (Excel 2016 program, Microsoft, Redmond, WA, USA). F-tests for mean fiber length, number of vascular bundles, basic density, volumetric swelling, and mold grades were performed using the IBM SPSS Statistics statistical software package, Version 24 (IBM Corporation, New York, USA). One-way analysis of variance (ANOVA) at a 0.05 significance level was applied to determine significant differences among rattan samples. When significant differences were found, Duncan's multiple-range test was performed.

3. Results and Discussion

3.1. Anatomical and Physical Properties

The anatomical characteristics of rattan samples were studied using light microscopy (Figure 2). The vascular bundles in rattan are scattered and embedded in the ground parenchymatous tissues. They are present just below the single layered epidermis and are small and incomplete, consisting of either clusters of fibers or a few phloem and vessel elements with fibers [28]. Vascular bundles present in the center are well developed. In both rattan species, vascular bundles consisted of two phloem fields and one metaxylem vessel and had horseshoe-shaped fibrous tissues. The ground tissue is parenchyma cells. According to Weiner and Liese [28], three forms of parenchyma cells are distinguished in cross sections, i.e., Type A: weakly branched with regular rounded intercellular spaces, Type B: small and rounded cells with irregular shaped intercellular spaces, and Type C: thin-walled, large, and round. The species showed the Type A parenchyma cell arrangement. During processing (polishing), the epidermis and some parts of the cortex were removed (Figure 1).

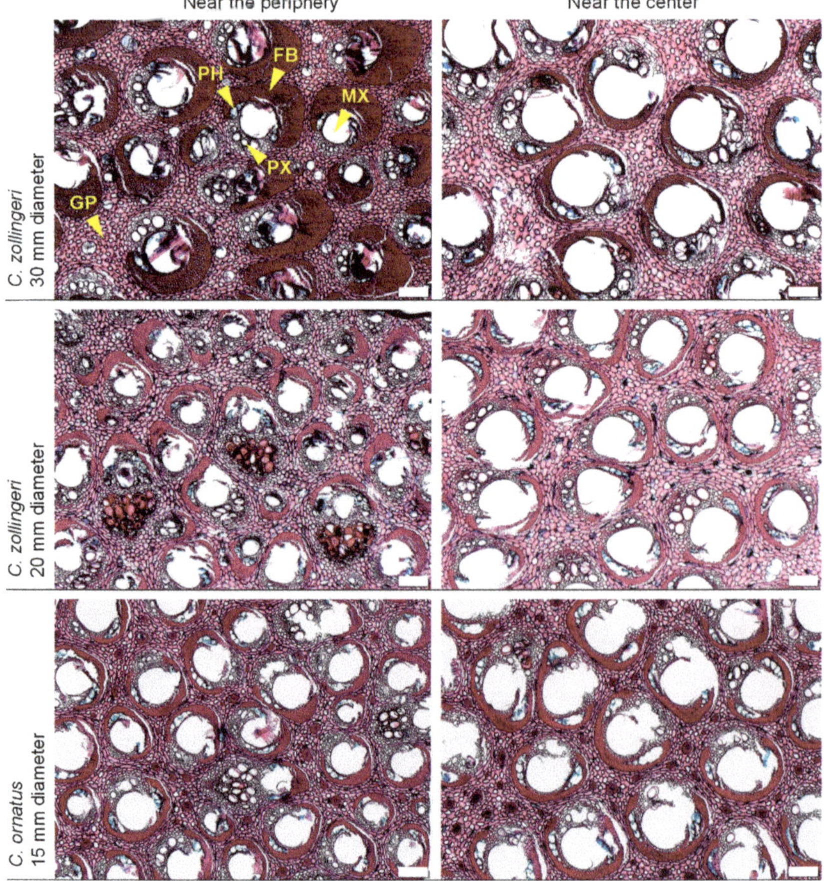

Figure 2. Light microscopic images of cross-sections of *C. zollingeri* and *C. ornatus* with different diameter classes. Arrow heads show MX = metaxylem, PX = protoxylem, PH = phloem, FB = fiber bundle, and GP = ground parenchyma. Scale bars = 200 μm.

The mean values of fiber length ranged from 1008 µm to 1329 µm, which were in accordance with a previous study [18]. Fiber length and number of vascular bundles of the rattan samples clearly decreased from the bottom to the top (Table 2). A similar trend was also reported in previous findings [5,16,18,29]. When all values were considered from the bottom and top of the canes, mean fiber length and number of vascular bundles were found to be significantly higher in *C. zollingeri* than in *C. ornatus*. Except for the *C. zollingeri* that had a 30 mm diameter, fiber length and number of vascular bundles were significantly higher in the bottom than in the top samples. The number of vascular bundles increased significantly with a decrease in stem diameter (Table 2). Similar results were reported by Bhat et al. [16] with a less consistent longitudinal variation.

Table 2. Mean values of fiber length and number of vascular bundles of different fully polished rattan samples. Mean values (± standard deviations) of n = 100 per sample type for fiber length.

Sample	Diameter Class	Fiber Length (µm)				Number of Vascular Bundles/mm^2			
		Bottom	Top	*t*-Value	Mean	Bottom	Top	*t*-Value	Mean
C. zollingeri	30	1221 (±261) b	1219 (±309) a	−0.14	1220 (±285) a	2.05 (±0.32) b	1.72 (±0.36) c	−1.22	1.89 (±0.37) b
C. zollingeri	20	1329 (±390) a	1102 (±254) b	−4.96 *	1216 (±348) a	3.40 (±0.63) a	3.17 (±0.60) a	−4.38 *	3.30 (±0.61) a
C. ornatus	15	1209 (±297) a	1008 (±289) c	5.00 *	1109 (±309) b	3.20 (±0.50) a	2.85 (±0.32) b	−2.62 *	3.02 (±0.44) a
F-value		26.31 *	5.06 *		7.99 *	76.10 *	90.64 *		148.31 *

Mean values followed by different letters within a column indicate that there is a significant difference ($p \leq 0.05$) as determined by ANOVA and Duncan's multiple range test. * Significant differences at the 0.05 level.

The higher frequency of vascular bundles could result in higher basic density. Besides, compression strength and the apparent Young's modulus increases with an increase in the number of vascular bundles [30,31]. It is thus expected that the outer part of rattan cane with densely distributed vascular bundles would have better mechanical strength than the core part with sparse distribution [15]. Anatomical investigations on *C. zollingeri* and *C. ornatus* with different diameter classes demonstrate that the fiber length and the number of vascular bundles per unit area varied significantly. Those properties decreased consistently in cane height from bottom to top. As mentioned above, the increase of basic density is dependent on the increase of cell wall thickness. The increase of fiber wall thickness with age is more pronounced than in cortical and ground parenchyma tissues, and the thickening of the fiber wall with increasing stem density results from fibers forming polylamellate walls [32]. Basic density values were positively correlated with the frequency of vascular bundles; however, the correlation was poor (r = 0.39). This could be attributed to the differences of share and thickness of fiber cells in rattan samples [16].

The physical properties, i.e., basic density and volumetric swelling, of rattan samples from different diameter classes are shown in Table 3. The top samples of fully polished *C. zollingeri* with a 20 mm diameter exhibited the highest basic density. This density was about 32% higher than in the bottom and top samples of *C. ornatus* with 15 mm diameters. Those differences were found to be statistically significant. No significant differences were observed in semi- and fully polished *C. zollingeri* samples between diameter classes (20 and 30 mm) and cane height (bottom and top). The density of semi-polished samples was higher than fully polished ones, except for the top samples of *C. zollingeri* with 20 mm diameters. However, such differences were not evaluated from a statistical point of view. Those density differences could be explained by the fact that semi-polished samples had some remaining cortex. The cortex between epidermis (single cell layer) and the vascular system consists of parenchyma cells, fibers, and incomplete vascular bundles. Higher frequency of vascular bundles, thicker fiber cell walls, and smaller cell sizes in the cortex contribute to higher basic density than in the core [11,30]. The basic density tends to decrease with cane height from the basal to the top portion of the cane [5,29,33]. This could be attributed to the steady decrease of fiber cells from base to top [15]. However, diverged results of density with cane heights are also reported [5,11]. In this study, for both semi- and fully polished samples, no statistical differences of basic density were found between the bottom and top parts. Basic density values are the result of various factors, such as the numerous fiber cells with thicker walls, higher vascular bundle frequency, narrower metaxylem vessel elements, and more lignified cells distributed in the cane. In particular, the higher percentage of fibers that constitute the total stem tissue considerably influences the density, and thus proportionally affects the strength properties of rattan cane [2,15].

Table 3. Mean basic density and volumetric swelling values of semi- and fully polished rattan samples in different diameter classes. Mean values ($\pm$ standard deviations) of n = 8 per sample type.

Sample	Diameter Class	Basic Density (g/cm^3)			Volumetric Swelling (%)		
		Fully-Polished					
		Bottom	Top	*t*-Value	Bottom	Top	*t*-Value
C. zollingeri	30	0.348 ($\pm$0.051) b	0.335 ($\pm$0.057) b	−0.76	17.39 ($\pm$2.40) b	19.06 ($\pm$6.55) b	−0.59
C. zollingeri	20	0.496 ($\pm$0.036) a	0.435 ($\pm$0.076) a	−1.92	15.80 ($\pm$2.30) b	13.79 ($\pm$1.13) b	1.78
C. ornatus	15	0.289 ($\pm$0.078) b	0.314 ($\pm$0.051) b	−0.61	27.30 ($\pm$6.13) a	27.39 ($\pm$5.64) a	-0.03
F-value		22.77 *	7.46 *		14.03 *	6.69 *	
		Semi-polished					
C. zollingeri	30	0.438 ($\pm$0.050)	0.412 ($\pm$0.052)	0.97	15.79 ($\pm$1.99)	17.82 ($\pm$3.32)	−1.36
C. zollingeri	20	0.458 ($\pm$0.059)	0.460 ($\pm$0.069)	−0.08	15.88 ($\pm$2.00)	16.22 ($\pm$1.99)	−0.26
t-value		−0.69	−1.49		−0.08	0.89	

Mean values followed by different letters within a column indicate that there is a significant difference ($p \leq 0.05$) as determined by ANOVA and Duncan's multiple range test. * Significant differences at the 0.05 level.

For fully polished samples, *C. ornatus* had the highest volumetric swelling, and differences were statistically significant from the other samples (Table 3). In bottom samples with 30 and 20 mm diameters, *C. ornatus* demonstrated about 46% and 42% higher volumetric swelling than *C. zollingeri*, respectively. These values were about 31% and 41% higher in the top samples of *C. ornatus* than in the top samples of *C. zollingeri*. No significant differences were observed in semi- and fully polished *C. zollingeri* samples between diameter classes (20 and 30 mm) and cane height (bottom and top). These results are in line with previous findings on other species of *Calamus* [34], although a significant increase of volumetric shrinkage from the bottom to the top of rattan cane is also reported [35]. The higher volumetric swelling values of *C. ornatus* could be caused by its lower basic density. Low basic density implies small cell wall thickness, and this in turn results in higher swelling [36].

Vascular bundles are scattered in the ground tissues (Figure 2). Just below the epidermis, vascular bundles are small and incomplete, consisting of either clusters of fibers or a few phloem and vessel elements with fibers. They are well-developed in the center of the stem cross-section. That implies a within-cane variation of density in the radial direction, i.e., a low-density core with large-diameter and low-frequency vascular bundles that is surrounded by a high-density outer zone with small-diameter and high-frequency vascular bundles [30].

Based on the anatomical and physical properties studied in rattan samples with different diameter classes, *C. zollingeri* with a 20 mm diameter showed longer fibers, higher basic density, and the lowest volumetric swelling in both the bottom and top of the canes. Therefore, we anticipate that *C. zollingeri* illustrates a more dimensional-stable character than the other rattan types.

3.2. Chemical Composition Analysis

The chemical composition of rattan samples in different diameter classes is shown in Table 4. The holocellulose content, which was about 78%, was identical in all rattan samples. This value agrees with previous studies [18,37]. The highest lignin content of 17.9% was obtained in *C. zollingeri* with a 30 mm diameter, while the *C. ornatus* exhibited the lowest lignin content of 9.1%. This might be related to the age of the rattan cane in that the cambial cells produce more mature cells with progressing cane age [17]. The content of alcohol-benzene solubles in different rattan species shows a non-specific trend from the bottom to top samples [18,34]. *C. ornatus* was found to have higher extractive and ash contents compared with other rattan types. *C. zollingeri* with a 20 mm diameter had higher extractive content than that of *C. zollingeri* with a 30 mm diameter. The high ash content in material like rattan is generally attributed to the high silica content, which can affect its working properties for some specific applications [34].

Table 4. Chemical composition of rattan samples in different diameter classes (wt%).

Sample	Diameter Class	Holocellulose	Lignin	Extractives	Ash
C. zollingeri	30	78.5	17.9	1.5	2.0
C. zollingeri	20	78.9	13.6	2.8	4.6
C. ornatus	15	78.0	9.1	5.1	7.7

3.3. Vapor Sorption Analysis

The MC of rattan is a very important factor as it affects the mechanical properties. Water is a good plasticizer and acts as a softener. Uptake of water loosens the bonds between the structural polymers within the lignocellulosic materials, i.e., lignin, cellulose, and hemicelluloses [38,39]. During vapor sorption analysis, the rattan samples showed the typical sigmoidal character of cellulosic materials (Figure 3). Different moisture sorption behavior was observed in rattan samples at various relative humidities (RHs) during adsorption and desorption processes.

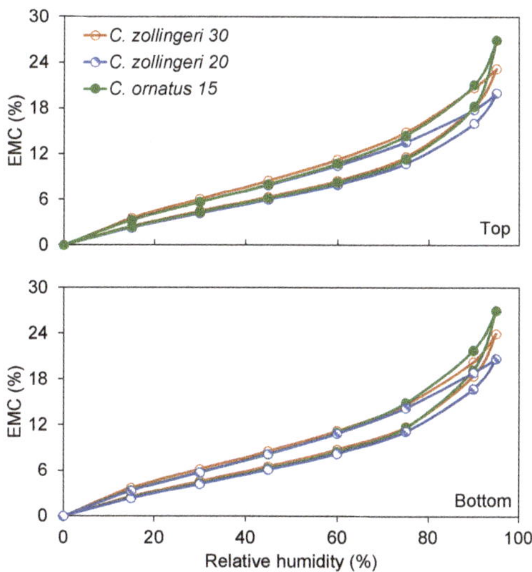

Figure 3. Equilibrium moisture content (EMC) of bottom/mature and top/juvenile rattan samples of different diameter classes (30, 20, and 15 mm) during the adsorption and desorption runs.

The EMC of rattan samples increased with increasing RH from 0 to 95%. Identical EMC values were observed in the top and bottom samples within the same species. There were no differences in the EMC values of the samples at RHs below 60% during both adsorption and desorption runs. The moisture adsorption isotherms are similar to the one reported by Yang et al. [36]. The highest EMC value was obtained at 95% RH for sample *C. ornatus*, which was ca. 27%. The sharp upward bend above 70% RH is generally accepted to be related to the softening point of hemicelluloses [40]. Removal of the lignin increased the moisture sorption capacity of holocellulose micro-veneers at RH above 75% [40]. Although the sorption mechanism of the rattan cell wall could be slightly different than wood, it is thus assumed that the higher EMC value of the *C. ornatus* samples could be mainly related to the lower lignin content as compared to other rattan samples, as indicated in chemical composition analysis (Table 4). As expected, at RHs above 75%, *C. zollingeri* with a 20 mm diameter demonstrated lower moisture sorption in comparison with *C. zollingeri* with a 30 mm diameter and *C. ornatus* in the top and bottom parts during both adsorption and desorption runs. The effect of high lignin content in rattan with a 30 mm diameter on moisture sorption might be offset by higher extractive and ash contents in *C. zollingeri* with a 20 mm diameter. The moisture sorption results are in accordance with the volumetric swelling values and confirmed a better hydrophobicity of *C. zollingeri* with a 20 mm diameter as compared with the other rattan samples.

The EMC values vary a lot in rattan age and the length of cane [2]. The results from dynamic moisture sorption analysis imply that the cane type and diameter should be considered when using rattan products for applications in high humidity conditions. The rattan-water vapor interaction study also helps to understand the material properties of rattan canes before harvesting. For instance, a young rattan could have similar or higher strength properties than a mature one because the plant produces an over-built stem that could withstand future load requirements [41]. In all aspects of properties, consideration of diameter class before harvesting is very important because it can lead to the shortening of the plantation and harvesting processes.

3.4. Mold and Decay Tests

After 10 days of incubation, mold growth was visually evaluated and graded on rattan samples of different diameter classes (Figure 4). The highest average mold rating was found on *C. ornatus* and the bottom sample of the 30 mm *C. zollingeri*. For *C. zollingeri*, statistically significant differences were neither observed between the bottom and top of the canes nor between the 30 mm and 20 mm diameter classes. Rattan hemicelluloses are partly composed of water-soluble polysaccharides, such as arabinogalactan, and partly of galactoglucomannan [18]. The sugar and acid units of hemicelluloses, such as arabinopyranose, arabinofuranose, galactose, methylglucuronic acid, and galacturonic acid are types of sugars and acids that are a source of fungal food. The chemical composition of the rattan samples illustrated a considerably lower lignin content in *C. ornatus*. This could be the reason for the higher mold infestation of this rattan type. However, differences in mold infestation between the bottom and top samples did not differ significantly.

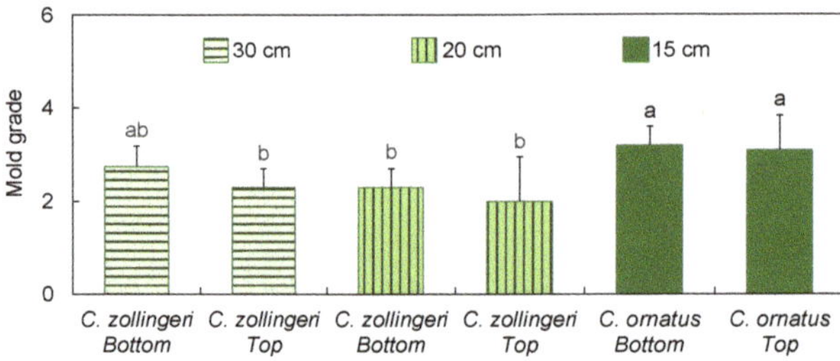

Figure 4. Mold grading of rattan samples for different diameter classes. Error bars show standard deviations. Bars with different letters indicate that there is a significant difference ($p \leq 0.05$), as determined by ANOVA and Duncan's multiple range test.

The decay resistance of rattan samples against the white-rot fungus *Trametes versicolor* is shown in Figure 5. To validate the results, beech and pine sapwood samples were selected as controls. Among all rattan samples of different diameter classes, the *C. ornatus* illustrated the highest mass loss. These results are in accordance with the ones from mold test and dynamic vapor sorption analysis. The differences of mass loss between the top and bottom samples of each diameter class were statistically insignificant. Among the three diameter classes, *C. zollingeri* of 20 mm diameter showed the lowest mass loss for both the bottom and top samples (Figure 5). However, no obvious trend was observed for mass loss of both the top and bottom samples in all diameter classes.

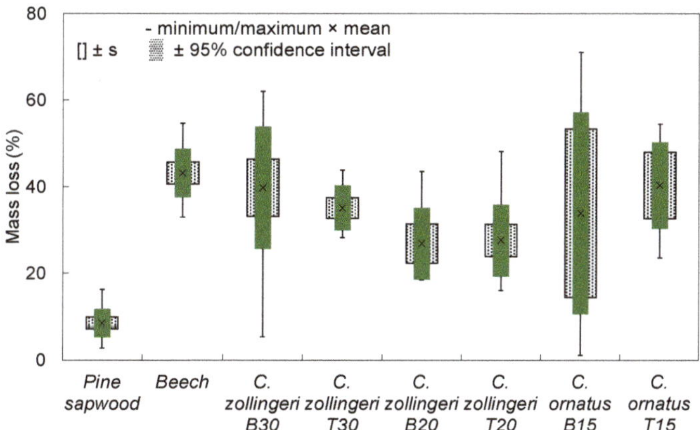

Figure 5. Mass loss of bottom/mature (B) and top/juvenile (T) rattan samples of different diameter classes (30, 20, and 15 mm) after decay test with the white-rot fungus *Trametes versicolor*.

As white-rot decay has a common feature of degrading lignin, cellulose, and hemicelluloses, the resistance of lignocellulosic material to this fungus highly depends on the relative proportion of their compositional polymers, i.e., it depends on the lignin content, carbohydrate ratio, and lignin monomer composition (mainly guiaicyl lignin) [42–44]. High mass loss of *C. ornatus* samples compared with the other rattan samples could be related to their considerably low lignin content, as demonstrated in chemical composition analysis (Table 4).

4. Conclusions

This study showed that *C. zollingeri* with a 20 mm diameter had the highest basic density in both bottom and top samples compared to *C. zollingeri* with a 30 mm diameter and *C. ornatus* with a 15 mm diameter. The reasoning for such higher density was supported by the anatomical features of higher vascular bundle frequency and longer thick-walled fibers. In addition, *C. zollingeri* with a 20 mm diameter exhibited the lowest volumetric swelling from an oven-dry condition to complete saturation. When water vapor sorption analysis was performed, this diameter class showed the lowest EMC at 95% RH, which verified its better dimensional stability even at high humidity levels. In addition, better resistance against mold and mass loss from white-rot fungal decay were also observed. Those properties followed a statistically similar pattern with sample height (bottom and top). From these findings, we conclude that *C. zollingeri* with a 20 mm diameter class exhibited better stability, moisture properties, and durability from older rattan with a 30 mm diameter class. On the contrary, *C. ornatus* with a 15 mm diameter showed inferior properties. Further research work needs to be extended to the evaluation of different mechanical properties.

Author Contributions: Conceptualization, S.A.A. and S.A.; methodology, S.A.A., R.H. and S.A.; software, S.A.A.; validation, S.A.A., R.H. and S.A.; formal analysis, S.A.A.; investigation, S.A.A.; R.H. and C.B.; visualization, S.A.A.; R.H. and C.B.; data curation, S.A.A. and S.A.; writing—original draft preparation, S.A.A; writing—review and editing, all authors. All authors have read and agreed to the published version of the manuscript.

Funding: The authors acknowledge funding provided by IKEA of Sweden.

Institutional Review Board Statement: Not applicable.

Informed Consent Statement: Not applicable.

Data Availability Statement: Not applicable.

Conflicts of Interest: The authors declare no conflict of interest.

References

1. Szczepanowska, H.M. Deconstructing rattan: Morphology of biogenic silica in rattan and its impact on preservation of Southeast Asian art and artifacts made of rattan. *Stud. Conserv.* **2018**, *63*, 356–374. [CrossRef]
2. Bhat, K.M.; Thulasidas, P.K.; Mohamed, C.P. Strength properties of ten south Indian canes. *J. Trop. For. Sci.* **1992**, *5*, 26–34.
3. Sastry, C.B. Rattan in the twenty-first century- an overview of major issues and needs for the global development of rattan. *Unasylva* **2001**, *205*, 3–12.
4. Olorunnisola, A.O.; Adefisan, O.O. Trial production and testing of cement-bonded particleboard from rattan furniture waste. *Wood Fiber Sci.* **2002**, *34*, 116–124.
5. Sharma, M.; Sharma, C.L.; Haokip, D. Anatomical and physical characteristics of some rattan species. *J. Indian Acad. Wood Sci.* **2018**, *15*, 132–139. [CrossRef]
6. Siebert, S.F. *The Nature and Culture of Rattan: Reflection of Vanishing Life in The Forests of Southeast Asia*; University of Hawai'i Press: Honolulu, HI, USA, 2012.
7. Krisdianto, K.; Jasni, J.; Tutiana, T. Anatomical properties of nine indigenous rattan species of Jambi, Indonesia. *Indonesian J. For. Res.* **2018**, *5*, 147–161. [CrossRef]
8. Ariffin, W.T.W.; Hisain, S.; Salleh, A.H. *Transfer of Technology Model: Rattan Furniture Making Unit*; International Network for Banboo and Rattan (INBAR), Forest Research Institute Malaysia: Kuala Lumpur, Malaysia, 2001.
9. Bhat, K.M.; Varghese, M. Anatomical basis for density and shrinkage behavior of rattan stem. *J. Inst. Wood Sci.* **1991**, *12*, 123–130.
10. Tomlinson, P.B.; Fisher, J.B.; Spangler, R.E.; Richer, R.A. Stem vascular architecture in the rattan palm *Calamus* (Arecaceae-Calamoideae-Calaminae). *Am. J. Bot.* **2001**, *88*, 797–809. [CrossRef] [PubMed]
11. Yang, S.; Xiang, E.; Shang, L.; Liu, X.; Tian, G.; Ma, J. Comparison of physical and mechanical properties of four rattan species grown in China. *J. Wood Sci.* **2020**, *66*, 3. [CrossRef]
12. Shang, L.; Jiang, Z.; Liu, X.; Tian, G.; Ma, J.; Yang, S. Effect of modification with methyl methacrylate on the mechanical properties of *Plectocomia kerrana* rattan. *BioResources* **2016**, *11*, 2071–2082. [CrossRef]
13. Abasolo, P.W.; Yoshida, M.; Yamamoto, H.; Okuyama, T. Microfibril angle determination of rattan fibers and its influence on the properties of the cane. *Holzforschung* **2000**, *54*, 437–442. [CrossRef]
14. Luo, Z.; Zhang, X.; Lu, B.; Pan, B.; Ruan, Z. Mechanical properties and test methods of rattan. *Furnit. Des. Room Dec.* **2012**, *7*, 108–110.
15. Abasolo, P.W.; Yamamoto, H.; Yoshida, M.; Mitsui, K.; Okuyama, T. Influence of heat and loading time on the mechanical properties of *Calamus merrillii* Becc. *Holzforschung* **2002**, *56*, 639–647. [CrossRef]
16. Bhat, K.M.; Nasser, K.M.M.; Thulasidas, P.K. Anatomy and identification of south Indian rattans (*Calamus* species). *IAWA J.* **1993**, *14*, 63–76. [CrossRef]
17. Abasolo, W.P. Properties of rattan cane as basis for determining optimum cutting cycle of cultivated *Calamus merrillii*. *J. Trop. For. Sci.* **2015**, *27*, 176–188.
18. Sanusi, D. *Rotan: Kekayaan Belantara Indonesia*; Brilian Internasional: Surabaya, Indonesia, 2012. (In Indonesian)
19. Hamid, N.H.; Hale, M. Decay threshold of acetylated rattan against white and brown rot fungi. *Int. Wood Prod. J.* **2012**, *3*, 96–106. [CrossRef]
20. Tiryaki, S.; Bardak, S.; Aydin, A.; Nemli, G. Analysis of volumetric swelling and shrinkage of heat treated woods: Experimental and artificial neural network modeling approach. *Maderas Ciencia. Tecnol.* **2016**, *18*, 477–492. [CrossRef]
21. Ghavidel, A.; Hosseinpourpia, R.; Gelbrich, J.; Bak, M.; Sandu, I. Microstructural and chemical characteristics of archaeological white elm (*Ulmus laevis* P.) and poplar (*Populus* spp.). *Forests* **2021**, *11*, 10271. [CrossRef]
22. EN 15403; Solid Recovered Fuels—Determination of Ash Content. European Committee for Standardization: Brussels, Belgium, 2011.
23. Hosseinpourpia, R.; Adamopoulos, S.; Walther, T.; Naydenov, V. Hydrophobic formulations based on tall oil distillation products for high-density fiberboards. *Materials* **2020**, *13*, 4025. [CrossRef]
24. Ahmed, S.A.; Sehlstedt-Persson, M.; Morén, T. Development of a new rapid method for mould testing in a climate chamber-Preliminary tests. *Eur. J. Wood Prod.* **2013**, *71*, 451–461. [CrossRef]
25. Sehlstedt-Persson, M.; Karlsson, O.; Wamming, T.; Morén, T. Mold growth on sapwood boards exposed outdoors: The impact of wood drying. *For. Prod. J.* **2011**, *61*, 170–179. [CrossRef]
26. EN 113-2; Durability of Wood and Wood-Based Products—Test Method against Wood Destroying Basidiomycetes—Part 2: Assessment of Inherent or Enhanced Durability. European Committee for Standardization: Brussels, Belgium, 2020.
27. Bravery, A.F. A miniaturised wood-block test for the rapid evaluation of wood preservative fungicides. In Proceedings of the 10th Annual Meeting, Peebles, Scotland, 17–22 September 1978.

28. Weiner, G.; Liese, W. Rattans- stem anatomy and taxonomic implications. *IAWA J.* **1990**, *11*, 61–70. [CrossRef]
29. Liu, X.; Tian, G.; Shang, L.; Yang, S.; Jiang, Z. Compression properties of vascular bundles and parenchyma of rattan (*Plectocomia assamica* Griff). *Holzforschung* **2014**, *68*, 927–932. [CrossRef]
30. Sulaiman, A.; Lim, S.C. Anatomical and physical features of 11-y-old cultivated *Calamus manan* in peninsular Malaysia. *J. Trop. For. Sci.* **1991**, *3*, 372–379.
31. Gu, Y.; Zhang, J. Tensile properties of natural and synthetic rattan strips used as furniture woven materials. *Forests* **2020**, *11*, 1299. [CrossRef]
32. Bhat, K.M.; Liese, W.; Schmitt, U. Structural variability of vascular bundles and cell wall in rattan stem. *Wood Sci. Technol.* **1990**, *24*, 211–224. [CrossRef]
33. Wahab, R.; Sulaiman, O.; Samsi, H.W. Basic density and strength properties of cultivated *Calamus manan*. *J. Bamboo Rattan* **2004**, *3*, 35–43. [CrossRef]
34. Ali, A.R.M.; Mohmod, A.L.; Khoo, K.C.; Kasim, J. Physical properties, fiber dimensions and proximate chemical analysis of Malaysian rattan. *Thai J. For.* **1995**, *14*, 59–70.
35. Chowdhury, M.Q. Assessment of some physical and mechanical properties of golla bet (*Daemonorops jenkinsiana*) from northeastern region of Bangladesh. *J. Bamboo Rattan* **2004**, *3*, 195–201. [CrossRef]
36. Yang, L.; Tian, G.; Yang, S.; Shang, L.; Liu, X.; Jiang, Z. Determination of fiber saturation point of rattan (*Calamus simplicifolius*) using the LF-NMR and two conventional methods. *Wood Sci. Technol.* **2020**, *54*, 667–682. [CrossRef]
37. Tellu, A.T. Chemical properties of different rattan species traded in Central Sulawesi Province. *Biodiversitas* **2008**, *9*, 108–111. (In Indonesian) [CrossRef]
38. Hillis, W.E. High temperature and chemical effects on wood stability. Part 1: General considerations. *Wood Sci. Technol.* **1984**, *18*, 281–293. [CrossRef]
39. Salmén, L. Viscoelastic properties of in situ lignin under water-saturated conditions. *J. Mater. Sci.* **1984**, *19*, 3090–3096. [CrossRef]
40. Hosseinpourpia, R.; Adamopoulos, S.; Mai, C. Dynamic vapour sorption of wood and holocellulose modified with thermosetting resins, *Wood Sci. Technol.* **2016**, *50*, 165–178.
41. Tomlinson, P.B. *The Structural Biology of Palms*; Clarendon Press: Oxford, UK, 1990.
42. Chen, M.; Wand, C.; Fei, B.; Ma, X.; Zhang, B.; Zhang, S.; Huang, A. Biological degradation of Chinese fir with *Trametes versicolor* (L.) Lloyd. *Materials* **2017**, *10*, 834. [CrossRef]
43. Kamperidou, V. The biological durability of thermally- and chemically-modified black pine and poplar wood against basidiomycetes and mold action. *Forests* **2019**, *10*, 1111. [CrossRef]
44. Bari, E.; Daniel, G.; Yilgor, N.; Kim, J.S.; Tajick-Ghanbary, M.A.; Singh, A.P.; Ribera, J. Comparison of the decay behavior of two white-rot fungi in relation to wood type and exposure conditions. *Microorganisms* **2020**, *8*, 1931. [CrossRef]

 *forests*

 MDPI

Article

Effect of Seasonal Variation on Leaf Cuticular Waxes' Composition in the Mediterranean Cork Oak (*Quercus suber* L.)

Rita Simões, Isabel Miranda * and Helena Pereira

Centro de Estudos Florestais (CEF), Instituto Superior de Agronomia, Universidade de Lisboa, 1349-017 Lisboa, Portugal
* Correspondence: imiranda@isa.ulisboa.pt

Abstract: *Quercus suber* L. (cork oak) leaves were analyzed along one annual cycle for cuticular wax content and chemical composition. This species, well adapted to the long dry summer conditions prevailing in the Mediterranean, has a leaf life span of about one year. The cuticular wax revealed a seasonal variation with a coverage increase from the newly expanded leaves (115.7 μg/cm^2 in spring) to a maximum value in fully expanded leaves (235.6 μg/cm^2 after summer). Triterpenoids dominated the wax composition throughout the leaf life cycle, corresponding in young leaves to 26 μg/cm^2 (22.6% of the total wax) and 116.0 μg/cm^2 (49% of the total wax) in mature leaves, with lupeol constituting about 70% of this fraction. The total aliphatic compounds increased from 39 μg/cm^2 (young leaves) to 71 μg/cm^2 (mature leaves) and then decreased to 22 μg/cm^2 and slightly increased during the remaining period. The major aliphatic compounds were fatty acids, mostly with C$_{16}$ (hexadecanoic acid) and C$_{28}$ (octacosanoic acid) chain lengths. Since pentacyclic triterpenoids are located almost exclusively within the cutin matrix (intracuticular wax), the increase in the cyclic-to-acyclic component ratio after summer shows an extensive deposition of intracuticular waxes in association with the establishment of mechanical and thermal stability and of water barrier properties in the mature leaf cuticle.

Keywords: *Quercus suber* L.; seasonal variation; cuticular waxes; leaves

Citation: Simões, R.; Miranda, I.; Pereira, H. Effect of Seasonal Variation on Leaf Cuticular Waxes' Composition in the Mediterranean Cork Oak (*Quercus suber* L.). *Forests* **2022**, *13*, 1236 . https://doi.org/10.3390/f13081236

Academic Editor: Adam M. Taylor

Received: 30 June 2022
Accepted: 2 August 2022
Published: 4 August 2022

Publisher's Note: MDPI stays neutral with regard to jurisdictional claims in published maps and institutional affiliations.

1. Introduction

The plant cuticle is a continuous extracellular hydrophobic membrane that forms a primary barrier between the plant's air-exposed surfaces and the external environment and also plays important functions in organ growth and development. Particularly in leaves, the cuticle plays a key role in limiting uncontrolled water loss and provides protection from biotic or abiotic stresses such as drought and high temperatures.

The cuticle is a heterogeneous membrane consisting of a matrix of polymeric cutin and cuticular waxes [1–3]. Cutin is a polyester formed by C$_{16}$ and C$_{18}$ hydroxy fatty acids and their derivatives and glycerol. Cuticular waxes predominantly comprise very long-chain fatty acids and their derivatives (including aldehydes, primary alcohols, and alkanes), as well as, in some species, cyclic molecules especially pentacyclic triterpenoids, sterols, and aromatics (e.g., *Ficus elastica* Roxb.) [1–3]. Cuticular waxes are distinguished according to their location in an intracuticular layer within the cutin matrix and an epicuticular layer deposited on the outer surface of the cutin [2].

The functional properties of the cuticle are largely related to the structural arrangement of the cuticular waxes' layers and their chemical compositions [1–3]. The limitation of the non-stomatal water loss is one of the most important features and has therefore been analyzed in terms of both the role of intracuticular and epicuticular waxes and the role of the different cuticular wax components. The functional barrier against water diffusion through the cuticle is preferentially established by the intracuticular wax, while the epicuticular wax does not contribute to the transpiration barrier [4]. Regarding the contribution of the chemical wax components, it was shown that it is the very long-chain aliphatic fraction of the wax that establishes the transpiration barrier [1,2,5], whereas the triterpenoids provide mechanical and thermal stability to the plant cuticle [6,7]. For instance, the triterpenoids deposited within the cutin matrix restrict the polymer's thermal expansion and thus prevent thermal damage to the highly ordered aliphatic wax barrier, even at high temperatures [7].

The cork oak (*Quercus suber* L.) is one of the most important evergreen species in the western Mediterranean Basin, with important economic value because of the production of cork that feeds a dedicated industrial chain [8]. This species is well-adapted to the adverse conditions of the Mediterranean, that is, to the long, dry summer conditions with high solar irradiances, air temperatures and vapor pressure deficit, and little or no precipitation, as well as moderately cold winters. The leaves are sclerophilic, oval in shape with a dark green color on the adaxial face and without epidermal hairs (trichomes), and the abaxial face is lighter with numerous stomata and densely covered with trichomes in the form of starry multicellular hairs [8,9]. The leaf's cuticular membrane contains substantial amounts of cuticular wax (154.3–235.1 $\mu g/cm^2$) composed largely of pentacyclic triterpenoid compounds (61%–72% of the identified compounds, with lupeol as the main component), while long-chain aliphatic components are mainly fatty acids (mainly in C_{30}, C_{28}, and C_{16}) (17%–23% of the identified compounds) that contribute to building a nearly impermeable membrane [8]. Cutin is present in high amounts (518 $\mu g/cm^2$ of leaf area) and has as major monomeric constituents 10,16-dihydroxyhexadecanoic and 9,10,18-trihydroxyoctadecanoic acids [10].

Cork oak has a specific foliage phenology, characterized by short-lived leaves that usually fall within one year concurrently with spring growth, with a cycle that is much shorter than that of other evergreen oaks, such as the Iberian holm oak (*Quercus rotundifolia* Lam.= *Q. ilex* L. subsp. *ballota*), whose leaves last 1–3 years, or the kermes oak (*Q. coccifera* L.), whose leaves can last 5–6 years [11–13]. In *Q. suber*, physiological activity starts in February/March with bud development and shoot growth, and the development of new foliage begins in April and is terminated by June. Most leaves emerge and expand within 1 month. The leaf life duration is approximately 14 months, with a range of 11–18 months [12,14,15]. Most of the older leaves (1-year-old) fall in spring during the early part of shoot growth [11].

The present work addresses the study of the seasonal variation of the chemical composition of cuticular waxes in cork oak (*Quercus suber* L.) leaves in the Mediterranean climate. It is hypothesized that the content and composition of the leaf cuticular waxes depend on the seasonally related leaf development stage and on the prevailing climatic variables. The results aim at understanding the role of cuticular wax components of sclerophilic leaves and their protective role regarding biotic and abiotic stresses in association with leaf development, thereby contributing to climate change adaptation measures.

2. Material and Methods

2.1. Sampling

The study was carried out on a provenance trial of 21-year-old *Quercus suber* L. trees at Herdade Monte Fava, Santiago do Cacém, in central Portugal (38°00′ N, 08°70′ W, altitude 79 m). This trial was established with seedlings of cork oak trees raised from seeds of different provenances: Portugal (PT35), Spain (ES11), Italy (IT13), France (FR3), Morocco (MA27), and Tunisia (TU32). A more detailed trial and site description is given in Sampaio et al. [16] and Varela [17]. The sampling included the collection of leaves from two trees from each of the six provenances. The first sampling was in May 2019 on fully expanded leaves from the current year's spring flushing followed by samplings, in the same trees, in September 2019, December 2019, January 2020, and March 2020. The leaves were collected randomly from different branches on the south-exposed crown side, in the lower part of the canopy up to a height of approximately 2 m, making up a total sample per tree of about 100 leaves. A composite leaf sample per provenance was prepared with the leaves of the two trees.

2.2. Morphological Variables

The morphological variables leaf area and specific leaf area were measured in 40 leaves that were randomly selected from the leaves sampled for each provenance. Leaf area was measured by digitalizing and calculated with Leica Qwin vs. 3.0 Image Analysis Software. The leaves were oven-dried at 70 °C until no change in mass was detected, and the total dry mass per leaf was determined. Specific leaf area (SLA, cm^2/g) was calculated as the ratio between the measured leaf area and the dry weight and used to describe the sclerophyllous nature of leaves.

2.3. Extraction of Cuticular Waxes

Cuticular wax was extracted from whole fresh leaves with dichloromethane over 6 h in a Soxhlet apparatus [9]. This extraction method yields a solution containing both the epicuticular and intracuticular waxes from both leaf sides. The amount of the soluble cuticular compounds was determined from the mass difference of the extracted leaves after drying at 105 °C and was expressed on a leaf surface area and dry weight basis (the ratio between wax in µg and the two-sided leaf surface area in cm^2, obtained by digitalization) [9].

2.4. Cuticular Wax Composition

The cuticular wax, obtained as dichloromethane extracts, was analyzed using gas chromatography–mass spectrometry (GC-MS). Two milligrams of each leaf extract was taken and derivatized in 120 µL of pyridine; the compounds with hydroxyl and carboxyl groups were trimethylsilylated into trimethylsilyl (TMS) ethers and esters, respectively, by adding 80 µL of bis(trimethylsily)-trifluoroacetamide (BSTFA) and reacting at 60 °C for 30 min. The derivatized extracts (1 µL) were immediately analyzed by GC-MS (EMIS, Agilent 5973 MSD, Palo Alto, CA, USA). The detailed experimental conditions are given elsewhere [9]. The compounds were identified as TMS derivatives by matching with GC-MS spectral libraries (Wiley, NIST) and published fragmentation profiles [18,19]. Two replicates were made per extract.

2.5. Statistical Analysis

Data are presented as the means $\pm$ standard deviation of the six independent provenances' samples analyzed in duplicate. To compare data along the leaf cycle, one-way analysis of variance (ANOVA) was performed. Duncan's post hoc tests were used to analyze pairwise differences between provenances. Statistical significance was set at $p < 0.05$. All statistical analyses were performed using the Sigmaplot® (Version 11.0, Systat Software, Inc., Chicago, IL, USA).

3. Results

The seasonal changes in *Q. suber* leaves regarding leaf size, specific leaf area, and cuticular wax content and composition were followed throughout the first year, beginning with the newly expanded leaves in late spring (May) and ending with one-year-old leaves in the early spring (March) of the following year.

3.1. Leaf Area and SLA

In May, the newly developed leaves had a leaf size of 5.8 ± 1.7 cm^2, which increased through summer to a maximum size of 6.5 ± 1.4 cm^2 in September (corresponding to the fully expanded mature leaves) that remained unchanged throughout the winter months (6.5 ± 2.3 cm^2 in December and 6.5 ± 1.4 cm^2 in January). In March, when the 1-year-old leaves began to fall, the average leaf size decreased to 5.7 ± 1.7 cm^2 (Figure 1). Although differences in leaf size occurred during the study period, they were not statistically significant ($p = 0.449$).

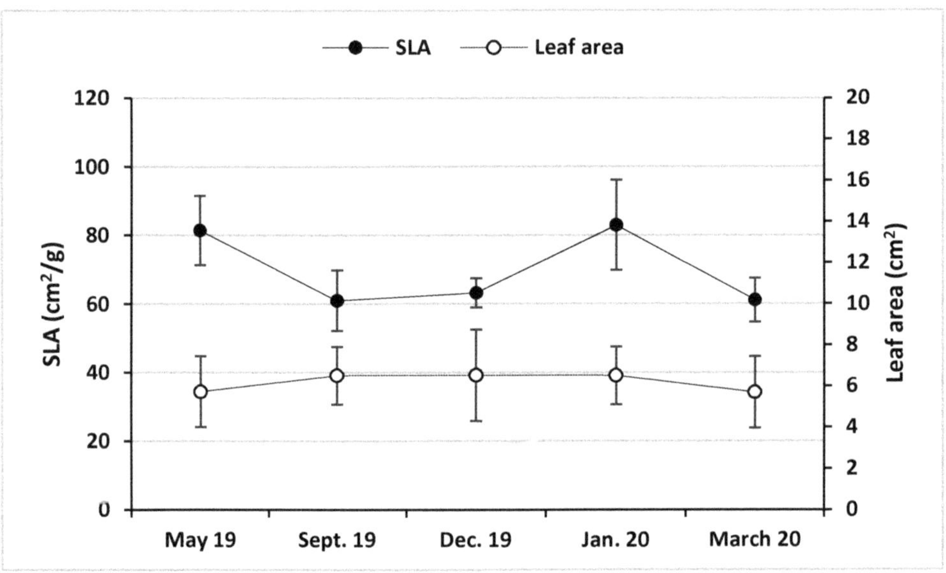

Figure 1. Seasonal variation in specific leaf area (SLA) and leaf area in *Quercus suber* L. (mean and standard deviation).

Specific leaf area (SLA), i.e., leaf surface per unit leaf dry mass, varied significantly throughout the study period ($p < 0.001$), with the values of May and January significantly different from those of September, December, and March (Figure 1). In May, the SLA of the young leaves was high (81.4 cm²/g) and declined with time to a mean value of 61.0 cm²/g, after the summer. SLA remained unchanged from September to December at a mean value of 63.2 cm²/g and increased in January (82.3 cm²/g). In March, the remaining one-year-old leaves were on average smaller, and the SLA was 61.1 cm²/g.

3.2. Cuticular Wax Content

Our results revealed a seasonal variation in total wax quantity deposited in the cuticle of the cork oak leaves ($p < 0.001$), with the May value significantly different from the other ones (Figure 2). In May, the newly developed leaves exhibited the lowest value of total wax coverage over the annual cycle (115.7 µg/cm² on average, ranging between 68.9 and 172.5 µg/cm²). In the next three months—within June, July and August —and until September, the biosynthesis of leaf wax lipids was very dynamic, doubling up to 235.6 µg/cm² (ranging from 173.7 to 267.4 µg/cm²). Wax content decreased afterwards to 182.1 µg/cm² in December (ranging 151.0 and 208.0 µg/cm²). The cuticular wax content in January and March (193.6 µg/cm² and 220.3 µg/cm², respectively) increased, although not significantly, due to sample variation.

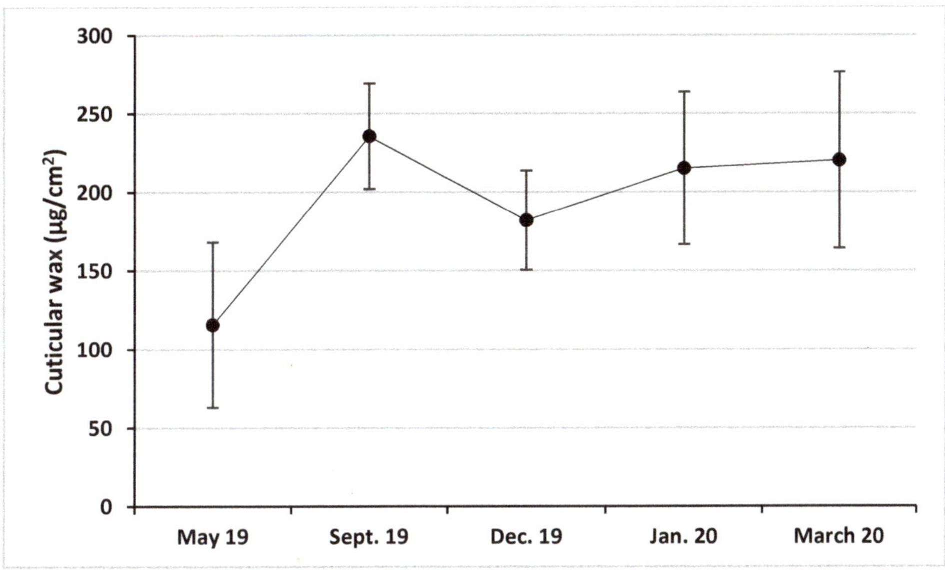

Figure 2. Seasonal variation of cuticular wax (μg/cm^2) in *Quercus suber* leaves (mean and standard deviation).

3.3. Cuticular Wax Composition

The seasonal changes in the composition of the leaf cuticular waxes grouped by major chemical families (μg/cm^2) are presented in Figure 3. The composition and the relative abundances of individual components are shown in Table 1.

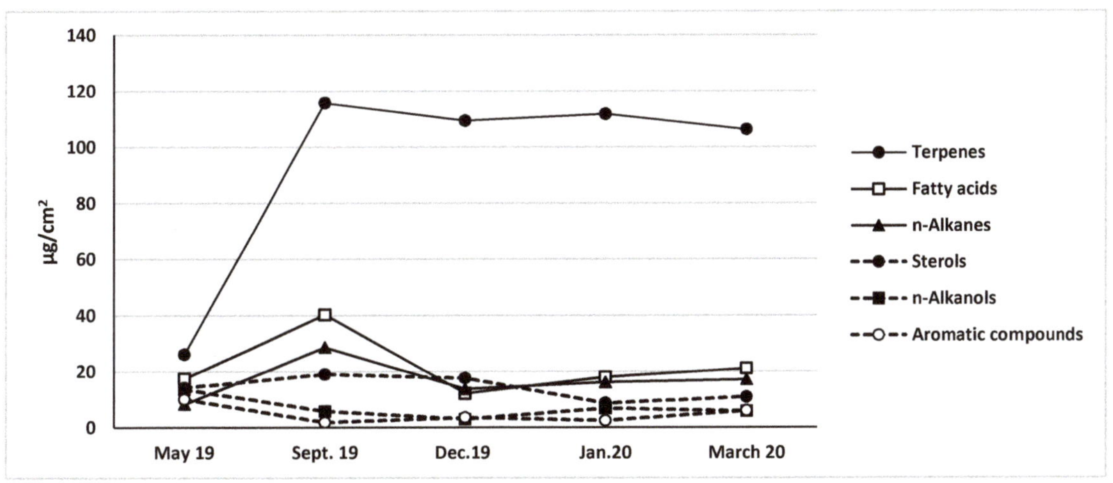

Figure 3. Seasonal variation in the cuticular wax chemical families (μg/cm^2) of *Quercus suber* leaves.

Table 1. Seasonal variation in the leaf cuticular waxes composition in cork oak (*Quercus suber*) (mean of six provenances), as determined by GC-MS, as a % of total peak area (only compounds with over 0.10% are shown).

	May 2019	September 2019	December 2019	January 2020	March 2020
Wax Content ($\mu g/cm^2$)	115.7 ± 52.6	235.6 ± 33.6	182.1 ± 31.6	215.1 ± 48.5	220.3 ± 56.0
n-Alkanes					
Hexacosane (C_{26})	-	0.16 ± 0.04	0.13 ± 0.15	0.13 ± 0.05	0.15 ± 0.17
Heptacosane (C_{27})	0.63 ± 0.24	1.06 ± 0.45	0.55 ± 0.45	0.68 ± 0.30	0.70 ± 0.17
Octacosane (C_{28})	0.21 ± 0.07	0.72 ± 0.20	0.46 ± 0.12	0.60 ± 0.19	0.46 ± 0.14
Nonacosane (C_{29})	4.75 ± 0.94	8.69 ± 3.38	7.23 ± 3.45	5.07 ± 2.61	5.24 ± 2.21
Triacontane (C_{30})	1.43 ± 0.54	1.38 ± 0.32	0.84 ± 0.40	0.93 ± 0.40	1.13 ± 0.29
n-Alkanols					
Hexadecan-1-ol ($C_{16}OH$)	0.10 ± 0.03	0.15 ± 0.05	0.14 ± 0.10	0.11 ± 0.09	0.12 ± 0.06
Docosan-1-ol ($C_{22}OH$)	0.89 ± 0.29	0.21 ± 0.10	0.12 ± 0.02	0.25 ± 0.10	0.18 ± 0.14
Tretracosan-1-ol ($C_{24}OH$)	9.92 ± 6.1	1.80 ± 0.89	1.14 ± 0.68	2.23 ± 1.05	1.84 ± 0.46
Pentacosan-1-ol ($C_{25}OH$	0.11 ± 0.07	-	-	-	-
Hexacosan-1-ol ($C_{26}OH$)	0.35 ± 0.30	-	-	-	-
Octacosn-1-ol ($C_{28}OH$)	-	-	-	0.32 ± 0.22	0.25 ± 0.26
Dotriacontan-1-ol ($C_{32}OH$)	-	0.11 ± 0.10	-	-	-
Fatty acids					
Saturated					
Decanoic acid ($C_{10:0}$)	-	-	-	-	0.21 ± 0.07
Dodecanoic acid ($C_{12:0}$)	0.10 ± 0.03	-	-	0.10 ± 0.07	0.22 ± 0.04
Tetradecanoic acid ($C_{14:0}$)	0.30 ± 0.16	0.12 ± 0.07	0.22 ± 0.11	0.15 ± 0.14	0.19 ± 0.23
Hexadecanoic acid ($C_{16:0}$)	4.97 ± 2.76	1.32 ± 0.65	4.31 ± 2.62	2.07 ± 0.60	2.02 ± 0.42
Octadecanoic acid ($C_{18:0}$)	0.51 ± 0.14	0.23 ± 0.09	0.17 ± 0.05	0.26 ± 0.14	0.20 ± 0.08
Eicosanoic acid ($C_{20:0}$)	0.20 ± 0.05	0.21 ± 0.11	-	0.18 ± 0.04	0.17 ± 0.08
Docosanoic acid ($C_{22:0}$)	0.32 ± 0.10	0.21 ± 0.17	-	0.26 ± 0.07	0.30 ± 0.17
Tetracosanoic acid ($C_{24:0}$)	0.80 ± 0.24	0.24 ± 0.15	-	0.24 ± 0.08	0.20 ± 0.17
Hexacosanoic acid ($C_{26:0}$)	0.71 ± 0.34	1.01 ± 0.43	-	0.56 ± 0.19	0.45 ± 0.16
Octacosanoic acid ($C_{28:0}$)	1.49 ± 1.09	4.65 ± 2.89	0.18 ± 0.09	1.46 ± 0.44	1.69 ± 0.36
Triacontanoic acid ($C_{30:0}$)	0.95 ± 1.01	8.15 ± 6.18	0.96 ± 0.74	2.28 ± 1.23	2.37 ± 0.61
Dotriacontanoic acid ($C_{32:0}$)	0.13 ± 0.09	0.41 ± 0.15	0.10 ± 0.7	0.32 ± 0.07	0.31 ± 0.12
Unsaturated					
9,12-Octadecadienoic acid ($C_{18:2}$)	1.42 ± 1.14	0.10 ± 0.05	-	-	0.15 ± 0.08
9,12,15-Octadecatrienoic acid ($C_{18:3}$)	2.54 ± 2.19	0.26 ± 0.16	0.24 ± 0.12	0.17 ± 0.12	0.51 ± 0.14
Glycerides					
Glycerol	3.52 ± 1.32	0.12 ± 0.05	0.16 ± 0.07	0.33 ± 0.38	1.03 ± 0.39
4-Hydroxyphenylglycolic acid	0.19 ± 0.21	0.35 ± 0.40	1.01 ± 0.45	1.18 ± 0.71	-
2-Palmitoglycerol	0.63 ± 0.09	0.21 ± 0.10	-	0.21 ± 0.26	-.
Glycerol monostearate	0.17 ± 0.07	-	-	-	-
1-Monolinoleate Glycerol	0.33 ± 0.15	-	-	-	-
Linolenoylglycerol	0.13 ± 0.07	-	-	-	0.18 ± 0.12
Sterols					
β-Systosterol	11.94 ± 5.28	6.29 ± 1.86	9.64 ± 4.26	4.01 ± 1.11	4.91 ± 1.03

Table 1. *Cont.*

	May 2019	September 2019	December 2019	January 2020	March 2020
Terpenes					
Diterpenes					
Phytol	4.00 ± 2.40	0.61 ± 0.28	0.35 ± 0.39	0.49 ± 0.36	1.01 ±0.23
Pentacylic triperpenes					
α-Amyrin	1.61 ± 0.53	1.34 ± 0.40	0.72 ± 0.47	0.76 ± 0.29	0.64 ± 0.21
β-Amyrin	2.33 ± 0.95	3.98 ± 1.38	4.90 ± 2.74	3.50 ± 1.32	3.97 ± 1.25
Germanicol	0.27 ± 0.42	8.16 ± 1.69	8.10 ± 6.50	8.40 ± 2.47	6.38 ± 1.49
Lupeol	7.38 ± 5.71	28.20 ± 10.41	36.59 ± 11.13	27.75 ± 7.06	24.11 ± 6.81
Epifriedelanol	0.43 ± 0.32	2.38 ± 2.26	3.75 ± 3.25	1.92 ± 2.22	3.12 ± 2.38
Erythrodiol	0.21 ± 0.17	-	0.14 ± 0.12	1.62 ± 1.27	0.39 ± 0.30
Friedelin	1.14 ±0.96	2.47 ± 2.73	2.82 ± 2.82	2.16 ± 2.00	2.38 ±1.85
Betulin	0.40 ± 1.09	0.36 ± 0.10	0.80 ± 0.63	1.09 ± 0.81	1.46 ± 0.71
Betulinic acid	0.46 ± 0.25	0.42 ± 0.15	0.75 ± 0.63-	0.98 ± 0.38	1.21 ± 0.53
Ursolic acid	0.44 ± 0.16	0.27 ± 0.10	0.24 ± 0.19	0.27 ± 0.16	0.28 ± 0.15
Aromatic compounds					
Benzoic acid	0.33 ± 0.29	0.16 ± 0.07	0.33 ± 0.09	0.27 ± 0.05	0.36 ± 0.06
Hexadecy-(E)-p-coumarate	4.93 ± 2.87	-	0.20 ± 0.09	-	-
Vanillin acid	0.47 ± 0.17	0.11 ± 0.06	0.11 ± 0.06	0.40 ± 0.49	0.16 ± 0.15
4-(Hydroxymethyl)phenol	0.66 ± 0.31	0.10 ± 0.08	0.51 ± 0.54	-	1.36 ± 0.24
Other compounds					
Myo-inositol	1.35 ± 0.99	0.34 ± 0.18	0.10 ± 0.09	1.35 ± 2.31	1.18 ±1.27
D-Fructose	0.67 ± 0.20	0.32 ± 0.09	-	3.31 ± 2.80	1.04 ±1.36
α-Tocopherol	1.62 ±1.11	-	-	-	1.04 ± 1.36
β-Tocopherol	0.33 ± 0.13	0.95 ± 0.50	-	-	-
γ-Tocopherol	-	-	0.55 ± 0.62	-	-
α-Tocopherolquinone	0.19 ± 0.17	0.13 ± 0.10	0.10 ± 0.16	0.40 ± 0.46	2.05 ± 2.1
Erythrono-1,4-lactone	0.33 ± 0.14	-	-	0.37 ± 0.22	-
Ribonic acid, 1,4-lactone	0.33 ± 0.13	0.10 ± 0.06	-	0.51 ± 0.34	0.31 ± 0.22
Quinic acid	-	-	-	1.93 ± 0.76	-

In the new expanding spring leaves in May, the cuticular wax was mainly composed of cyclic terpene compounds and linear long-chain aliphatic molecules. Pentacyclic triterpenoids were abundant, accounting for 26 µg/cm^2 of the leaf wax coverage (22.6% of the total wax). The major component was lupeol (7.4% of all compounds, 32% of the triterpenic fraction), with β-amyrin (2.3% of all compounds) and betulin (2% of all compounds) as other important constituents. Sterols amounted to 14 µg/cm^2 (12.2% of the wax), mainly β-systosterol (12.0% of the compounds). Aromatics comprised 8.7% of all compounds (10 µg/cm^2), with hexadecy-(E)-p-coumarate as the major compound. The linear long aliphatic molecules of acids, alcohols, and alkanes together represented 39.1 µg/cm^2 (33.9% of the total wax) and were the most abundant wax components. Fatty acids were the main compounds (17.3 µg/cm^2, 15% of total wax), mostly C$_{16}$ (hexadecanoic acid) and C$_{28}$ (octacosanoic acid) fatty acids, which together accounted for 59% of the total fatty acids. The aliphatic alcohols contributed to 11.4 µg/cm^2 of leaf surface coverage (12% of total wax), mostly C$_{24}$ (tetracosan-1-ol), with 85% of total aliphatic alcohols. Alkanes accounted for 7.2% of the wax, with C$_{29}$ (nonacosane, 4.8% of all compounds) and C$_{30}$ (triacontane, 1.4%) as predominant molecular species.

After the summer, the cuticular wax on the mature leaves increased significantly, mainly by pentacylic triperpenes, which reached 116.0 µg/cm^2 (49% of the total wax) (five times more than in spring leaves). Lupeol was the major constituent, followed by smaller amounts of epifriedelanol and friedelin. Pentacylic triterpenes were always the main wax components during the following months until the end of the annual cycle (March, with 12-month-old leaves), with a constant amount and composition (109.4 µg/cm^2, 111.8 µg/cm^2, and 106.3 µg/cm^2, respectively, in December, January, and March). The statistical analysis showed a significant difference ($p < 0.001$) in the time-related chemical composition in the amount of terpenes (Figure 3), with the May value different from the other ones.

Sterols were relatively constant over time (8% of the wax compounds in September and 5% in March) and composed exclusively of β-systosterol. Aromatic compounds were present in minor amounts.

Fatty acids showed the highest variability in concentration, with a maximum in September (75.6 µg/cm^2, 17.1% of all compounds), decreasing to 28.7 µg/cm^2 in December and slightly increasing during the remaining period. The statistical analysis showed a significant difference ($p < 0.001$) in the amount of fatty acids (Figure 3), with the September value different from the other ones. Fatty acids were composed of a homologous series of even and saturated components from C_{16} to C_{30}, mostly including octacosanoic (C_{28}) and triacontanoic (C_{30}) acids, followed by hexadecanoic (C_{16}) acid.

The aliphatic alcohols, including mainly tretracosan-1-ol, were highest in spring but remained rather constant throughout the period.

n-Alkanes, including only nonacosane and triacontane, were highest in September (28.5 µg/cm^2) and then decreased and remained relatively constant (between 13.6 and 17.0 µg/cm^2).

4. Discussion

4.1. Leaf Area and SLA

In *Q. suber*, the shoot growth and the development of new foliage occur during the late spring and early summer, most likely reaching a maximum enlargement rate in summer (June/August) [11,20,21]. This was confirmed in the present study. The leaf size increased through summer to a maximum in September corresponding to the fully expanded mature leaves, which remained unchanged throughout the winter. The smaller leaf size measured in March is probably related to a canopy change related to leaf fall. In fact, the longevity of cork oak leaves is about one year, much shorter than other evergreen oaks; e.g., leaves last 1–3 years in *Quercus rotundifolia* and 5–6 years in *Q. coccifera* [13,14,22]. The leaf size measured in the present study matches the few available reports for mature *Q. suber* leaves. Mediavilla et al. [23] reported values between 5.5 ± 0.3 and 7.4 ± 0.5 cm^2 in leaves taken from different orientations in the canopy, and Prats et al. [21] observed 7.1 ± 1.5 cm^2 for full expanded leaves.

Specific leaf area (SLA) was the highest for the young leaves in May (Figure 1), when leaves were expanding and growing quickly, with the tree investing less in dry matter per leaf [24]. SLA declined after the summer as a result of the increase in leaf mass by the retention of the photosynthetic compounds produced during the late spring and early summer period, when the mature leaves were totally expanded. SLA changes resulting from leaf structure modification are an underlying mechanism facilitating acclimatization to a drought that increases with declining rainfall [25,26]. SLA increased in January when low temperatures limited photosynthesis and reflect the remobilization or loss of resources, namely, before leaf abcission [22]. In March, the remaining one-year-old leaves were on average smaller and maintained their loss of photosynthetic capacity. Photosynthetic activity has been shown to decrease with leaf age in other sclerophylls, and this strategy fits expectations for drought-tolerant species such as evergreen Mediterranean oaks [27].

The seasonal leaf variation in *Q. suber* was previously reported by Passarinho et al. [28] for three periods, spring to early summer, summer, and autumn–winter; the dry matter content of new leaves was low, increased with age to the maximum level in August/September, with SLA values of 85.0 ± 39 cm^2/g (April–July), 74.4 ± 41 cm^2/g (July–September), and 89.9 ± 37 cm^2/g (September–February). The findings of the present study accord with these reported results.

4.2. Cuticular Wax Content

The occurrence of cuticular waxes in the leaves revealed a significant seasonal variation in total wax quantity, as expressed by the unit area of the leaf surface, as shown in Figure 2.

The lowest value was found in May, in the newly developed cork oak leaves, suggesting that leaf growth was more efficient than the wax lipids biosynthesis during this spring period, when the environmental conditions are mild in terms of temperature and water availability, and therefore, the leaves did not require special protection. This changed drastically during the summer months, when biosynthesis of leaf wax lipids was very dynamic, doubling from May to September. This is in line with the need to provide efficient protection to minimize water loss through the cuticular layer when all stomata are closed due to the summer abiotic stress [24,29–32]. In fact, one of the leaf-related strategies that plants may adopt to cope with drought conditions is increasing wax accumulation on the leaf surface [33,34]. In winter, wax content decreased, likely because of natural wax erosion and evaporation due to environmental factors such as light, temperature, rain, and wind [29,31,35,36] and then increased slightly during the remaining period. There are no previous studies on the variation in the cuticular waxes along the development of *Q. suber* leaves, and the results reported here are the first for this species. They accord with the few available studies on leaves of several species that showed that the cuticular wax content in the early development stage is lower than that observed in mature leaves, reaching maximum values after full leaf expansion, e.g., on the leaves of *Quercus robur* L. [30], *Fagus sylvatica* L. [29], three *Hosta* genotypes (*Hosta plantaginea* (Lam) Asch., *H. lancifolia* Engl. and *H.* 'Krossa Regal') [31], *Hedera helix* L. [35], and *Actinidia deliciosa* (A.Chev.) C.F.Liang & A.R.Ferguson [36].

The importance of a steadily present continuous outer leaf coverage as a protective strategy is shown by the self-healing of voids in the epicuticular wax layer on living plants, which is a dynamic process, at least in the early stages of leaf development [37–39]. Wax regeneration after removing the original epicuticular wax layer by peeling with a water-based glue applied to the leaf surface was observed on young leaves of *Prunus laurocerasus* L. [40] and on leaves with different ages of diverse species [37,39,41].

4.3. Cuticular Wax Composition

To our knowledge, no previous studies have been made on the seasonal changes in the composition of the cuticular waxes of *Q. suber* leaves. In this work, we have found a clear compositional variation, namely between the young and the mature leaves, i.e., between May and September (Figure 3), regarding the proportion of each chemical family. The cuticular wax composition of the young leaves (Table 1) included terpenes, sterols, linear long aliphatics (fatty acids, alcohols, and alkanes), and aromatics representing a proportion in the wax of 22.6%, 12.2%, 33.9%, and 8.7%, respectively. After the summer, the cuticular wax on the mature leaves increased significantly, mainly by pentacylic triperpenes, which represented nearly half of the total wax and about five times more than in spring leaves. The proportion of the four wax chemical classes of terpenes, sterols, linear long aliphatics, and aromatics in the mature leaves was, respectively, 60.5%, 9.6%, 15.7%, and 1.9%. Within each chemical family, the composition did not vary over time (Table 1).

In general, the *Q. suber* cuticular wax compositions found in this study were similar to that previously identified in 1-year-old *Q. suber* mature leaves sampled in March showing predominantly pentacyclic triterpenoids (61%–72% of the identified compounds, mainly lupeol) and aliphatic compounds (17%–23% of the identified compounds, mainly fatty acids (C_{30}, C_{28}, and C_{16})) [9].

A similar wax composition is found in other oaks. For instance, *Quercus coccifera* leaf wax contains pentacyclic triterpenoids as the most abundant class (49% and 61%), with germanicol and lupeol as major constituents, and very-long-chain aliphatic compounds from C_{20} to C_{51} (25%–34%) with *n*-alkanes as the main class comprising a homologous series from C_{25} to C_{32} [42]. The cuticular wax composition from the adaxial surface of fully expanded olive tree leaves (*Olea europaea* L.) was also mainly composed of pentacyclic triterpenoids (83% of total wax, with oleanolic acid as the major triterpenoid) with a ratio between the very long chain acyclic and the cyclic wax of 0.04 with *n*-alkanes as the main compound class (3% of the total wax) of the long-chain aliphatic fraction [43]. A high proportion of pentacyclic triterpenoids is found in the leaf cuticles of desert plants, for instance, in *Rhazya stricta* Decne., for which pentacyclic triterpenoids make up 85.2% of the total cuticle wax, while long-chain aliphatics with chain lengths ranging from C_{20} to C_{33} represent only 3.4% with alkanes as the main class (2.5% of the total wax, 73.6% of the aliphatics) [7].

In the present study, the solubilized cuticular waxes were not differentiated into the epicuticular waxes, deposited on cutin, and the intracuticular waxes found within the cutin matrix, since the extraction method was aimed at the total removal of the lipophilic soluble components [9,10].

Recent studies revealed a substantial compositional gradient between the epicuticular wax and the intracuticular wax: while epicuticular wax is only composed of very-long-chain fatty acid derivatives (e.g., alkanes, primary alcohols, fatty acids), intracuticular wax is composed of alicyclic compounds (e.g., triterpenoids and steroids) and long-chain aliphatic molecules [4,44]. The wax compositional differences between the cork oak leaves throughout their development cycle may be linked to the proportional occurrence of epicuticular and intracuticular waxes over time, which can be estimated by analyzing the ratio between cyclic compounds and linear-chain compounds. In the young leaves from May, the pentacyclic triterpenoids (26 $\mu g/cm^2$) in the wax mixture were less than aliphatic compounds (39 $\mu g/cm^2$), with a cyclic-to-acyclic component ratio of 0.7, while in the mature leaves, this ratio was 1.6 in September, 3.8 in December, and 2.4 in March. Since the pentacyclic triterpenoids are located almost exclusively in the intracuticular wax compartment, the increase in the ratio after summer shows an extensive deposition of intracuticular waxes. This is in association with the establishment of mechanical and thermal stability given by the triterpenoids [6] and the water loss barrier in the mature leaf cuticle as given by the intracuticular waxes [7]; however, a natural effect of wax weathering with leaf age and with environmental factors was superimposed on the seasonal pattern of variation. This was detected by comparing the lowest ratio value from September with the highest value from December.

5. Conclusions

The seasonal dynamics of cuticular waxes in the cuticle of cork oak leaves were characterized for the first time, and the initial investigation hypothesis of an effect of seasonality on leaf wax coverage and composition was confirmed. There was an increase in wax coverage on the newly expanded leaves to a maximum value in the fully expanded leaves, and a decline after the summer as a result of the natural weathering of the cuticular waxes with leaf age and with environmental factors. A subsequent renewal of the external layer followed, thereby reflecting the importance of a steadily present continuous outer leaf coverage in such harsh climatic conditions. The leaf cuticular wax layer was mainly composed of cyclic terpene compounds and linear long-chain aliphatic molecules. Triterpenoids dominated the wax mixture throughout the leaf life cycle, with a high proportion of lupeol. In the very long-chain aliphatic wax fraction, fatty acids showed the highest concentration variability along the leaf-development cycle, with a decrease in autumn but an increasing trend in the next spring. The results on the chemical dynamics of the cuticular waxes along the leaf cycle support the role of the intracuticular layer and the long-chain lipids as a transpiration barrier during the summer drought in the Mediterranean climate.

Author Contributions: Conceptualization, H.P. and I.M.; methodology, I.M. and R.S.; validation, H.P., I.M. and R.S.; experimental analysis, R.S.; writing—original draft preparation, I.M. and R.S.; writing—review and editing, H.P. and I.M. All authors have read and agreed to the published version of the manuscript.

Funding: This research was funded by Fundação para a Ciência e a Tecnologia (FCT) through funding from the Forest Research Centre (UIDB/00239/2020). Funding for this work was also provided by a doctoral scholarship from the FCT SUSFOR Doctoral Programme (PD/BD/128259/2016).

Institutional Review Board Statement: Not applicable.

Informed Consent Statement: Not applicable.

Data Availability Statement: The raw data are available upon request to the corresponding author.

Acknowledgments: Rita Simões acknowledges a doctoral scholarship from FCT with the SUSFOR Doctoral Programme (PD/BD/128259/2016). The cork oak provenance field trials were funded by the European Commission (FAIR1-CT-95-0202) and national programs (PBIC/AGR/2282/95, PAMAF 4027, PRAXIS/3/3.2/Flor/2110/95). We thank Ana Rodrigues for her technical assistance in field sampling.

Conflicts of Interest: The authors declare no conflict of interest.

References

1. Jetter, R.; Kunst, L.; Samuels, A.L. Composition of plant cuticular waxes. In *Biology of the Plant Puticle*; Riederer, M., Müller, C., Eds.; Blackwell: Oxford, MS, USA, 2006; pp. 145–181.
2. Buschhaus, C.; Jetter, R. Composition differences between epicuticular and intracuticular wax substructures: How do plants seal their epidermal surfaces? *J. Exp. Bot.* **2011**, *62*, 841–853. [CrossRef] [PubMed]
3. Yeats, T.H.; Rose, J.K.C. The formation and function of plant cuticles. *Plant Physiol.* **2013**, *163*, 5–20. [CrossRef] [PubMed]
4. Zeisler-Diehl, V.; Müller, Y.; Schreiber, L. Epicuticular wax on leaf cuticles does not establish the transpiration barrier, which is essentially formed by intracuticular wax. *J. Plant Physiol.* **2018**, *227*, 66–74. [CrossRef] [PubMed]
5. Seufert, P.; Staiger, S.; Arand, K.; Bueno, A.; Burghardt, M.; Riederer, M. Building a barrier: The influence of different wax fractions on the water transpiration barrier of leaf cuticles. *Front. Plant Sci.* **2022**, *12*, 766602. [CrossRef]
6. Tsubaki, S.; Sugimura, K.; Teramoto, Y.; Yonemori, K.; Azuma, J.I. Cuticular membrane of Fuyu persimmon fruit is strengthened by triterpenoid nano-fillers. *PLoS ONE* **2013**, *8*, e75275. [CrossRef]
7. Schuster, A.-C.; Burghardt, M.; Alfarhan, A.; Bueno, A.; Hedrich, R.; Leide, J.; Thomas, J.; Riederer, M. Effectiveness of cuticular transpiration barriers in a desert plant at controlling water loss at high temperatures. *AoB Plants* **2016**, *8*, plw027. [CrossRef]
8. Pereira, H. *Cork: Biology, Production and Uses*; Elsevier: Amsterdam, The Netherlands, 2007.
9. Simões, R.; Rodrigues, A.; Ferreira-Dias, S.; Miranda, I.; Pereira, H. Chemical composition of cuticular waxes, pigments and morphology of leaves of *Quercus suber* trees from different provenances. *Plants* **2020**, *9*, 1165. [CrossRef]
10. Simões, R.; Miranda, I.; Pereira, H. Chemical composition of leaf cutin in six *Quercus suber* provenances. *Phytochemistry* **2021**, *181*, 112570. [CrossRef]
11. Pereira, J.S.; Beyschlag, G.; Lange, O.L.; Beyschlag, W.; Tenhunen, J.D. Comparative phenology of four mediterranean shrub species growing in Portugal. In *Plant Response to Stress*; Tenhunen, J.D., Catarino, F.M., Lange, O.L., Oechel, W.C., Eds.; Springer: Berlin/Heidelberg, Germany, 1987; pp. 503–514.
12. Escudero, A.; Del Arco, J.M.; Sanz, I.C.; Ayala, J. Effects of leaf longevity and retranslocation efficiency on the retention time of nutrients in the leaf biomass of different woody species. *Oecologia* **1992**, *90*, 80–87. [CrossRef]
13. Pausas, J.G.; Pereira, J.S.; Aronson, J. The tree. In *Cork Oak Woodlands on the Edge: Conservation, Adaptive Management, and Restoration*; Aronson, J., Pereira, J.S., Pausas, J.G., Eds.; Island Press: Washington, DC, USA, 2009; pp. 11–21.
14. Oliveira, G.; Correia, O.; Martins-Loução, M.A.; Catarino, F. Phenological and growth patterns of the Mediterranean oak *Quercus suber* L. *Trees* **1994**, *9*, 41–46. [CrossRef]
15. Fialho, C.; Lopes, F.; Perreira, H. The effect of cork removal on the radial growth and phenology of young cork oak trees. *For. Ecol. Manag.* **2001**, *141*, 251–258. [CrossRef]
16. Sampaio, T.; Gonçalves, E.; Patrício, M.S.; Cota, T.M.; Almeida, M.H. Seed origin drives differences in survival and growth traits of cork oak (*Quercus suber* L.) populations. *For. Ecol. Manag.* **2019**, *448*, 267–277. [CrossRef]
17. Varela, M.C. *European Network for the Evaluation of Genetic Resources of Cork Oak for Appropriate Use in Breeding and Gene Conservation Strategies*; INIA: Lisbon, Portugal, 2020.
18. Kolattukudy, P.E.; Agrawal, V.P. Structure and composition of aliphatic constituents of potato tuber skin (suberin). *Lipids* **1974**, *9*, 682–691. [CrossRef]
19. Ferreira, J.P.; Miranda, I.; Sen, A.; Pereira, H. Chemical and cellular features of virgin and reproduction cork from *Quercus variabilis*. *Ind. Crops Prod.* **2016**, *94*, 638–648. [CrossRef]
20. Grant, O.M.; Tronina, L.; Ramalho, J.C.; Besson, C.K.; Lobo-Do-Vale, R.; Pereira, J.S.; Jones, H.G.; Chaves, M.M. The impact of drought on leaf physiology of *Quercus suber* L. trees: Comparison of an extreme drought event with chronic rainfall reduction. *J. Exp. Bot.* **2010**, *61*, 4361–4371. [CrossRef]
21. Prats, K.A.; Brodersen, C.R.; Ashton, M.S. Influence of dry season on *Quercus suber* L. leaf traits in the Iberian Peninsula. *Am. J. Bot.* **2019**, *106*, 656–666. [CrossRef]
22. Garcia-Plazaola, J.I.; Faria, T.; Abadía, J.; Abadía, A.; Chaves, M.M.; Pereira, J.S. Seasonal changes in xanthophyll composition and photosynthesis of cork oak (*Quercus suber* L.) leaves under Mediterranean climate. *J. Exp. Bot.* **1997**, *48*, 1667–1674. [CrossRef]
23. Mediavilla, S.; Martín, I.; Babiano, J.; Escudero, A. Foliar plasticity related to gradients of heat and drought stress across crown orientations in three Mediterranean *Quercus* species. *PLoS ONE* **2019**, *14*, e0224462. [CrossRef]

24. Dwyer, J.M.; Hobbs, R.J.; Mayfield, M.M. Specific leaf area responses to environmental gradients through space and time. *Ecology* **2014**, *95*, 399–410. [CrossRef]

25. Aranda, I.; Pardos, M.; Puértolas, J.; Jiménez, M.D.; Pardos, J.A. Water-use efficiency in cork oak (*Quercus suber*) is modified by the interaction of water and light availabilities. *Tree Physiol.* **2007**, *27*, 671–677. [CrossRef]

26. Gouveia, A.C.; Freitas, H. Modulation of leaf attributes and water use efficiency in *Quercus suber* along a rainfall gradient. *Trees* **2009**, *23*, 267–275. [CrossRef]

27. Ramírez-Valiente, J.A.; Sánchez-Gómez, D.; Aranda, I.; Valladares, F. Phenotypic plasticity and local adaptation in leaf ecophysiological traitsof 13 contrasting cork oak populations under different water availabilities. *Tree Physiol.* **2010**, *30*, 618–627. [CrossRef] [PubMed]

28. Passarinho, J.A.P.; Lamosa, P.; Baeta, J.P.; Santos, H.; Ricardo, C.P.P. Annual changes in the concentration of minerals and organic compounds of *Quercus suber* leaves. *Physiol. Plant.* **2006**, *127*, 100–110. [CrossRef]

29. Prasad, R.B.N.; Gülz, P.G. Surface structure and chemical composition of leaf waxes from *Quercus robur* L., *Acer pseudoplatanus* L. and *Juglans regia* L. *Z. Naturforsch. C* **1990**, *45*, 813–817. [CrossRef]

30. Gülz, P.-G.; Müller, E. Seasonal variation in the composition of epicuticular waxes of *Quercus robur* leaves. *Z. Naturforsch. C* **1992**, *47*, 800–806. [CrossRef]

31. Jenks, M.A.; Gaston, C.H.; Goodwin, M.S.; Keith, J.A.; Teusink, R.S. Seasonal variation in cuticular waxes on hosta genotypes differing in leaf surface glaucousness. *HortScience* **2002**, *37*, 637–677. [CrossRef]

32. Fang, Y.; Xiong, L. General mechanisms of drought response and their applicationin drought resistance improvement in plants. *Cell. Mol. Life Sci.* **2015**, *72*, 673–689. [CrossRef]

33. Zhang, J.Y.; Broeckling, C.D.; Blancaflor, E.B.; Sledge, M.K.; Sumner, L.W.; Wang, Z.Y. Overexpression of WXP1, a putativeMedicago truncatula AP2 domain-containing transcription fac-tor gene, increases cuticular wax accumulation and enhancesdrought tolerance in transgenic alfalfa (*Medicago sativa*). *Plant J.* **2005**, *42*, 689–707. [CrossRef]

34. Cameron, K.D.; Teece, M.A.; Smart, L.B. Increased accumu- lation of cuticular wax and expression of lipid transfer protein in response to periodic drying events in leaves of tree tobacco. *Plant Physiol.* **2006**, *140*, 176–183. [CrossRef]

35. Hauke, V.; Schreiber, L. Ontogenetic and seasonal development of wax composition and cuticular transpiration of ivy (*Hedera helix* L.) sun and shade leaves. *Planta* **1998**, *207*, 67–75. [CrossRef]

36. Celano, G.; D'Auria, M.; Xiloyannis, C.; Mauriello, G.; Baldassarre, M. Composition and seasonal variation of soluble cuticular waxes in *Actinidia deliciosa* leaves. *Nat. Prod. Res.* **2006**, *20*, 701–709. [CrossRef] [PubMed]

37. Neinhuis, C.; Koch, K.; Barthlott, W. Movement and regeneration of epicuticlar waxes through plant cuticles. *Planta* **2001**, *213*, 427–434. [CrossRef] [PubMed]

38. Koch, K.; Neinhuis, C.; Ensikat, H.-J.; Barthlott, W. Self assembly of epicuticular waxes on living plant surfaces imaged by atomic force microscopy (AFM). *J. Exp. Bot.* **2004**, *55*, 711–718. [CrossRef] [PubMed]

39. Koch, K.; Bhushan, B.; Ensikat, H.-J.; Barthlott, W. Self-healing of voids in the wax coating on plant surfaces. *Philos. Trans. Royal Society A* **2009**, *367*, 1673–1688. [CrossRef] [PubMed]

40. Jetter, R.; Schäffer, S. Chemical composition of the *Prunus laurocerasus* leaf surface. dynamic changes of the epicuticular wax film during leaf development. *Plant Physiol.* **2001**, *126*, 1725–1737. [CrossRef]

41. Huth, M.A.; Huth, A.; Koch, K. Morphological diversity of β-diketone wax tubules on *Eucalyptus gunnii* leaves and real time observation of self-healing of defects in the wax layer. *Aust. J. Bot.* **2018**, *66*, 313–324. [CrossRef]

42. Bueno, A.; Sancho-Knapik, D.; Gil-Pelegrín, E.; Leide, J.; Peguero-Pina, J.J.; Burghardt, M.; Riederer, M. Cuticular wax coverage and its transpiration barrier properties in *Quercus coccifera* L. leaves: Does the environment matter? *Tree Physiol.* **2020**, *40*, 827–840. [CrossRef]

43. Huang, H.; Burghardt, M.; Schuster, A.; Leide, J.; Lara, I.; Riederer, M. Chemical compositions and water permeabilities of fruit and leaf cuticles of *Olea europaea* L. *J. Agric. Food Chem.* **2017**, *65*, 8790–8797. [CrossRef]

44. Jetter, R.; Riederer, M. Localization of the transpiration barrier in the epi- and intracuticular waxes of eight plant species: Water transport resistances are associated with fatty acyl rather than alicyclic components. *Plant Physiol.* **2016**, *170*, 921–934. [CrossRef]

Article

Highlighting a New Morphospecies within the *Dialium* Genus Using Leaves and Wood Traits

Robin Doucet [1,*], Gaël Bibang Bengono [1,2,*], Marius Ruwet [1,*], Isabelle Van De Vreken [1], Brieuc Lecart [1], Jean-Louis Doucet [1], Juan Antonio Fernandez Pierna [3], Philippe Lejeune [1], Benoit Jourez [3], Alain Souza [2] and Aurore Richel [1]

[1] Forest Is Life, TERRA Teaching and Research Centre, Gembloux Agro-Bio Tech, University of Liege, B-5030 Gembloux, Belgium

[2] Laboratoire Electrophysiologie-Pharmacologie, Unité de recherche Agrobiologie, Université des Sciences et Techniques de Masuku, Franceville B-901, Gabon

[3] Centre Wallon de Recherche Agronomique (CRA-W), B-5030 Gembloux, Belgium

* Correspondence: robin.doucet@uliege.be (R.D.); gael.bibangbengono@student.uliege.be (G.B.B.); marius.ruwet@hotmail.com (M.R.)

Abstract: During inventories of lesser-known timber species in eastern Gabon, a new *Dialium* morphospecies was discovered. To discriminate it from the two other 2–5 leaflets *Dialium* species, 25 leaf traits were measured on 45 trees (16 *Dialium pachyphyllum*, 14 *Dialium lopense*, 15 *Dialium* sp. nov.). Nine wood chemical traits, as well as infrared spectra, were also examined on harvestable trees (four *Dialium pachyphyllum* and four *Dialium* sp. nov.). This study revealed seven discriminant leaf traits that allowed to create a field identification key. Nine significant differences (five in sapwood and four in heartwood) in terms of wood composition were highlighted. The use of the PLS-DA technique on FT-IR wood spectra allowed to accurately identify the new morphospecies. These results provide strong support for describing a new species in this genus. Implications for sustainable management of its populations are also discussed.

Keywords: new species; leaves; morphology; wood; properties; chemical composition; spectroscopy; chemometrics; unconventional species

Citation: Doucet, R.; Bibang Bengono, G.; Ruwet, M.; Van De Vreken, I.; Lecart, B.; Doucet, J.-L.; Fernandez Pierna, J.A.; Lejeune, P.; Jourez, B.; Souza, A.; et al. Highlighting a New Morphospecies within the *Dialium* Genus Using Leaves and Wood Traits. *Forests* **2022**, *13*, 1339. https://doi.org/10.3390/f13081339

Academic Editors: Vicelina Sousa, Helena Pereira, Teresa Quilhó and Isabel Miranda

Received: 1 July 2022
Accepted: 16 August 2022
Published: 22 August 2022

Publisher's Note: MDPI stays neutral with regard to jurisdictional claims in published maps and institutional affiliations.

1. Introduction

Despite efforts implemented by Central African countries in sustainable forest management, a progressive depletion of some timber flagship species has been observed [1,2]. This is mainly due to an extremely selective exploitation that focuses on a dozen species [3]. Most of them are light-demanding and logging impacts (canopy openness of less than 10% every cutting cycle of 25 years) are too small to positively influence their regeneration [4,5]. This depletion of species jeopardizes forest biodiversity and the future of productive forests.

The timber sector that covers 53 million ha (29 million with sustainable management plan) contributes significantly to employment and to the Gross National Product (GNP), 4% on average [3]. Apart from conservation areas, legal selective logging appears to be the least degrading use of forest ecosystems [6,7]. Moreover, some certified companies (totaling five million ha for the label Forest Stewardship Council, FSC [3]) go further by bringing significant improvement to life condition of workers, implementing Reduce Impact Logging (RIL) techniques, regulating hunting or enriching the forest [8–10]. By assigning an economic value to a forest, while having a limited impact on it, production forests can strengthen the network of 18 million hectares of protected area [6,7,11] and thus prevent land-use conversion.

Central African states plan to increase areas allocated to timber production and the intensity of logging by 50% by 2030 [3], with a an average logging intensity of 8.7–9.5 m^3/ha. In order to avoid depleting those forests, it is more necessary than ever to reduce the pressure on traditional species and to redirect the exploitation towards lesser-known timber species [2,10,12] with: (i) a high wood quality, (ii) lower light requirements, and thus, a better regeneration than traditional light-demanding timber species.

Some species of the genus *Dialium* could meet such requirements [13]. They reach high densities in evergreen forests, have sustained regeneration, and very durable wood. However, the systematics of the genus remains unclear [14]. In the forests of central and eastern Gabon, two species of *Dialium* can reach large dimensions and could interest the logging industry: *Dialium pachyphyllum* Harms and *Dialium lopense* Breteler [14]. In the FSC-certified concession of Precious Woods operating in this region, a group of similar individuals appears to have differences in morphological traits from previously known species, especially on the leaves. [15]. Those morphological distinctions suggest the presence of an undescribed *Dialium* morphospecies [16].

The species identification issue is not a recent topic and many tree species continue to be discovered in Central Africa [17–19]. Recently, a species from the same genus, *Dialium heterophyllum* M. J. Falção & Mansano, was discovered in the South Amazonian Basin [20]. An accurate identification of species is crucial to understand their ecology and their timber properties. Therefore, the aim of this study is to evaluate the feasibility of differentiation between closely related species based on vegetative characteristics. More specifically, it aims to: (i) objectify morphological leave traits that discriminate the new morphospecies from the others, (ii) to determine chemical composition variation in the wood of harvestable species, and (iii) to verify whether this chemical composition, using Fourier Transformed Infrared (FT-IR) Spectra and chemometrics, can allow discriminating this new morphospecies.

2. Materials and Methods

2.1. Study Genus

Although there is no consensus among botanists, the *Dialium* genus (*Fabaceae, Dialioideae*) could include 44 species, 22 of which are endemic to the Guinean-Congolese region. The differences between species are sometimes tenuous, and descriptions are based on a limited number of individuals. All are hermaphroditic and dispersed by animals with abundant regeneration in the understory of evergreen forests [14]. Some species can reach 40 m in height and a meter in diameter. These species are generally grouped in Gabon under the name of "omvong" and typically have 3–5 leaflets. They have durable woods, resistant to fungi, termites, marine borers, and insects [13,21]. The wood can be used in parquetry, cabinetry, for exterior cladding, bridge construction and other heavy industrial uses [14].

2.2. Study Area

The study was performed in in the logging concession granted to Precious Woods-Compagnie Equatoriale des Bois A.S. (PW-CEB). This FSC-certified logging company is located in Bambidie, Lastoursville (Gabon) (0°41.65′ S–12°59.01′ E). The average temperature and precipitation are 25 °C and 1700 mm, respectively [22]. The forest is evergreen and its canopy is dominated by *Aucoumea klaineana* Pierre, *Scyphocephalium mannii* (Benth.) Warb., and *Julbernardia pellegriniana* Troupin. PW-CEB wishes to diversify its production to guarantee the maintenance of these activities in the long term. From this perspective, the possibility of valorizing *Dialium* wood has been analyzed. During field inventories, 3 morphospecies of "omvong" have been observed: *Dialium pachyphyllum, Dialium lopense*, and a new one *Dialium* sp. nov. All are present in the dynamic observation permanent plots set up by the Dynafac network (https://www.dynafac.org (accessed on 1 July 2022)).

2.3. Herbarium Collection and Description

Samples (Figure 1) of *D. pachyphyllum*, *D. lopense*, and *D.* sp. nov. were collected in the field. In addition, reference herbaria from the botanical garden of Meise (Belgium) were studied (Appendix A). They were selected based on the quality of their identification by botanists specialized in the *Fabaceae* family. A total of 45 individuals were analyzed, including 29 individuals (diameter measured at 1.3 m, height >10 cm) collected in the field and 16 individuals from the Meise collection. For each sample, 25 leaf morphological traits were observed (Table 1). All traits were not systematically visible and/or available (e.g., absence of basal leaflet).

Figure 1. Field herbarium examples of *Dialium pachyphullum* Harms (**a**), *Dialium* sp. nov. (**b**) and *Dialium lopense* Breteler (**c**).

Table 1. Leaf morphological traits investigated.

Trait	Units/Modalities
Petiole type	"cylindrical"; "bulged"; "canaliculated"
Petiole length	mm
Rachis length	mm
Foliole number	-
Pilosity on the upper face	"pubescent"; "glabrous"
Pilosity on the above face	"pubescent"; "glabrous"
Curved margin	"yes"; "no"
Acumen shape	"oblong"; "elliptical"; "lanceolate"; "ovate"; "obovate"; "oblanceolate"; "falciform"
Asymmetric leaflet base	"yes"; "no"
Tertiary venation	"tight and prominent"; "discrete"
Translucent dots	"yes"; "no"
Terminal leaflet lamina shape	"no acumen"; "sharp"; "tapered"; "retuse"; "obtuse"; "truncated"; "emarginated"
Terminal leaflet lamina length	Mm
Terminal leaflet lamina width	mm
Basal leaflet lamina shape	"no acumen"; "sharp"; "tapered"; "retuse"; "obtuse"; "truncated"; "emarginated"
Basal leaflet lamina length	Mm
Basal leaflet lamina width	Mm
Coriaceous lamina	"yes"; "no"
Twisted lamina	"yes"; "no"
Petiolule length	mm
Ratio petiole/rachis	-
Ratio length/width of terminal leaflet	-
Ratio length/width of basal leaflet	-
Ratio basal leaflet length/terminal leaflet length	-
Ratio terminal leaflet length/petiolule length	-

2.4. Wood Material

For each of the studied species, *D. pachyphyllum* and *D.* sp. nov., four trees were cut following national regulations. Observed stems of *D. lopense* did not reach the legal minimum cutting diameter of 70 cm. Its wood properties were therefore not investigated. For each tree, altitude, floor slope, stem perimeter at the base of the log (measured at 1.3 m height when possible or above buttresses), position in the stand, and tree height were recorded (Table 2). Tree position in the stand was assessed from the ground. A tree with a completely shaded crown was considered as "dominated", a crown situated in the canopy and partially lit as "canopy", and a fully sunny crown above the canopy as "emergent". Wood disk samples, 10 cm thick, were collected at two different heights: at the bottom of the stem, corresponding to base of the log after felling and removing buttresses which are too important for the sawing or rotten heart (h1), and at the top of the stem corresponding to the first living branch (h2). To avoid sample contamination by fuel and oil of the chainsaw, wooden disk faces were planned in the society carpentry. Finally, 100 g of both sapwood and heartwood (taken at least 15 cm away from the pith and 10 cm away from sapwood, to avoid both juvenile wood and potential transition wood) were cut in each disk.

Table 2. Dendrometric and environmental parameters of selected trees. Position: Em = "Emergent" and Can = "Canopy"; altitude (m); slope (°); p is stem perimeter (m); hp is the height at which the perimeter was measured (m); h1 is the sampling height at the base of the stem (m); h2 is the sampling height at the top of the stem (m); ht is the total tree height (m).

Species	Tree	Position	Altitude	Slope	p	hp	h1	h2	ht
	1	Em	369	0	3.30	2.2	1.3	12.0	43.0
D. pachyphyllum	2	Can	358	5	3.10	3.1	1.6	10.2	41.0
	3	Em	360	0	2.90	6.1	2.0	17.1	42.0
	4	Em	365	0	2.86	1.3	1.7	14.0	46.5
	1	Can	348	0	3.60	5.1	2.7	18.9	43.0
D. sp. nov.	2	Em	343	0	2.60	3.3	1.9	19.6	39.0
	3	Em	347	30	2.65	2.4	2.5	21.0	42.0
	4	Can	345	0	2.55	2.8	1.8	16.7	44.5

2.5. Wood Sample Preparation

2.5.1. For Chemical Analyses and Moisture Content

Wood samples were first planed using an electric hand planer Black et Decker© DN 710 and shavings were collected by a dust collector. Shavings were then ground into <1 mm powder using a Fritsch *Pulverisette* universal cutting mill and stored at −18 °C before extraction. Moisture Content (MC) was assessed before each experiment by measuring the mass loss of 1 g of powder drying at 105 °C; this loss of mass was expressed as a percentage of initial mass.

2.5.2. For FTIR Measurements

For each of the 8 trees (4 *D. pachyphyllum* and 4 *D.* sp. nov.) described in Table 1, thin shavings were collected on 5 radially distributed heartwood strips (section 15 × 25 mm). Those strips were cut at the lower sampling height (h1). Shavings were produced using a Veritas© low-angle jack plane and a shooting board. Those thin shavings were then stored in a standard atmosphere of 20 ± 2 °C and 65 ± 5% air relative moisture.

2.6. Wood Primary Metabolites Content and Mineral

For each morphospecies, the variation of primary metabolites was investigated between heartwood and sapwood. Samples were collected at two heights h1 and h2 (Table 2). For each species, wood and sampling height combinations, experimental repetitions are presented in Table 3. All metabolites are expressed as a percentage of wood dry matter (DM). DM was calculated by subtraction of MC.

Table 3. Experimental iterations on the four selected trees for each species and each modality combination of the studied variables, i.e., wood and sampling height.

	Experimental Repetition (n)				
Tree	**Hemicellulose**	**Cellulose**	**Lignin**	**Nitrogen**	**Ash**
1	2	2	2	3	3
2	2	2	2	3	3
3	2	2	2	-	-
4	2	2	2	-	-

2.6.1. Cellulose, Hemicellulose and Lignin Content

Fibers and lignin content were assessed following the Van Soest method [23–25] using a FOSS© FT 122 FiberTech tm. This method consists in successive matter treatments to remove sequentially soluble hemicellulose, cellulose, and lignin.

2.6.2. Ash Content

Ash content was determined as the mass residue of 1 g wood anhydrous powder after calcining. The calcining was processed according to [26]. It consisted in a first ramp from 21 °C to 575 °C within 2 h, including two temperature stages (102 °C for 12 min and 250 °C for 30 min). Then, a constant heating was processed at the reached temperature during 3 h.

2.6.3. Nitrogen Content

Wood nitrogen content was measured with the Kjeldahl method [27]. A hundred mg of wood powder was weighted on a Nitrogen-free paper and placed into the test tube FOSS Tecator digester for the mineralization. The digestion took place during 2 h, at 360 °C with 7.2 mL of concentrate H_2SO_4 and a Kjeltabs tablette (1.5 g K_2SO_4 + 0.045 g $CuSO_4.5H_2O$ + 0.045 g TiO_2). Tubes were then cooled and titrated with H_2SO_4 0.02 N in a FOSS© Kjeltec 2300.

2.6.4. Silica Content

The silica content was measured on 1 g of wood ash (see the Ash content protocol) by the gravimetric measure after hydrofluoric acid (HF) attack on $HClO_4/HNO_3$ acid-insoluble compounds [28]. Due to economic issues, this test was only applied on 3 samples (from 3 different trees) of sapwood and heartwood, at the lowest sample height (h1), for both *D. sp. nov.* and *D. pachyphyllum.*

2.7. Ethanol-Water Extracts Charaterisation

Extracts were prepared by maceration. After a preliminary test based on the Oreopoulou et al., (2019) review [29], the following extraction parameters were used: five grams of wood powder (W_M) (with a known MC) were incorporated into 100 mL of an ethanol-water (70–30, *v/v*) solution. Extraction was processed at 50 °C, using a rotary table for 2 h. Extracts were then vacuum filtered using a Büchner funnel and a Whatman© cellulose filter, 11 µm pore size. The ethanol was then evaporated using a rotary evaporator and the filtrate was finally freeze-dried and the dry mass of extracts was recorded (E_{DM}). The extraction yields were finally measured using E_{DM}. All extractions were processed once for each combination of variables modalities (species, heartwood and sapwood, and sampling height) for each tree.

2.7.1. Phenolic Content

The Phenolic Content (PC) was measured using the Folin–Ciocalteu method [30] with some modifications. The tested solution was prepared by dissolving 50 µg of dry extracts into 10 mL of a methanol-water (70–30, *v/v*) solution. A volume of 100 µL of the tested solution was added into 500 μL of the Folin–Ciocalteau diluted in water (1:10, *v/v*). After 2 min, 2 mL of a 20% Na_2CO_3 solution was added. The mixture was then stirred and kept in a dark place at local temperature for 30 min. Finally, the mixture absorbance was recorded at 750 nm. The PC is expressed as mg gallic acid equivalent per g of dry extract (GAE). The GAE was obtained from a calibration curve of standard gallic acid dilution (from 0.5 g/L to 0.05 g/L). The PC determination was processed in triplicate for each extract.

2.7.2. Condensed Tannins Content

The Condensed Tannins Content (CTC) was measured using the Vanillin assay as mentioned by [31] with some modifications. The tested solution was prepared by dissolving 50 μg of dry extract into 10 mL of a methanol-water solution (70–30, *v/v*). In a cuvette, 500 μL of tested solution was added to 1500 μL of a vanillin-methanol solution (4–96, m–v). Then, 750 μL of HCl 37% was added. After stirring the mixture, the reaction took place in a dark place at 20 °C for 20 min. Finally, the absorbance of the solution was recorded at 550 nm. The CTC is expressed as mg catechin Equivalent per g of dry extract (CE). The CE was obtained from a calibration curve of standard catechin dilution (from 0.5 g/L to 0.05 g/L). The CTC determination was processed in triplicate for each extract.

2.8. FT-IR Acquisition

FT-IR spectra acquisition was processed using a Bruker VERTEX 70 paired to a Bruker Platinum Attenuated Total Reflectance (ATR) accessory. ATR principle is based on a reflectance measurement; then, roughly prepared samples such as wood transversal shavings can be used. These have been therefore directly pressed under the crystal diamond of the device for the scans. Spectra were obtained by averaging 32 scans with a resolution of 2 cm^{-1} in a range of 4000 cm^{-1} to 400 cm^{-1}. For each tested heartwood strip, 3 different wood shavings were scanned and considered as 3 repetitions. Spectra were then preprocessed using first a standard normal variate and, then, the first derivative with the Savitzky–Golay algorithm (polynomic order = 1; window = 7) using the *mdatools* package [32]. Regions of 800–1775 and 2810–3000 cm^{-1} were chosen to focus the analyses on relevant spectral information [33,34]. Average preprocessed spectra were calculated for each species and, then, *D.* sp. nov. spectra were subtracted from that of *D. pachyphyllum* to highlight the average differences between species.

2.9. Statistical Analyses

All statistical analyses and graphics were produced using RStudio (v 1.2.5001) [35]. Before the analyses, aberrant values (5 for lignin and 2 for hemicellulose) due to manipulation errors were removed from the datasets.

2.9.1. Conditional Inference Tree

To highlight most discriminant leaf morphological traits that allow morphospecies identification, the Conditional Inference Tree (CIT) was used (ctree function from the *partykit* package [36]). This recursive partitioning method consists in testing the independence of all explicative variables (25 measured leaf traits) and the response variable (species determination) at each partitioning step. The variable that had the strongest association to the response value, and an association test *p*-value lower than the selected alpha (0.05), was selected and the population was segmented. Each subpopulation was submitted to the same procedure until getting homogenous populations (null hypothesis of the independence test accepted for all variables in each subpopulation). To counteract the negative effect of the "multiple comparisons problem" due to inferences, the Bonferroni correction was systematically applied.

2.9.2. PLS-DA

The Partial Least Square Regression Discriminant Analysis (PLS-DA) was used to discriminate *D. pachyphyllum* and *D.* sp. nov. based on their wood shaving FT-IR spectra.

The outlier detection was carried out using Principal Component Analysis (PCA). Spectra were projected on PCs, score distance, and orthogonal distance were measured on the 5 first PCs accounting for 81.5% of the total variance. The region of acceptance proposed by [37] was used on distances and no outliers were found. Then, the dataset was divided into a training set (containing all preprocessed spectra from 3 trees of each species) and a validation set (containing the remaining spectra from the last trees of each species). Each set was composed by an explicative X matrix (FT-IR spectra) and a response Y matrix (species affiliation). To process PLS on a categorical Y, the matrix was first coded as dummy blocks. Then, using *plsda* and *perf* functions from the *mixomics* package [38], a PLS-DA model was fitted. The PLS-DA procedure aims to create a linear combination of variable from X, called Latent Variables (LVs), that maximize the covariance with Y. Scores of samples are calculated for those new variables and projected on the new LVs n-dimensional space, where n corresponds to the number of LVs. For each Y modality (*D.* sp. nov. and *D. pachyphyllum*) parameter (such as centroids) of the n-dimensional scores, a distribution is calculated. Using those parameters and a distance metric, it is possible to measure the distribution proximity of a new sample projected in the LVs space. Those distances to distributions are finally used to classify the samples. In this study, the Mahalanobis distance metric, which takes into account the shape of the distribution [39], was chosen. The optimal number of LVs to select (that explain most of the Y variance while avoiding overfitting) was assessed using a k-fold (fold = 5) cross-validation repeated a hundred times. This cross-validation processed tree operations for a hundred times: (i) five samples were randomly selected in the training set, (ii) the PLS-DA model was fitted on the remaining dataset for 20 LVs, (iii) the five samples were predicted by the model and the Overall Classification Error rate (OCE) is computed for each LVs of the model. After all repetitions, mean and standard deviation of OCE were measured for each LV. The lowest LV number that minimized the OCE was selected. During this procedure, the Variable Importance in Projection score (VIP score) confidence interval ($\alpha = 0.05$) was calculated from its distribution for each variable on each selected LV. This interval was used to verify whether variable VIP scores were significantly higher than 1, the commonly used threshold to assess the importance of a variable in prediction [40]. Within LVs, VIP scores' 95th percentile were also measured to identify variables that were situated among the 5% most important in the considered LV. After fitting the model on the training dataset, the validation set was predicted by it and a confusion matrix has been built. A confusion matrix allows to measure model accuracy and classes sensitivity/specificity.

2.9.3. Variance Analyses

Analysis of variance (ANOVA) was used to investigate leave trait differences between morphospecies but also the influence of wood (sapwood and heartwood) and sampling height (h1 and h2) on wood chemical properties. To confirm ANOVA preconditions, normality and homoscedasticity, the Ryan–Joiner and the Levene tests were used, respectively. If means were unequal, they were compared by using the post-HOC *t*-test (t). If the homoscedasticity precondition was not met, the non-parametric test of Kruskall–Wallis (χ^2) was used and medians were compared by using the Wilcoxson–Mann–Whitney (W) test. The selected α was 0.05 for all analyses excepted when variable modalities to compare were superior to two. In this case, the Bonferroni correction was applied to limit the family-wise error rate. Those analyses were conducted using the *rstatix* package [41].

3. Results
3.1. Leaf Morphological Traits
3.1.1. Conditional Inference Tree

The algorithm only selected qualitative variables for the CIT construction (Figure 2). The first partition, based on venation type (variable independence test: $p < 0.001$) allowed discriminating all 15 *D. pachyphyllum* individuals that had a discrete tertiary venation from the 25 other trees. The second partition, based on terminal leaflet's acumen shape (variable independence test: $p = 0.024$) produced a pure node with only *D. lopense* and a node with 68% of *D.* sp. and 32% of *D. lopense*. All individuals that presented leaves with a tight, prominent tertiary nervation and an obtuse/no acumen can therefore be predicted as *D. lopense*. However, trees which presented leaves with tight, prominent tertiary venation and a sharp, tapered or retuse acumen had 68% probability to belong to the *D.* sp. and 32% to *D. lopense*.

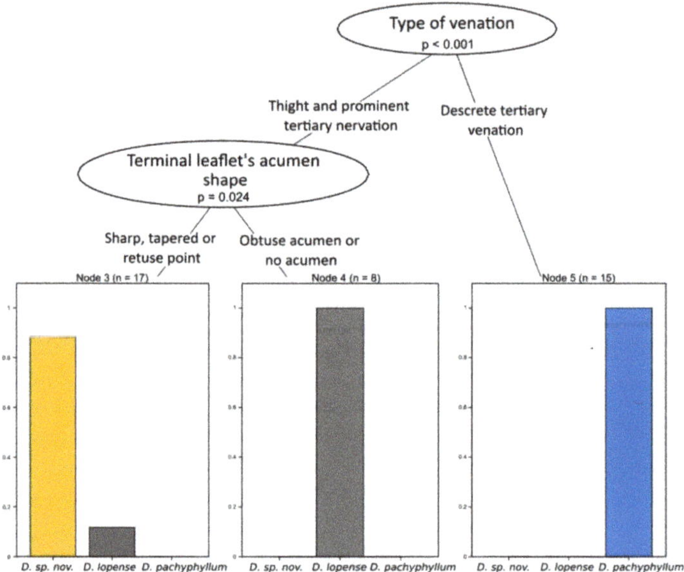

Figure 2. Conditional inference tree using the 25 leaf morphological variables to discriminate the 40 individuals for which all leaf traits could be measured.

3.1.2. Variance Analyses

Variance analyses on quantitative variables highlighted many significant differences between *D. lopense* and *D.* sp. nov. (Figure 3). The mean width/length ratio of the basal leaflet was the most significant (t = 3.31, $p = 0.0043$), with 0.54 ± 0.08 and 0.45 ± 0.06 for *D. lopense* and *D.* sp. nov., respectively. The mean terminal leaflet width also differed (t = 2.96, $p = 0.0073$) between the two morphospecies, 51.93 ± 13.48 mm and 39.4 ± 8.63 mm respectively. The mean width/length ratio on terminal leaflet is higher (t = 2.64, $p = 0.015$) for *D. lopense* (0.44 ± 0.084) than for *D.* sp. nov. (0.37 ± 0.05). The median petiole/rachis ratio is lower (W = 38, $p = 0.01$) for *D. lopense* (0.25 ± 0.24) than for *D.* sp. nov. (0.49 ± 0.44). The mean length of terminal petiolule/length of terminal leaflet ratio is lower (t= −2.68, $p = 0.012$) for *D. lopense* (0.06 ± 0.02) than for *D.* sp. nov. (0.07 ± 0.02).

D. pachyphyllum only differed significantly (W = 59.5, $p = 0.018$) from *D.* sp. nov. based on a lower petiole/rachis ratio median (0.22 ±0.18 for *D. pachyphyllum*). It however varied from *D. lopense* in two traits. *D. pachyphyllum* had a higher (t = −2.24, $p = 0.034$) mean length of the terminal petiolule/length of terminal leaflet ratio (0.072 ± 0.022 and 0.057 ± 0.016, respectively, for *D. pachyphyllum* and *D. lopense*) and a higher (W = 35, $p = 0.013$) basal leaflet length/terminal leaflet length ratio median (0.67 ± 0.11 and 0.58 ± 0.08, respectively, for *D. pachyphyllum* and *D. lopense*).

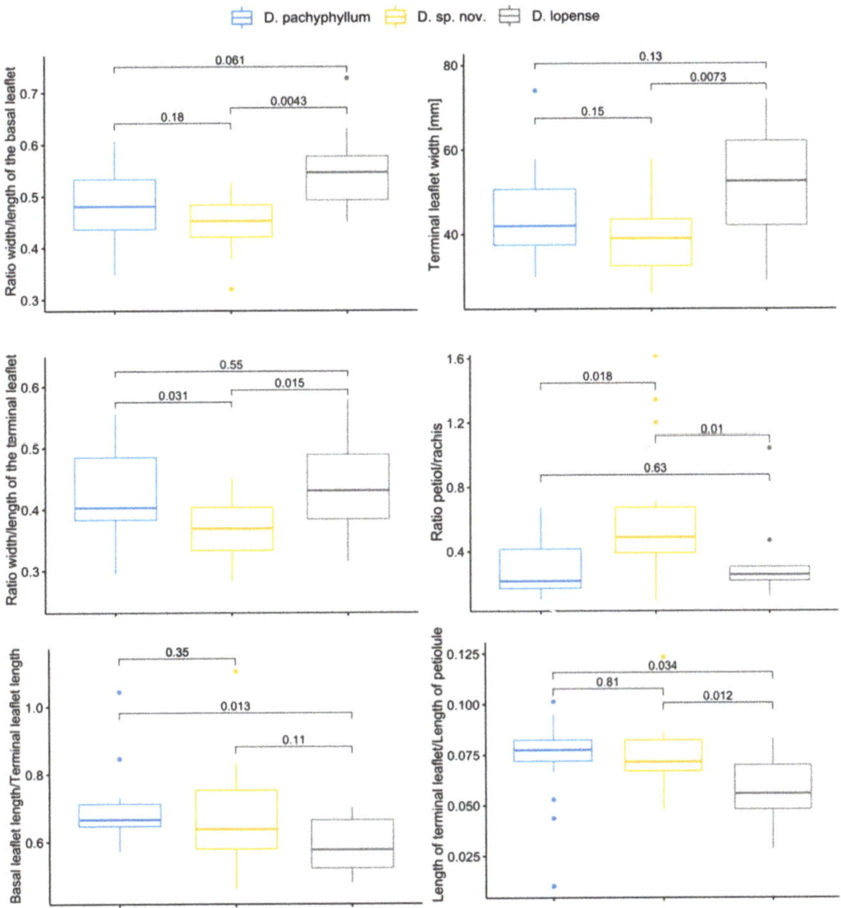

Figure 3. Significant leaf morphological trait variations between *Dialium pachyphyllum*, *Dialium* sp. nov., and *Dialium lopense*. *p*-values for species comparison are related to *t*-tests (ratio *w*/*l* of the basal leaflet, terminal leaflet width, ratio length of terminal leaflet/length of petiolule) and Wilcoxon tests (ratio petiol/rachis, ratio basal leaflet length/terminal leaflet length).

3.2. Chemical Composition

3.2.1. Primary Metabolites and Minerals

No significant effect of the sampling height was observed on primary metabolites. The mean cellulose content significantly differed (F = 3.56, p = 0.0014) between sapwood of *D. pachyphyllum* (55.4 ± 2.98%DM) and *D.* sp. nov. (52.0 ± 2.52%DM). The mean lignin content was significantly different between heartwood and sapwood (F = 72.3, $p < 0.001$) with a higher content in heartwood (27.9 ± 2.01%DM) than sapwood (23.5 ± 1.93%DM). As for lignin, the mean hemicellulose content was only significantly different between heartwood and sapwood (F = 54.4, $p < 0.001$). However, heartwood had a lower hemicellulose content (12.4 ± 1.14%DM) than sapwood (14.9 ± 1.4%DM).

The median ash content was significantly different (χ^2 = 25.9, $p < 0.001$) between species (2.46 ± 0.78%DM and 0.90 ± 0.282%DM for *D.* sp. nov. and *D. pachyphyllum*, respectively). The median nitrogen content was significantly (χ^2 = 25.9, $p < 0.001$) different within sapwood (0.28 ± 0.04%DM) and heartwood (0.35 ± 0.05%DM). The median of this metabolite also significantly differed (χ^2 = 6.65, p= 0.009) between species: 0.31 ± 0.04%DM for *D. pachyphyllum* and 0.35 ± 0.07 for *D.* sp. nov. However, this difference between species was due to samples from heartwood for which species differentiation (0.32 ± 0.03%DM for *D. pachyphyllum* and 0.39 ± 0.05%DM for *D.* sp. nov.) was very significant (W = −4.53, *p* < 0.001) unlike sapwood (Figure 4).

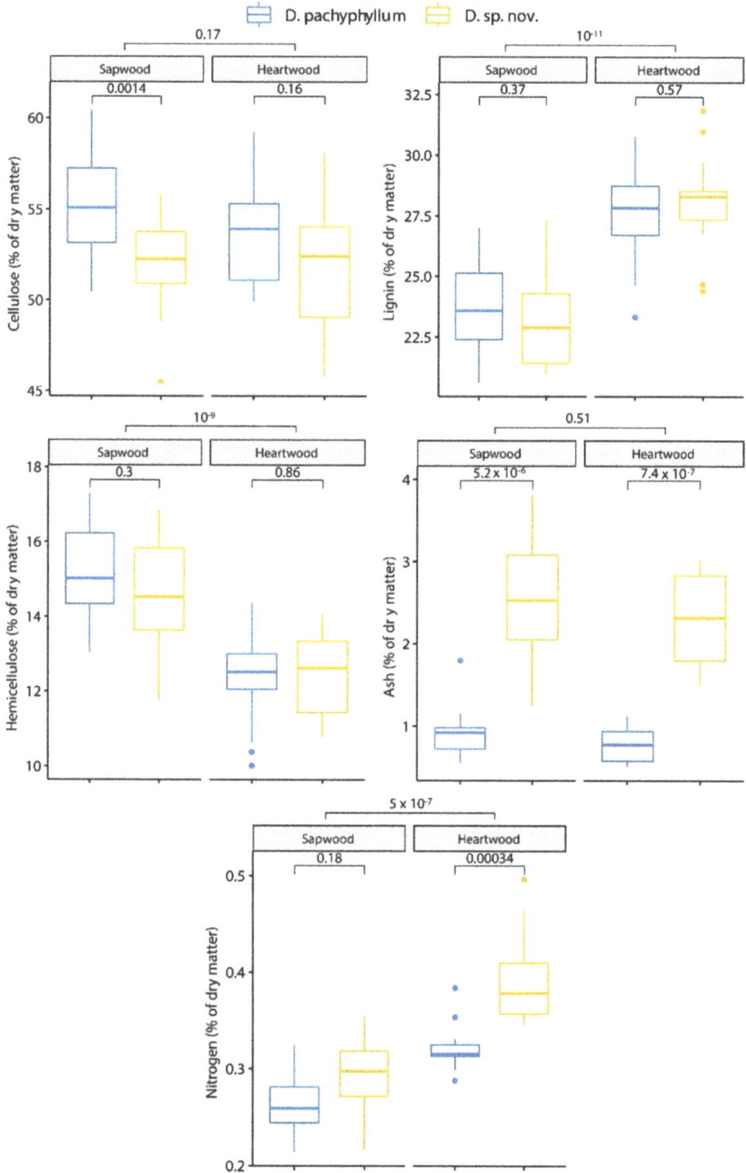

Figure 4. Primary metabolite's variation in heartwood and sapwood among the different studied species; *p*-values concerning heartwood and sapwood are related to ANOVA test (Cellulose, Lignin and Hemicellulose) and Kruskal–Wallis test (Ash and Nitrogen); *p*-values that compare species are related to the *t*-test (Cellulose, Lignin and Hemicellulose) and Wilcoxon's test (Ash and Nitrogen).

The mean silicate content was higher in *D.* sp. nov. (2.42 ± 0.57%DM and 2.64 ± 0.59%DM for sapwood and heartwood, respectively) than *D. pachyphyllum* (1.23 ± 0.39%DM and 0.92 ± 0.22%DM for sapwood and heartwood, respectively).

3.2.2. Ethanol Extracts

Median extraction yields differed significantly ($W = 698$, $p < 0.001$) with the sampling height. Samples from the top of the stem had higher extraction yields (2.6 ± 1.0%DM) than samples at the bottom (2.2 ± 0.6%DM). *D.* sp. nov. sapwood (2.7 ± 0.7%DM) had a higher extraction yield ($W = 158$, $p = 0.007$) than *D. pachyphyllum* sapwood (2.13 ± 0.36%DM).

The mean PC was significantly affected by the three studied variables: for sampling height ($F = 5.0$, $p = 0.028$), for heartwood and sapwood ($F = 77.4$, $p < 0.001$), and for the species ($F = 37.5$, $p < 0.001$). The mean PC was lower at the lower sampling height (330 ± 84 GAE) than top of the stem (359 ± 11 GAE). The mean PC was higher for heartwood (402 ± 82 GAE) than sapwood (280 ± 72 GAE). Finally, it was lower for *D. pachyphyllum* (304 ± 93 GAE) than *D.* sp. nov. (384 ± 82 GAE). Figure 5 emphasizes the significant difference of mean PC between the two species for both sapwood (253 ± 72 GAE for *D. pachyphyllum*; 320 ± 57 GAE for *D.* sp. nov.) and heartwood (355 ± 83 GAE for *D. pachyphyllum*; 448 ± 4.6 GAE for *D.* sp. nov.).

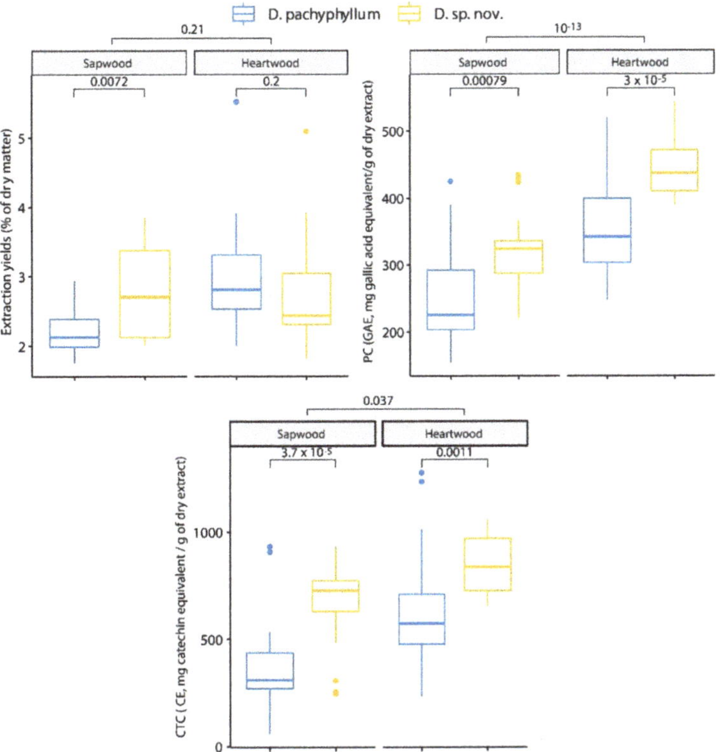

Figure 5. Extract yields and extract composition variation in heartwood and sapwood among the different studied species. *p*-values concerning heartwood and sapwood are related to ANOVA test (Phenolic content and Condensed tannin content) and Kruskal–Wallis test (Extraction yields); *p*-values that compare species are related to the *t*-test (Phenolic content and Condensed tannin content) and Wilcoxon's test (Extraction yields).

The three variables also induced a significant variation of mean CTC: for sampling height ($F = 4.5$, $p = 0.037$), for heartwood and sapwood ($F = 22.3$, $p < 0.001$), and for species

(F = 35.3, *p* < 0.001). The same PC variation pattern was observed: the mean CTC was lower at the lower sampling height (585 ± 265 CE and 677 ± 275 CE for lower and higher sampling height, respectively), higher for the heartwood (733 ± 248 CE and 528 ± 259 CE for heartwood and sapwood, respectively), and lower for *D. pachyphyllum* (502 ± 287 CE and 760 ± 185 CE for *D. pachyphyllum* and *D.* sp. nov., respectively). Figure 5 highlights the difference of mean CTC between the two species for both sapwood (385 ± 241 CE for *D. pachyphyllum*; 672 ± 190 CE for *D.* sp. nov.) and heartwood (619 ± 285 CE for *D. pachyphyllum*; 848 ± 132 CE for *D.* sp. nov.).

3.2.3. Wood FT-IR Distinction

Average preprocessed spectra for each species are presented in Figure 6. The spectral differences (Δ) highlight 15 wavenumbers that present high peaks. The PLS-DA model was fitted using 7 LVs which reached an average OCE of 2% during the cross-validation procedure (Figure 7). After predicting the validation set, composed with an independent tree of each species, the accuracy (29/30) of 96.6% is close to cross-validation (Table 4). The sensitivity and specificity of *D. pachyphyllum* are, respectively, 93.8% and 100%. For *D.* sp. nov., specificity and sensitivity are 100% and 87.5%, respectively. Those results highlight a high accuracy of the model and a low risk of predicting *D.* sp. nov. as *D. pachyphyllum*.

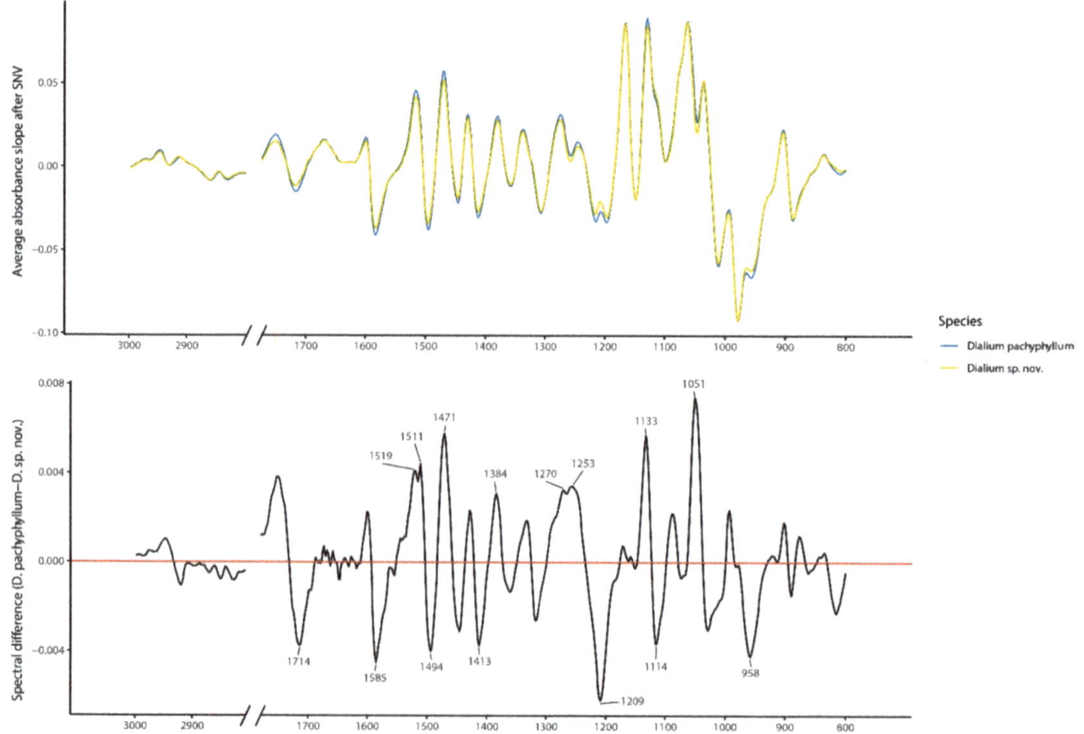

Figure 6. Upper plot presents average preprocessed (SNV + first derivatitive) spectrum of each species. The lower graph represents the difference (*D. pachyphyllum-D.* sp. nov.) between the average spectra. The 15 wavenumbers correspond to 15 high peaks of difference between the species.

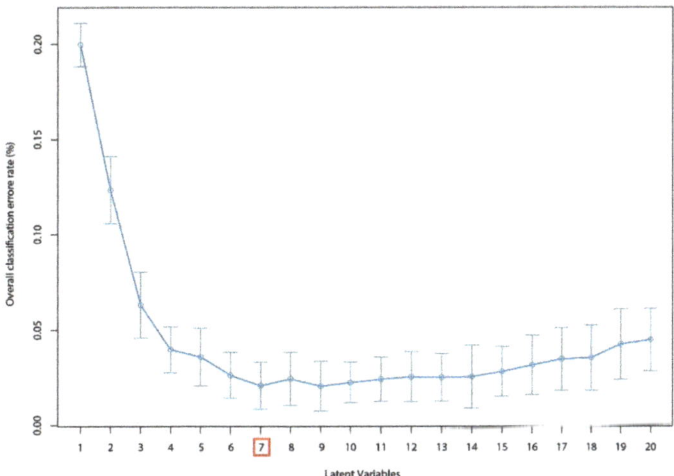

Figure 7. Overall classification error rate from the cross-validation procedure. The number of LVs selected is red-circled.

Table 4. Confusion matrix obtained by predicting the independent validation dataset.

		D. pachyphyllum	*D.* **sp. nov.**
Predicted	*D. pachyphyllum*	15	0
	D. **sp. nov.**	1	14

Table 5 compares the wavenumbers identified by the spectral difference (Δ) and their importance in the prediction model. The 1051 cm^{-1} shows the largest peak difference and reaches the top 5% of VIP scores for each LV. This variable therefore highly participated to species distinction. According to the literature, this band can be attributed to the C-H stretching bond from tannins (*Terminalia chebula* (Gaertner) Retz., *Caesalpinia spinosa* (Molina) Kuntze, and *Schinopsis lorentzii* Engl. tannin extracts in their study [42]). The 1114 cm^{-1} peak, although not the highest Δ, reaches top 5% VIP scores of 6 LVs and strongly contributes to the predictive model. As for the previous band, this wavenumber can be attributed to tannins (C-H bending observed in *S. lorentzii* and *Castanea sativa* Mill. extracts [42]). In opposition, the 1253 cm^{-1} and 1270 cm^{-1} peaks did not show VIP scores significantly higher than 1, except for LV1 and 2, respectively. Those wavenumbers, assigned to lignin bonds, are not important in the discriminant model. Wavenumbers from 1714 cm^{-1} to 1511 cm^{-1} were assigned to bonds in lignin or phenols. Except the 1511 cm^{-1} band for LV5 and LV6, this range presents a significant importance in species differentiation. Peaks at 1413, 1471, and 1494 cm^{-1} had VIP scores higher than 1 and reached the top 5% for 1413 cm^{-1} on LV1 and 1471 cm^{-1} on LV1 and LV2. Those bands are associated to lignin bonds. The 1209 band, assigned to a cellulose bond, has a high Δ and a significant importance in all LVs. The two last bands investigated, 1133 and 958 cm^{-1}, are less important in the model (VIP scores around 1 or not higher) while presenting a medium Δ.

Table 5. Wavenumbers (W), expressed in cm^{-1}, that present 15 large peak differences ($\times 10^{-3}$) between average species spectrum (Δ); Mean VIP scores from of PLS-DA model, shaded scores are upper to the 95th percentile of vip score within the LV, scores with [n] are not significantly higher than 1 according to the confidence interval from the cross-validation process; Bond: chemical bond assignement to the wavenumber; Mol corresponds to the molecular class in which the chemical bonds are found (L = Lignin, H = Hemicellulose, C = Cellulose, P = Polyphenol, T = Tannin); Ref. corresponds to references that mention the bond assignment.

W	Δ	VIP Scores							Chemical Bond Assignment	Mol	Ref.
		LV1	LV2	LV3	LV4	LV5	LV6	LV7			
1714	38	1.47	1.19	1.14	1.12	1.10	1.08	1.09	**1712 cm^{-1}**: C=O groups	L	[43]
1585	45	1.62	1.28	1.25	1.22	1.21	1.20	1.18	**1585 cm^{-1}**: Aromatic ring stretching	L	[44]
1519	41	1.56	1.29	1.22	1.20	1.19	1.16	1.15	**1517 cm^{-1}**: Aromatic skeletal vibration	L	[43]
1511	44	1.37	1.13	1.07	1.05	1.04 [n]	1.02 [n]	1.05	**1508–1513 cm^{-1}**: Aromatic skeletal vibration	L, T	[42,43]
1494	40	1.60	1.31	1.24	1.23	1.22	1.19	1.19	**1495 cm^{-1}**: aromatic vibration	P, T	[45]
1471	58	1.77	1.40	1.33	1.32	1.30	1.28	1.26	**1460–1470 cm^{-1}**: C-H vibration	L	[46,47]
1413	83	1.93	1.53	1.46	1.43	1.41	1.38	1.37	**1413 cm^{-1}**: C=C aromatic	L	[48]
1384	31	1.49	1.20	1.14	1.13	1.14	1.11	1.10	**1384 cm^{-1}**: -OH bending of phenolic bond	L	[49]
1270	32	1.09 [n]	1.14	1.09	1.07	1.05 [n]	1.03 [n]	1.01 [n]	**1265 cm^{-1}**: C-O vibration	L	[50,51]
1253	33	1.07	0.89 [n]	0.85 [n]	0.84 [n]	0.83 [n]	0.82 [n]	0.81 [n]	**1250 cm^{-1}**: C-O-C asymmetric stretch	L	[34]
1209	62	1.50	1.21	1.15	1.13	1.11	1.10	1.09	**1203–1210 cm^{-1}**: O-H bending	C	[52]
1133	57	1.18	0.94 [n]	1.07 [n]	1.11	1.10	1.09	1.08	**1134 cm^{-1}**: C-O-C glycosidic vibration of xylan	H	[50]
1114	37	1.54	1.64	1.63	1.62	1.60	1.56	1.54	**1112–1113 cm^{-1}**: C-H bending in plane	T	[42]
1051	98	1.73	1.64	1.56	1.53	1.51	1.50	1.48	**1050 cm^{-1}**: C-H stretching in plane	T	[42]
958	42	1.33	1.12 [n]	1.12	1.13	1.11	1.09	1.08 [n]	**957–961 cm^{-1}**: C-H aromatic out of plane deformation	-	[45]

4. Discussion

4.1. Species Distinction in Forest Inventories

Species discrimination criteria based on leaf morphology is frequently used in the *Fabaceae* family [53,54] and, particularly in the *Dialium* genus [55]. Furthermore, during field surveys, reproductive characteristics are generally not visible and only trunks and leaves can be observed. Leaves in *Dialium* can contribute to distinguishing species from each other [56]. For example, distinction of *Dialium heterophyllum* from other species in the Amazonian Basin is done by its reduced rachis and unifoliate to trifoliate leaves [20]. Using the 7 significant leaf traits to classify this species seems therefore relevant. Table 6 presents the identification key for the 3–5 leaflets *Dialium* species that occur in the study area.

4.2. Wood Chemical Properties and Their Variation

4.2.1. Primary Metabolites and Minerals

As Mbagou (2017) already observed, sampling height has no influence on structural primary metabolites [57]. However, many differences can be observed considering the radial position of the sample. Indeed, considering both species, the lignin content was 4.4%DM higher in heartwood than sapwood. This pattern was already observed by many authors [58–60]. Hemicellulose varies oppositely and, considering both species, is on average 2.5%DM lower in heartwood than sapwood. The difference is less pronounced but remains encountered in the literature [60].

Table 6. Identification key for 3–5 leaflets *Dialium* species present in PW-CEB forests; on the left side of the table, the sections with the observable traits are presented; on the right side is either the species classification corresponding to the observed trait, or the reference of the section to continue the identification key; the range of values presented for the quantitative variables, in the third section, corresponds to the interval in which 50% of the observations made during this study occurs for each species.

1.	-	Tight and prominent tertiary venation	2.
	-	Discrete tertiary venation	*D. pachyphyllum*
2.	-	Sharp, tapered or retuse terminal leaflet acumen	3.
	-	No terminal leaflet acumen or obtuse acumen	*D. lopense*
3.	-	Ratio width/length of the basal leaflet: [0.42–0.48]	*D.* sp. nov.
	-	Terminal leaflet width: [33–44] mm	
	-	Ratio width/length of the terminal leaflet: [0.33–0.40]	
	-	Ratio petiole/rachis: [0.39–0.67]	
	-	Length of terminal leaflet/length of petiolule: [0.06–0.08]	
	-	Ratio width/length of the basal leaflet: [0.49–0.57]	*D. lopense*
	-	Terminal leaflet width: [42–62] mm	
	-	Ratio width/length of the terminal leaflet: [0.38–0.49]	
	-	Ratio petiole/rachis: [0.22–0.30]	
	-	Length of terminal leaflet/length of petiolule: [0.05–0.07]	

The highest cellulose content (55.4%DM) was obtained from the sapwood of *D. pachyphyllum* which appears to reach higher values than generally observed in African tropical woods. According to Gérard et al., (2019), the average is 42.2%DM, and the maximum observed is 58.1%DM [28]. For instance, *Aucoumea klaineana* present a cellulose content of 46.1%DM [61], *Lophira alata* Banks ex C.F. Gaertn. of 40.0%DM [28]. This difference could be explained by the cellulose determination method. Heartwood lignin content (27.9%DM in average) did not significantly differ between species and was a bit lower than in other tropical woods for both species. Gérard et al., (2019) measured an average lignin content of 29.2%DM for 549 tropical species tested [28]. Nuopponen et al., (2006) reported that the lignin content of tropical hardwoods can exceed that of softwoods and reach between 29%DM and 41%DM [62]. Hemicellulose did not differ between species. Heartwood hemicellulose content seems comparable to the average pentosane content (major constituent of hemicellulose) for tropical hardwood of 15.9%DM [28]. Ash content was significantly higher in *D.* sp. nov. which had a high value (2.46%DM) compared to other tropical species that, on average, have 1.1%DM [28]. Concerning nitrogen content, *D.* sp. nov. heartwood has the highest value which is almost twice the average of 0.24%DM observed for 59 Panamanian tree species [63]. Results about silica content follow the same pattern. *D.* sp. nov. has twice more silica content in sapwood (2.42%DM) and nearly three times more in heartwood (2.64%DM) than *D. pachyphyllum*. Both species had a very high silica content according to the 0.1%DM average silica content of 599 hardwoods [28].

4.2.2. Ethanol-Water Extracts

Extract yields were higher at the top of the stem for both species. Gérardin et al., (2020) also obtained higher yields for water-ethanol extracts at the top of the trees [64]. Concerning species distinction, extracts yields were higher in *D.* sp. nov. sapwood. However, using the same extraction method, several authors obtained higher yields than observed for both species. Saha Tchinda (2015), using hot water as solvent, obtained 8.4%DM from *Baillonella toxisperma* Pierre heartwood and 3.8%DM for *Distemonanthus benthamianus* Baill. heartwood. PC and CTC are significantly higher in the heartwood than sapwood. Those results are probably species-dependent since Engozogho et al., (2020) reported higher polyphenol and tannin levels in sapwood (2 ± 0.8%DM) than heartwood (0.7 ± 0.1%DM) of *Aucoumea klaineana*, whereas Mbagou et al., (2017) bring out significant higher PC in heartwood than sapwood for *Tessmania Africana* Harms, and Bikoro et al., (2018) had significantly more PC in sapwood but significantly more CTC in heartwood of *Khaya ivorensis* A. Chev. [65]. Considering species variation, *D.* sp. nov. has higher PC and CTC than *D. pachyphyllum*. According to Nisca et al., (2021), those metabolite are interesting to discriminate species [66]. *D.* sp. nov. heartwood PC is comparable to *Pterocarpus soyauxii* Hooker acetone extract (430 GAE) and both species have higher PC than *Baillonella toxisperma* for the acetone extract [67].

4.3. FT-IR Wood Distinction

The high accuracy of the PLS-DA model (96.6%) on two independent trees of the training set added to the significant differences in analytical metabolite measurements, and proves the difference in wood composition between the two species. As already highlighted by Wang et al., (2016), spectroscopy is an adapted tool to discriminate closely related species as *Dalbergia* spp. [34]. However, the interpretation of the spectra without the use of statistical techniques can be hazardous. Indeed, as it has been shown, the peak differences observed on the average spectra are not always associated to discriminant bands. This can come from a high intra-species variation being greater than inter-species variation. Knowing this, it is likely that the peaks at 1270 and 1253 cm^{-1}, despite a high Δ, present a large inter-species variation that do not allow discrimination. The peak at 1114 cm^{-1} assigned to tannin bonds, has the same delta but its importance in the predictive model might be due to low intra-species variations. The higher Δ peak at 1051 cm^{-1} is the most interesting because it represents the greatest average difference between the species and participates very strongly in the PLS-DA model. Added to the significant difference in CTC between *Dialium* species, those results are in lines with Nisca et al., (2021) to highlight the efficiency of this class of molecule in the distinction of closely related species.

4.4. Impacts on Tree Ecology, Forest Management, and Wood Production

Leaf traits are highly correlated to tree ecology [68]. For instance, venation density significantly influences water management [69] and, therefore, tree light strategy. According to Martin A. et al., (2014), wood nitrogen content is negatively correlated to the relative growth rate and higher values could correspond to late successional species [63]. As those two traits significantly discriminate the studied *Dialium* species, their ecology might differ. Consequently, *D.* sp. nov. should be discriminated in annual monitoring plots on the study site and its population dynamic should be separately modelled from the other species. Once population dynamic parameters will be known, management practices should be adapted. Moreover, before any valuation of the wood, the conservation status of the species, in relation to its distribution area, should be specified.

Wood properties are directly related to chemical composition [70]. The cellulose proportion variation between *D.* sp. nov. and *D. pachyphyllum* could therefore lead to differences in mechanical properties [71]. The silica content increases resistance to marine borers [72] while PC, CTC, and suberin increase durability against fungi [73–75]. However, silica content is highly abrasive and increases sawing difficulties [14]. To know if the species should be promoted to a different timber sector, it will be interesting to check whether the difference in chemical properties induces a significant difference in wood properties and processing.

Dialium pachyphyllum has a wide distribution in Central Africa while *Dialium lopense* is restricted to Gabon [14]. As this new species has not yet been described, its area of occurrence as well as its exploitation potential are not known. If those are restricted, it may be interesting to control the production of this species. This control can be done through customs checks before export. In view of the ease of preparation of shavings chips and the accuracy of the model, these techniques could be used routinely to identify wood packages. However, this requires a large increase in the number of individuals to train the model to capture inter-specific variation. In addition, the model should favor a sensibility of 100% for *D.* sp. nov. to avoid classifying the samples into another species.

5. Conclusions

This study brought out many results that support the existence of a new *Dialium* species. Using 7 significant leaf traits to discriminate *D.* sp. nov. from other 3–5 *Dialium* leaflets, an identification key was proposed for forest inventories. A total of nine significant differences in chemicals properties were highlighted (five in sapwood and four in heartwood). The PLS-DA model trained on heartwood shavings FT-IR spectra accurately discriminate *D.* sp. nov. from *D. pachyphyllum*. Those outcomes may have consequences on the newly discovered morphospecies ecology, population dynamic, wood properties, and timber production potential. It is therefore advisable to investigate those aspects along with the study of its reproductive system before considering its large-scale logging.

Author Contributions: Conceptualization, R.D., G.B.B., M.R. and J.-L.D.; methodology, A.R., J.-L.D. and I.V.D.V.; FT-IR data acquisition, R.D.; FT-IR data statistical analysis, R.D.; chemical data collection, R.D. and I.V.D.V.; chemical data statistical analysis, R.D.; leaves traits data collection, M.R., R.D., J.-L.D. and G.B.B.; leaves traits statistical analysis, M.R.; validation, A.R., B.L., J.-L.D., J.A.F.P. and A.S.; formal analysis, R.D.; writing—original draft preparation, R.D., G.B.B. and M.R.; writing—review and editing, A.R., B.L., J.-L.D., J.A.F.P., A.S., B.J. and P.L.; supervision, A.R., P.L., B.J. and J.-L.D.; funding acquisition, J.-L.D. All authors have read and agreed to the published version of the manuscript.

Funding: The research was supported by the Identification d'Essences à Haut Potential de Valorisation (EHPval) project, funded by the Programme de Promotion de l'Exploitation Certifiée des Forêts (PPECF).

Data Availability Statement: Datasets will be deposited on DRYAD after paper acceptance.

Acknowledgments: We are grateful to PW–CEB, especially Monnet Vincent, Nna Ekome Stévy, Assam Augustin, Bokomba Jean-Bosco, Mboulamab Jean Arnisse for harvesting/sending the wood material and for logistical support during herbarium collection. We are grateful to the *Ministère des Eaux et Forêts du Gabon* (MINEF) for delivering the export license, and to the *Centre national de la Recherche Scientifique* (CENAREST) for giving authorization to do research in Gabon. We are also grateful to the botanical garden of Meise for lending their herbarium and particularly to Janssens Steven, De Smedt Sofie, and Bogaerts Ann for their help. We thank the 1st master students of Gembloux Agro-Bio Tech (academic year 2020–2021) for their help in primary metabolites data collection. We more specifically thank Groignet Eric, Kammoun Maroua, Werrie Pierre-Yves, De Crane d'Heysselaer Simon, Berchem Thomas, Istasse Thibaut, Morin Sophie, and Dumoulin Lionel for their advices in chemical data collection and interpretation. We thank Maesen Philippe and the BEAGx lab for the silica dosage. We would also like to thank Kayoka Mukendi Nicaise, Baetten Vincent, Vincke Damien, Arnould Quentin, Stevens François, Plasman Lisa from CRA-W for their help in spectroscopy understanding and their support in spectral acquisition.

Conflicts of Interest: The work is all original research carried out by the authors. All authors agree with the content of the manuscript and its submission to the journal. No part of the research has been submitted nor published in any form in this journal or elsewhere. The manuscript is not being considered for publication elsewhere while it is being considered for publication in this journal. All sources of funding are acknowledged in the manuscript, and authors have no financial benefits that could result from publication.

Appendix A

Table A1. Herbarium information. Herbarium reference that mentions the "BR00000" prefix comes from the botanical garden of Meise (Belgium), "field" corresponds to the herbarium collected in PW-CEB forest for this study.

Harvester	Herbaria Reference	Date	Identification	Country
de Wilde	BR0000016564503	19-12-96	*Dialium lopense*	Gabon
Breteler	BR0000016564534	23-01-99	*Dialium lopense*	Gabon
Wieringa	BR0000016564589	17-11-94	*Dialium lopense*	Gabon
Wieringa	BR0000016564572	02-04-04	*Dialium lopense*	Gabon
Breteler	BR0000016564527	05-02-99	*Dialium lopense*	Gabon
Breteler	BR0000016564541	17-02-99	*Dialium lopense*	Gabon
Doucet J-L	BR0000016564558	01-05-96	*Dialium lopense*	Gabon
Wilk	BR0000016564565	17-05-87	*Dialium lopense*	Gabon
McPherson	BR0000016564596	22-11-93	*Dialium lopense*	Gabon
McPherson	BR0000016564602	18-05-92	*Dialium lopense*	Gabon
McPherson	BR0000016564626	15-05-92	*Dialium lopense*	Gabon
Bibang & Doucet J-L.	field	30-03-21	*Dialium lopense*	Gabon
Doucet R.	field	17-06-21	*Dialium lopense*	Gabon
Doucet R.	field	May 2019	*Dialium lopense*	Gabon
McPherson	BR0000013636012	04-05-92	*Dialium pachyphyllum*	DRC
Tshibamba	BR0000013639617	01-07-11	*Dialium pachyphyllum*	DRC
van de Burgt	BR0000013639631	12-09-11	*Dialium pachyphyllum*	Gabon
Wieringa	BR0000013635824	16-11-94	*Dialium pachyphyllum*	Gabon
Wieringa	BR0000013635916	31-10-03	*Dialium pachyphyllum*	Gabon
Doucet R.	field	May 2019	*Dialium pachyphyllum*	Gabon
Doucet R.	field	May 2019	*Dialium pachyphyllum*	Gabon
Doucet R.	field	16-06-21	*Dialium pachyphyllum*	Gabon
Doucet R.	field	16-06-21	*Dialium pachyphyllum*	Gabon
Doucet R.	field	16-06-21	*Dialium pachyphyllum*	Gabon
Doucet R.	field	16-06-21	*Dialium pachyphyllum*	Gabon
Doucet R.	field	17-06-21	*Dialium pachyphyllum*	Gabon
Doucet R.	field	17-06-21	*Dialium pachyphyllum*	Gabon
Doucet R.	field	17-06-21	*Dialium pachyphyllum*	Gabon
Doucet R.	field	19-06-21	*Dialium pachyphyllum*	Gabon
Doucet R.	field	19-06-21	*Dialium pachyphyllum*	Gabon
Bibang & Doucet J-L.	field	30-03-21	*Dialium* sp. nov.	Gabon
Bibang & Doucet J-L.	field	30-03-21	*Dialium* sp. nov.	Gabon
Bibang & Doucet J-L.	field	31-03-21	*Dialium* sp. nov.	Gabon
Bibang & Doucet J-L.	field	31-03-21	*Dialium* sp. nov.	Gabon
Bibang & Doucet J-L.	field	31-03-21	*Dialium* sp. nov.	Gabon

Table A1. *Cont.*

Harvester	Herbaria Reference	Date	Identification	Country
Bibang & Doucet J.-L.	field	31-03-21	*Dialium* sp. nov.	Gabon
Bibang & Doucet J.-L.	field	31-03-21	*Dialium* sp. nov.	Gabon
Doucet R.	field	May 2019	*Dialium* sp. nov.	Gabon
Doucet R.	field	May 2019	*Dialium* sp. nov.	Gabon
Doucet R.	field	May 2019	*Dialium* sp. nov.	Gabon
Doucet R.	field	May 2019	*Dialium* sp. nov.	Gabon
Doucet R.	field	17-06-21	*Dialium* sp. nov.	Gabon
Doucet R.	field	19-06-21	*Dialium* sp. nov.	Gabon
Doucet R.	field	19-06-21	*Dialium* sp. nov.	Gabon
Doucet R.	field	19-06-21	*Dialium* sp. nov.	Gabon

References

1. Biwolé, A.B.; Ouédraogo, D.-Y.; Betti, J.L.; Picard, N.; Rossi, V.; Delion, S.; Lagoute, P.; Gourlet-Fleury, S.; Lejeune, P.; Doucet, J.-L. Dynamique des populations d'azobé, Lophira alata Banks ex C. F. Gaertn., et implications pour sa gestion durable au Cameroun. *Bois Forets Des Trop.* **2019**, *342*, 55–68. [CrossRef]
2. Karsenty, A.; Gourlet-Fleury, S. Assessing Sustainability of Logging Practices in the Congo Basin's Managed Forests: The Issue of Commercial Species Recovery. *Ecol. Soc.* **2006**, *11*, art26. [CrossRef]
3. FRM. *Vision Stratégique et Industrialisation de la Filière Bois dans les 6 Pays du Bassin du Congo, Horizon 2030*; FRM: Montpellier, France, 2018.
4. Doucet, J.-L. L'alliance Délicate de la Gestion Forestière et de la Biodiversité dans les Forêts du Centre du Gabon. Ph.D. Thesis, Université de Liège, Gembloux Agro-Bio Tech, Gembloux, Belgium, 2003.
5. Kleinschroth, F.; Gourlet-Fleury, S.; Sist, P.; Mortier, F.; Healey, J.R. Legacy of logging roads in the Congo Basin: How persistent are the scars in forest cover? *Ecosphere* **2015**, *6*, 64. [CrossRef]
6. Edwards, D.P.; Magrach, A.; Woodcock, P.; Ji, Y.; Lim, N.T.L.; Edwards, F.A.; Larsen, T.H.; Hsu, W.W.; Benedick, S.; Khen, C.V.; et al. Selective-logging and oil palm: Multitaxon impacts, biodiversity indicators, and trade-offs for conservation planning. *Ecol. Appl.* **2014**, *24*, 2029–2049. [CrossRef]
7. Edwards, D.P.; Laurance, W.F. Biodiversity Despite Selective Logging. *Science* **2013**, *339*, 646–647. [CrossRef]
8. Cerutti, P.O.; Lescuyer, G.; Tacconi, L.; Eba'a Atyi, R.; Essiane, E.; Nasi, R.; Tabi Eckbil, P.P.; Tsanga, R. Social impacts of the Forest Stewardship Council certification: An assessment in the Congo basin. *Int. For. Rev.* **2017**, *19*, 50–63. [CrossRef]
9. Burivalova, Z.; Hua, F.; Koh, L.P.; Garcia, C.; Putz, F. A Critical Comparison of Conventional, Certified, and Community Management of Tropical Forests for Timber in Terms of Environmental, Economic, and Social Variables. *Conserv. Lett.* **2017**, *10*, 4–14. [CrossRef]
10. FSC. *FSC Forest Stewardship Council Standard for the Congo Basin*; Forest Stewardship Council A.C.: Bonn, Germany, 2012.
11. Putz, F.E.; Zuidema, P.A.; Synnott, T.; Peña-Claros, M.; Pinard, M.A.; Sheil, D.; Vanclay, J.K.; Sist, P.; Gourlet-Fleury, S.; Griscom, B.; et al. Sustaining conservation values in selectively logged tropical forests: The attained and the attainable. *Conserv. Lett.* **2012**, *5*, 296–303. [CrossRef]
12. Sist, P.; Piponiot, C.; Kanashiro, M.; Pena-Claros, M.; Putz, F.E.; Schulze, M.; Verissimo, A.; Vidal, E. Sustainability of Brazilian forest concessions. *For. Ecol. Manag.* **2021**, *496*, 119440. [CrossRef]
13. Gérard, J.; Guibal, D.; Paradis, S.; Cerre, J.-C. *Tropical Timber Atlas*; Quae: Versailles, France, 2017.
14. Bengono, G.B.; Souza, A.; Tosso, F.; Doucet, R.; Richel, A.; Doucet, J.-L. Les Dialium de la région guinéo-congolaise (synthèse bibliographique). *BASE* **2021**, *25*, 172–191. [CrossRef]
15. Ruwet, M. Caractérisations Morphologiques, Spectrales et Génétiques des Eyoum (*Dialium* spp.). Master's Thesis, Université de Liège, Gembloux Agro-Bio Tech, Gembloux, Belgium, 2021.
16. Allaby, M. *A Dictionary of Ecology*; Oxford University Press: Oxford, UK, 2010.
17. Wieringa, J.J.; MacKinder, B.A. Novitates Gabonensis 79: *Hymenostegia elegans* and *H. robusta* spp. nov. (*Leguminosae-Caesalpinioideae*) from Gabon. *Nord. J. Bot.* **2012**, *30*, 144–152. [CrossRef]
18. Kenfack, D.; Sainge, M.N.; Chuyong, G.B.; Thomas, D.W. The genus cola (Malvaceae) in Cameroon's Korup national park, with two novelties. *Plant Ecol. Evol.* **2018**, *151*, 241–251. [CrossRef]
19. Bissiengou, P.; Chatrou, L.W.; Wieringa, J.J.; Sosef, M.S.M. Taxonomic novelties in the genus Campylospermum (Ochnaceae). *Blumea J. Plant Taxon. Plant Geogr.* **2013**, *58*, 1–7. [CrossRef]
20. De Azevedo Falcão, M.J.; De Freitas Mansano, V. Dialium heterophyllum (Fabaceae: Dialioideae), a new tree species from the Amazon. *Phytotaxa* **2020**, *477*, 47–59. [CrossRef]

21. Gérard, J.; Guibal, D.; Paradis, S.; Vernay, M.; Beauchêne, J.; Brancheriau, L.; Châlon, I.; Daigremont, C.; Détienne, P.; Fouquet, D.; et al. *Tropix 7*; CIRAD: Montpelier, France, 2011.

22. TerEA. *Résumé du Plan D'aménagement. Compagnie Equatoriale des Bois. Precious Wood Gabon*; TerEA (Terre Environnement Aménagement): Marseille, France, 2007.

23. Van Soest, P.J.; Wine, R.H. Use of Detergents in the Analysis of Fibrous Feeds. IV. Determination of Plant Cell-Wall Constituents. *J. AOAC Int.* **1967**, *50*, 50–55. [CrossRef]

24. Van Soest, P.J. Use of Detergents in the Analysis of Fibrous Feeds. II. A Rapid Method for the Determination of Fiber and Lignin. *J. AOAC Int.* **1990**, *73*, 491–497. [CrossRef]

25. Chen, H.; Ferrari, C.; Angiuli, M.; Yao, J.; Raspi, C.; Bramanti, E. Qualitative and quantitative analysis of wood samples by Fourier transform infrared spectroscopy and multivariate analysis. *Carbohydr. Polym.* **2010**, *82*, 772–778. [CrossRef]

26. Sluiter, A.; Hames, B.; Ruiz, R.O.; Scarlata, C.; Sluiter, J.; Templeton, D. *Determination of Ash in Biomass*; Renew; Energy Lab.: Golden, CO, USA; Laboratory Analytical Procedure (LAP): Golden, CO, USA, 2005.

27. Sheppard, P.R.; Thompson, T.L. Effect of Extraction Pretreatment on Radial Variation of Nitrogen Concentration in Tree Rings. *J. Environ. Qual.* **2000**, *29*, 2037–2042. [CrossRef]

28. Gérard, J.; Paradis, S.; Thibaut, B. Survey on the chemical composition of several tropical wood species. *Bois Forets Des Trop.* **2019**, *342*, 79–91. [CrossRef]

29. Oreopoulou, A.; Tsimogiannis, D.; Oreopoulou, V. Extraction of Polyphenols from Aromatic and Medicinal Plants: An Overview of the Methods and the Effect of Extraction Parameters. In *Polyphenols in Plants*; Elsevier: Amsterdam, The Netherlands, 2019; pp. 243–259.

30. Ainsworth, E.A.; Gillespie, K.M. Estimation of total phenolic content and other oxidation substrates in plant tissues using Folin-Ciocalteu reagent. *Nat. Protoc.* **2007**, *2*, 875–877. [CrossRef]

31. Engozogho Anris, S.P.; Bi Athomo, A.B.; Safou Tchiama, R.; Santiago-Medina, F.J.; Cabaret, T.; Pizzi, A.; Charrier, B. The condensed tannins of Okoume (Aucoumea klaineana Pierre): A molecular structure and thermal stability study. *Sci. Rep.* **2020**, *10*, 1773. [CrossRef]

32. Kucheryavskiy, S. mdatools—R package for chemometrics. *Chemom. Intell. Lab. Syst.* **2020**, *198*, 103937. [CrossRef]

33. Hobro, A.J.; Kuligowski, J.; Döll, M.; Lendl, B. Differentiation of walnut wood species and steam treatment using ATR-FTIR and partial least squares discriminant analysis (PLS-DA). *Anal. Bioanal. Chem.* **2010**, *398*, 2713–2722. [CrossRef]

34. Wang, S.N.; Da Zhang, F.; Huang, A.M.; Zhou, Q. Distinction of four Dalbergia species by FTIR, 2nd derivative IR, and 2D-IR spectroscopy of their ethanol-benzene extractives. *Holzforschung* **2016**, *70*, 503–510. [CrossRef]

35. R Studio Team. *RStudio: Integrated Development for R. RStudio*; PBC: Boston, MA, USA, 2020.

36. Hothorn, T.; Zeileis, A. Partykit: A modular toolkit for recursive partitioning in R. *J. Mach. Learn. Res.* **2015**, *16*, 3905–3909.

37. Pomerantsev, A.L. Acceptance areas for multivariate classification derived by projection methods. *J. Chemom.* **2008**, *22*, 601–609. [CrossRef]

38. Rohart, F.; Gauthier, B.; Signh, A.; Lê Cao, K.A. mixOmics: An R package for omics feature selection and multiple data integration. *PLoS Comput Biol.* **2017**, *13*, e1005752. [CrossRef]

39. Mahalanobis, P.C. On the Generalised Distance in Statistics. *Proc. Natl. Inst. Sci. India* **1936**, *2*, 49–55.

40. Chong, I.G.; Jun, C.H. Performance of some variable selection methods when multicollinearity is present. *Chemom. Intell. Lab. Syst.* **2005**, *78*, 103–112. [CrossRef]

41. Kassambara, A. *rstatix: Pipe-Friendly Framework for Basic Statistical Tests*; (v 0.7.0); R Project for Statistical Computing: Vienna, Austria, 2021.

42. Grasel, F.D.S.; Ferrão, M.F.; Wolf, C.R. Development of methodology for identification the nature of the polyphenolic extracts by FTIR associated with multivariate analysis. *Spectrochim. Acta—Part A Mol. Biomol. Spectrosc.* **2016**, *153*, 94–101. [CrossRef] [PubMed]

43. González-Peña, M.M.; Hale, M.D.C. Rapid assessment of physical properties and chemical composition of thermally modified wood by mid-infrared spectroscopy. *Wood Sci. Technol.* **2011**, *45*, 83–102. [CrossRef]

44. Yap, M.G.S.; Que, Y.T.; Chia, L.H.L. FTIR characterization of tropical wood–polymer composites. *J. Appl. Polym. Sci.* **1991**, *43*, 2083–2090. [CrossRef]

45. Popescu, C.-M.; Jones, D.; Kržišnik, D.; Humar, M. Determination of the effectiveness of a combined thermal/chemical wood modification by the use of FT–IR spectroscopy and chemometric methods. *J. Mol. Struct.* **2020**, *1200*, 127133. [CrossRef]

46. Backa, S.; Brolin, A.; Nilsson, T. Characterisation of fungal degraded birch wood by FTIR and Py-GC. *Holzforschung* **2001**, *55*, 225–232. [CrossRef]

47. Shen, D.; Liu, G.; Zhao, J.; Xue, J.; Guan, S.; Xiao, R. Thermo-chemical conversion of lignin to aromatic compounds: Effect of lignin source and reaction temperature. *J. Anal. Appl. Pyrolysis* **2015**, *112*, 56–65. [CrossRef]

48. Trilokesh, C.; Uppuluri, K.B. Isolation and characterization of cellulose nanocrystals from jackfruit peel. *Sci. Rep.* **2019**, *9*, 16709. [CrossRef]

49. Abdul Latif, M.H.; Attiya, H.G.; Al–Abayaji, M.A. Lignin FT-IR study of Iraqi date palm Phoenix dactylifera frond bases wood. *Plant Arch.* **2019**, *19*, 327–332.

50. Nuopponen, M.H.; Wikberg, H.I.; Birch, G.M.; Jääskeläinen, A.S.; Maunu, S.L.; Vuorinen, T.; Stewart, D. Characterization of 25 tropical hardwoods with fourier transform infrared, ultraviolet resonance raman, and13C-NMR cross-polarization/magic- angle spinning spectroscopy. *J. Appl. Polym. Sci.* **2006**, *102*, 810–819. [CrossRef]
51. Huang, A.; Zhou, Q.; Liu, J.; Fei, B. Distinction of three wood species by Fourier transform infrared spectroscopy and two-dimensional correlation IR spectroscopy. *J. Mol. Struct.* **2008**, *883–884*, 160–166. [CrossRef]
52. Özgenç, Ö.; Durmaz, S.; Boyaci, I.H.; Eksi-Kocak, H. Determination of chemical changes in heat-treated wood using ATR-FTIR and FT Raman spectrometry. *Spectrochim. Acta—Part A Mol. Biomol. Spectrosc.* **2017**, *171*, 395–400. [CrossRef]
53. Pasquet, R.S. Classification infraspecifique des formes spontanees de *Vigna unguiculata* (L.) Walp. (Fabaceae) a partir de donnees morphologiques. *Bull. Du Jard. Bot. Natl. Belg.* **1993**, *62*, 127. [CrossRef]
54. Vandrot, H. Nomenclature of the new caledonian genera arthroclianthus Baill. and nephrodesmus schindl. (Fabaceae-Desmodieae). *Adansonia* **2018**, *40*, 103–129. [CrossRef]
55. Rojo, J.P. Studies in the Genus Dialium (Cassieae-Caesalpinioideae). Ph.D. Thesis, Oxford University, Oxford, UK, 1982.
56. Zimmerman, E.; Prenner, G.; Bruneau, A. Floral ontogeny in Dialiinae (Caesalpinioideae: Cassieae), a study in organ loss and instability. *S. Afr. J. Bot.* **2013**, *89*, 188–209. [CrossRef]
57. Bosco Mbagou, J. Variabilité Intra-Arbre des Propriétés Physico-Mécaniques Et Chimiques du Tessmania Africana en Provenance du Gabon. Ph.D. Thesis, Université de Laval, Québec, QC, Canada, 2017.
58. Lourenço, A.; Neiva, D.M.; Gominho, J.; Marques, A.V.; Pereira, H. Characterization of lignin in heartwood, sapwood and bark from Tectona grandis using Py–GC–MS/FID. *Wood Sci. Technol.* **2015**, *49*, 159–175. [CrossRef]
59. Adamopoulos, S.; Voulgaridis, E.; Passialis, C. Variation of certain chemical properties within the stemwood of black locust (*Robinia pseudoacacia* L.). *Eur. J. Wood Wood Prod.* **2005**, *63*, 327–333. [CrossRef]
60. Lhate, I.; Cuvilas, C.; Terziev, N.; Jirjis, R. Chemical composition of traditionally and lesser used wood species from Mozambique. *Wood Mater. Sci. Eng.* **2010**, *5*, 143–150. [CrossRef]
61. Doat, J. Etude papetière de l'Okoumé. Essais de laboratoire, semi-industriels et industriels (1re partie). *Bois Forets Des Trop.* **1972**, *146*, 31–52.
62. Nuopponen, M.H.; Birch, G.M.; Sykes, R.J.; Lee, S.J.; Stewart, D. Estimation of wood density and chemical composition by means of diffuse reflectance mid-infrared fourier transform (DRIFT-MIR) spectroscopy. *J. Agric. Food Chem.* **2006**, *54*, 34–40. [CrossRef]
63. Martin, A.R.; Erickson, D.L.; Kress, W.J.; Thomas, S.C. Wood nitrogen concentrations in tropical trees: Phylogenetic patterns and ecological correlates. *N. Phytol.* **2014**, *204*, 484–495. [CrossRef]
64. Gérardin, P.; Fritsch, C.; Cosgun, S.; Brennan, M.; Dumarçay, S.; Colin, F.; Gérardin, P. Effet de la hauteur de prélèvement sur la composition quantitative et qualitative des polyphénols de l'écorce d'Abies alba Mill. *Rev. For. Française* **2020**, *72*, 411–423. [CrossRef]
65. Bikoro Bi Athomo, A.; Engozogho Anris, S.P.; Safou-Tchiama, R.; Santiago-Medina, F.J.; Cabaret, T.; Pizzi, A.; Charrier, B. Chemical composition of African mahogany (K. ivorensis A. Chev) extractive and tannin structures of the bark by MALDI-TOF. *Ind. Crops Prod.* **2018**, *113*, 167–178. [CrossRef]
66. Nisca, A.; Ștefănescu, R.; Stegăruș, D.I.; Mare, A.D.; Farczadi, L.; Tanase, C. Comparative Study Regarding the Chemical Composition and Biological Activity of Pine (*Pinus nigra* and *P. sylvestris*) Bark Extracts. *Antioxidants* **2021**, *10*, 327. [CrossRef]
67. Tchinda Saha, J.-B.; Abia, D.; Dumarç Ay, S.; Kor Ndikontar, M.; Gérardin, P.; Noah, J.N.; Perrin, D. Antioxidant activities, total phenolic contents and chemical compositions of extracts from four Cameroonian woods: Padouk (Pterocarpus soyauxii Taubb), tali (*Erythrophleum suaveolens*), moabi (*Baillonella toxisperma*), and movingui (*Distemonanthus benthamianus*). *Ind. Crops Prod.* **2012**, *41*, 71–77.
68. Poorter, L. Leaf traits show different relationships with shade tolerance in moist versus dry tropical forests. *N. Phytol.* **2009**, *181*, 890–900. [CrossRef]
69. Sack, L.; Cowan, P.D.; Jaikumar, N.; Holbrook, N.M. The 'hydrology' of leaves: Co-ordination of structure and function in temperate woody species. *Plant. Cell Environ.* **2003**, *26*, 1343–1356. [CrossRef]
70. Barnett, J.; Jeronimidis, G. *Wood Quality and Its Biological Basis*; Blackwell Publishing Ltd.: Carlton, VIC, Australia, 2003.
71. Cruz, N.; Bustos, C.; Aguayo, M.G.; Cloutier, A.; Castillo, R. THM densification of wood. *BioResources* **2018**, *13*, 2268–2282.
72. Dirol, D.; Deglise, X. *Durabilité des Bois et Problèmes Associés*; Hermès Science: Paris, France, 2001.
73. Anouhe, B.J.S.; Niamké, F.B.; Faustin, M.; Virieux, D.; Pirat, J.L.; Adima, A.A.; Kati-Coulibaly, S.; Amusant, N. The role of extractives in the natural durability of the heartwood of Dicorynia guianensis Amsh: New insights in antioxydant and antifungal properties. *Ann. For. Sci.* **2018**, *75*, 15. [CrossRef]
74. Scalbert, A. Tannins in Woods and Their Contribution to Microbial Decay Prevention. In *Plant Polyphenols*; Springer: Boston, MA, USA, 1992; pp. 935–952.
75. Akai, S.; Fukutomi, M. Preformed Internal Physical Defenses. In *How Plants Defend Themselves*; Horsfall, J.G., Cowling, E.B., Eds.; Elsevier: Amsterdam, The Netherlands, 1980; pp. 139–159.

Article

Phytochemical Composition of Extractives in the Inner Cork Layer of Cork Oaks with Low and Moderate *Coraebus undatus* Attack

Rita Simões [1], Manuela Branco [1], Carla Nogueira [1], Carolina Carvalho [1], Conceição Santos-Silva [2], Suzana Ferreira-Dias [3], Isabel Miranda [1,*] and Helena Pereira [1]

[1] Centro de Estudos Florestais (CEF), Associate Laboratory TERRA, Instituto Superior de Agronomia, Universidade de Lisboa, Tapada da Ajuda, 1349-017 Lisboa, Portugal
[2] UNAC, União da Floresta Mediterrânica, Rua Mestre Lima de Freitas 1, 1549-012 Lisboa, Portugal
[3] Linking Landscape, Environment, Agriculture and Food (LEAF), Associate Laboratory TERRA, Instituto Superior de Agronomia, Universidade de Lisboa, Tapada da Ajuda, 1349-017 Lisboa, Portugal
* Correspondence: imiranda@isa.ulisboa.pt

Abstract: The beetle *Coraebus undatus*, during its larval stage feeds, and excavates galleries on the cork-generating layer of *Quercus suber* L. trees, seriously affecting the cork quality with significant economic losses for the cork industry. This work compared the composition of the extracts present in the innermost cork layers (the belly) of cork planks from *Q. suber* trees with low and moderate *C. undatus* attack in one stand. The total extractives in the inner cork layer from trees with moderate and low *C. undatus* attacks were similar (on average 22% of the cork mass) with a high proportion of polar compounds (91% of the total extractives). The chemical composition of the inner cork lipophilic extractives was the same in trees infested and free of larvae, with triterpenes as the most abundant family accounting for 77% of all the compounds, predominantly friedeline. The hydrophilic extractives differed on the levels of phenolic compounds, with higher levels in the inner cork extracts of samples from trees with low attack (90.0 mg GAE g^{-1} vs. 59.0 mg GAE g^{-1} of inner cork mass) The potential toxic activity of phenolic compounds may have a role in decreasing the larval feeding.

Keywords: *Quercus suber*; cork borer; cork; chemical composition; phenolics

Citation: Simões, R.; Branco, M.; Nogueira, C.; Carvalho, C.; Santos-Silva, C.; Ferreira-Dias, S.; Miranda, I.; Pereira, H. Phytochemical Composition of Extractives in the Inner Cork Layer of Cork Oaks with Low and Moderate *Coraebus undatus* Attack. *Forests* **2022**, *13*, 1517. https://doi.org/10.3390/f13091517

Academic Editor: Hans Beeckman

Received: 26 July 2022
Accepted: 15 September 2022
Published: 19 September 2022

Publisher's Note: MDPI stays neutral with regard to jurisdictional claims in published maps and institutional affiliations.

1. Introduction

The cork oak (*Quercus suber* L.) is an evergreen oak species adapted to dry and warm regions, distributed along the western Mediterranean basin, covering an area of approximately 2.1 million ha, in agro-forestry systems with high ecological and socio-economic importance [1,2]. Cork oak management is oriented towards the production of cork. This is one of the most important non-wood forest products that feeds a dedicated industrial chain, with wine cork stoppers as its worldwide famous product [3]. The world cork leader is Portugal, with an annual production of 100 thousand tonnes of raw cork, and second is Spain, with 62 thousand tonnes in 2010 [4].

Biotic threats to cork oak forests are a major concern since they may affect tree vitality and survival, as well as cork quality. One such biotic threat is related to infestations by the cork beetle *Coraebus undatus* Fabricius (Coleoptera, Buprestidae), named "cobrilha da cortiça" in Portugal and "culebrilla" in Spain, which negatively impacts the cork value for industrial processing [3] and, to a smaller extent, may induce stem wounds. Several reports indicated high infestation intensity in several regions of southern Spain and southern France [5,6]. *Coraebus. undatus* females lay their eggs on the surface or in cracks of the cork back (i.e., the external surface of the cork layer covering the stem) about 4 years after the previous cork extraction [7,8]. The hatched larvae perforate the layers of the cork tissue underneath the cork back to the phellogen, where they feed and grow, excavating long, sinuous galleries reaching 2 m in length and 3–4 mm in width, with a higher incidence in the stem region at a height between 0.5 m and 1.5 m [9]. Larval development lasts from

1 to 2 years, after which, the larva builds a pupal chamber within the cork layers, where it matures until the adult insect is formed and perforates the cork, emerging to the outside.

The occurrence of *C. undatus* attacks is rather elusive since the galleries are only visible when the cork is removed at the phellogen region at the end of one production cycle (on average 9 years) and appear as ribbon-like scars marked on the stem and on the inner part (i.e., the belly) of the cork plank, often showing dark fillings of larval feed excrements (Figure 1). Galleries may also be detected within the cork planks during cork processing and quality evaluation since they become embedded in the cork tissue during growth, and this constitutes a major devaluation of the cork plank [3,10,11]. Given the *C. undatus* cycle (egg laying starting in 4-year-old cork back and 2-year larval development) and the most common 9-year cork production cycle that prevails in the most important cork regions [12], it is to be expected that two to four attacks may occur during one cork production cycle. In addition to the decrease in cork quality, *C. undatus* attacks may also negatively impact tree health since, during the cork extraction, a localized phloem tearing may originate in the regions of the galleries, causing irreversible damage to the tree in that area with response to wounding and the risk for potential biological attacks.

Figure 1. Galleries of *Coraebus undatus* larvae at the time of cork extraction (**left**) on *Q. suber* trunk (**right**) showing a larva and an excrement-filled gallery on the inner side (belly) of cork planks.

Although a few studies on *C. undatus* attacks have tried to link infestation and the intensity of attacks to stand characteristics, e.g., tree density, the presence of understory, solar orientation, drought stress, or tree parameters, e.g., age, diameter, height, or health status [8,9], a clear pattern has not been obtained, in part due to a high degree between tree variation regarding attack levels: For instance, in one study in eight cork oak forest plots exploited for cork production in southern Spain (Natural Park "Los Alcornocales"; Sierra Morena, Huelva), the infestation index ranged from 0.40 to 2.32 [8].

To the best of our knowledge, no studies have related the presence of *C. undatus* with the chemical characteristics of the cork in the phellogenic region. The chemical composition of cork has been studied in detail, given its importance on the material's properties that are at the base of its applications [13], and it is known that there is a large natural variability regarding its most relevant components, namely in the content of extractives, suberin, and lignin [13,14].

Extractives are small molecules non-linked to the cell wall structural components that are soluble in the appropriate solvents. They are of particular importance since resistance against pests and pathogens usually relies on high concentrations of a diverse array of plant secondary metabolites [15,16]. Secondary metabolites are essential for reducing plant palatability and affect pest growth, development, and digestion, as shown in various examples. The activity of triterpenic compounds (betulinic and ursolic acids) on insect feeding was observed against the third instar larvae of castor semi-looper (*Achoea janata* (Linnaeus) (Lepidoptera: Noctuidae)) [17]. The triterpenoid fridelin isolated from *Azima tetracantha* Lam. (Salvadoraceae) leaves showed anti-food, larvicidal, and pupicidal activities against *Helicoverpa armigera* (Hübner) (Lepidoptera: Noctuidae) and *Spodoptera litura* (Fabricius) (Lepidoptera: Noctuidae) [18]. Phenolics can also be toxic to insects, for example, to the larvae of *Spodoptera litura* and its parasitoid *Bracon hebetor* (Say) (Hymenoptera: Braconidae) [19], and to the southern armyworm *Spodoptera eridania* (Cramer) (Lepidoptera: Noctuidae) [20].

Cork possesses significant amounts of secondary metabolites that are soluble compounds not chemically linked to the structural polymers of the cell walls, mainly composed of aliphatic, triterpenic, and phenolic compounds. The relative abundance and chemical composition of these compounds show a very high natural variability, even among trees of the same species and forest [21–28]. This work focuses on the content and chemical composition of the extractives present in the innermost cork layers of the cork planks, where larvae feed and excavate their galleries, taken from trees with very low *C. undatus* attack and from trees with moderate attack, as assessed at the time of cork extraction. The objective was to analyze the chemical profile of these secondary metabolites in cork in relation to the presence of the cork borer.

2. Material and Methods

2.1. Sampling

Cork planks were collected from *Q. suber* mature trees under exploitation (i.e., cork extraction) in one cork oak forest (montado), located in central western Portugal in the Coruche municipality (38°57′ N, 8°37′ W). The region has a Mediterranean-type climate with Atlantic influence, with its highest temperatures in summer (June to September) when precipitation is lowest: a mean average annual rainfall of 775 mm, 83.0% of precipitation concentrated from October to April, an average annual maximum temperature of 21 °C, and an average annual minimum temperature of 14 °C. The sampling took place in July and August during the period of cork stripping. A total of 97 cork oak trees were randomly selected from 11 five-tree plots scattered and covering all of the area of the cork oak stand and measured regarding d.b.h. (diameter at 1.30 m from the ground and cork stripping height). The mean d.b.h. was 39.4 ± 6.2 cm, and the debarking height of the cork stripping ranged from 1.2 m to 2.7 m with an average of 1.9 m ± 0.51 m.

The presence of *C. undatus* galleries was identified by visual observation of the decorked stems and the inner side (belly) of the cork planks. The classification of the intensity of attack (damage index) was based on the visual observation of the debarked stem after cork removal following the methodology used by Du Merle and Attié [29] and applied in *C. undatus* assessments [8,9]. On each tree, four vertical lines oriented to the north, east, south, and west sides of the stripped part of the stem were divided into 50 cm long sections from the soil surface to a maximum height of four levels (200 cm). The crosses between galleries and each of these lines were counted and the tree damage intensity (AI) was calculated as AI = total number of gallery intersections/4 × number of vertical levels occupied (4 is the number of orientation sections; N, S, E, and W). The plot infestation index was calculated as IP = $\sum$ AI/N, where N is the number of trees per plot. The cork samples (20 cm × 20 cm) for chemical analysis were taken from a subset of 22 *Q. suber* trees with different attack intensities that were grouped as (a) trees with very low *C. undatus* attack, corresponding to trees with a damage index between 0 and 0.06 (mean 0.03); and (b) infested trees with *C. undatus* attack corresponding to a damage index between 0.13 and 0.38 (mean 0.25). For the sake of simplicity, the two groups of samples were coded as "low attack" and "moderate attack".

The innermost cork layers of the cork plank (the belly side) were manually separated with a chisel corresponding to a removal of approximately 2 mm of thickness. This sample comprises the first few layers of phellem (cork tissue) adjacent to the phellogen and some remains of the phellogen tissue. The samples were ground individually in a Retsch (SM2000, Retsch GmbH, Haan, Germany) cutting mill, passing through a 1 mm × 1 mm sieve. The milled cork samples were dried in an oven at 60 °C and kept for analysis.

2.2. Lipophilic Compounds Extraction

The milled cork samples were Soxhlet extracted for 6 h with dichloromethane (Sigma-Aldrich, ≥99.8% purity, St. Louis, MO, USA) to recover the soluble lipophilic compounds. The extraction yield was determined by the mass difference of the solid residue after drying at 60 °C overnight and for 1 h at 105 °C and reported as a percent of the original sample. The extract was used for gas chromatography–mass spectrometry (GC–MS) qualitative and quantitative analysis.

2.3. Hydrophilic Compounds Extraction

The dichloromethane extracted samples were suspended in a 50:50 (v/v) ethanol/water mixture in an ultrasonic bath at room temperature for 16 h. The suspension was filtered through glass filter crucible G3 (pore size 15 to 30 μm) and the resulting solution was used for the quantitative analysis of total phenols, flavonoids, proanthocyanidins, and antioxidant activity. The yield of extractives solubilized by the ethanol–water mixture was determined by the mass difference of the solid residue after drying at 60 °C overnight and for 1 h at 105 °C and reported as a percent of the original sample.

2.4. Chemical Characterization of the Lipophilic Extract

The lipophilic extracts solubilized by dichloromethane were recovered as a solid residue after solvent evaporation under N_2 flow and dried overnight under vacuum at room temperature. A 2 mg extract sample was derivatized in 100 μL of pyridine (Sigma-Aldrich, ≥99.8% purity, St. Louis, MO, USA) by adding 100 μL of bis(trimethylsilyl)-trifluoroacetamide (BSTFA, Sigma-Aldrich, ≥99.0% purity, St. Louis, MO, USA) and kept for 30 min in an oven at 60 °C, by which the compounds with hydroxyl and carboxyl groups were trimethylsilylated into trimethylsilyl (TMS) ethers and esters, respectively.

The derivatized extracts were immediately injected in a GC–MS Agilent 5973 MSD with the following GC conditions: Zebron 7HG-G015-02 column (30 m, 0.25 mm; ID, 0.1 μm film thickness), flow 1 mL/min, injector 280 °C, oven temperature program, 100 °C (1 min), rate of 10 °C/min up to 150 °C, rate of 4 °C/min up to 300 °C, rate of 5 °C/min up to 370 °C, rate of 8 °C/min up to 380 °C (5 min). The MS source was kept at 220 °C, and the electron impact mass spectra (EIMS) were taken at 70 eV of energy. The compounds were identified as TMS derivatives by comparing their mass spectra with a GC–MS spectral library database (Wiley, NIST Mass Spectral Library) with over 90% similarity and by comparing their fragmentation profiles with published data [30–32]. For semi-quantitative analysis, the area of peaks in the total ion chromatograms of the GC–MS analysis was integrated, and their relative proportions were expressed as the area proportion of the total chromatogram area. Each aliquot was injected in triplicate, and the mean results are given (only a standard deviation inferior to 5% was considered).

2.5. Ethanol-Water Extract Composition

The ethanol–water extracts obtained from the dichloromethane-extracted samples were analyzed in terms of total phenolics, flavonoids, and condensed tannins. The total phenolic content (TPC) was determined according to the Folin–Ciocalteu method [33]. Briefly, 100 μL of the extract sample was added to 4 mL of diluted Folin–Ciocalteu reagent (Sigma-Aldrich, ≥99.8% purity, St. Louis, MO, USA) (1:10 v/v) and afterwards to 4 mL of aqueous sodium carbonate (Sigma-Aldrich, ≥99.9% purity, St. Louis, MO, USA) (7.5 g/L). The mixture was incubated for 30 min in the dark and was recorded with a spectrophotometer UV/Vis V-530 spectrophotometer (Jasco, Tokyo, Japan) at 750 nm against a blank containing only water. The TPC was calculated from a standard curve with gallic acid (Sigma-Aldrich, ≥99% purity, St. Louis, MO, USA) (range 0.014–0.762 mg/mL) and expressed as the mg of gallic acid equivalents (GAE) per gram of extract. The analyses were carried out in triplicate, and the average value was calculated.

The total flavonoid content was determined using the aluminum chloride colorimetric assay with catechin (CA) as standard [34]. Briefly, 1 mL of the extract sample was added with 0.3 mL $NaNO_2$ (Sigma-Aldrich, ≥99% purity, St. Louis, MO, USA) solution (5% w/v) and 0.3 mL $AlCl_3$ (Sigma-Aldrich, ≥ 99% purity, St. Louis, MO, USA) solution (10% w/v). The mixture was then allowed to stand for 6 min. Afterwards, 2 mL of sodium hydroxide (1 M) and 2.4 mL of water were added sequentially and vigorously shaken. The absorbance was recorded at 510 nm after 30 min of incubation against water (UV/Vis V-530 spectrophotometer). The results were calculated according to the calibration curve for catechin (Sigma-Aldrich, ≥99% purity, St. Louis, MO, USA) (0.10–1.0 mg/mL). The total flavonoid content was expressed as the mg of catechin (CE) equivalent/g of the extract. Triplicate measurements were carried out.

The total proanthocyanidin content (condensed tannins) was determined according to the vanillin-sulfuric acid method [35]. Briefly, a volume of 100 μL of the extract was mixed with 2.5 mL of 1.0% (w/v) vanillin (Sigma-Aldrich, ≥99% purity, St. Louis, MO, USA) methanolic solution and 2.5 mL of 25% (v/v) sulfuric acid in absolute methanol. The blank solution was prepared with the same procedure without vanillin. The absorbances of the extract samples and blanks were recorded at 500 nm after 15 min. Catechin standard solutions (0.1–0.6 mg/mL) were used for constructing the calibration curve, and the amount of total condensed tannins was expressed as the mg of catechin (CE) equivalent/g of the extract. Triplicate measurements were carried out.

2.6. Antioxidant Activity of Ethanol–Water Extracts

The antioxidant activity of the ethanol–water extracts was determined by the 2,2-diphenyl-1-picrylhydrazyl (DPPH) (DPPH, Sigma-Aldrich, ≥99.0% purity, St. Louis, MO, USA), which measures the free radical scavenging capacity [35]. Different dilutions of the extract were prepared and an aliquot of 100 μL of each solution was added to 3.9 mL of a DPPH methanolic solution (24 μg/mL). A control was prepared by adding 100 μL of methanol to 3.9 mL of a DPPH methanolic solution (24 μg/mL). The mixtures were shaken vigorously and left to stand in the dark for 30 min. The absorbance at 517 nm was measured using a UV/Vis V-530 spectrophotometer and compared to the initial absorbance of the DPPH solution using methanol as blank. The scavenging activity was estimated based on the percentage of the DPPH radical scavenged. Trolox (Sigma-Aldrich, ≥97% purity, St. Louis, MO, USA), and (+)-catechin were used as reference compounds. The IC_{50} values were determined from the plotted graphs of scavenging activity against the concentration extracts and represent the amount of extract necessary to decrease the initial DPPH concentration by 50%. Low IC_{50} values indicate high free-radical scavenging activities. The scavenging effect on the DPPH radical of the extract was also expressed as the mg of Trolox equivalent/g of the extract. All analyses were run with three replicates and averaged.

2.7. Statistical Analysis

The data were analyzed using the Sigmaplot statistical software (version 11.0, Systat Software Inc., San Jose, CA, USA). The normality of each distribution was analyzed by the Shapiro–Wilk test and the homogeneity of variances by Levene's test. Student's t test or the Mann–Whitney U test, depending, respectively, on the presence or absence of a normal distribution with equal variances. Principal component analysis (PCA) and cluster analysis (CA) were performed in order to evaluate the presence of an eventual relationship between chemical compounds present in the inner cork layer and the *C. undatus* intensity attack level and the presence of sample groups [36,37]. Samples 1 to 10 correspond to very low attack samples; sample codes from 11 to 22 correspond to moderate attack. In CA, the

Euclidean distance and single linkage methods were used. PCA and CA were performed with the Statistica™ software, version 7, from Statsoft (Tulsa, OK, USA).

3. Results and Discussion

3.1. Incidence of C. undatus

Previous studies showed that *C. undatus* is a very common phytophagous insect in cork oak forests in the Mediterranean region, in some sites affecting more than 90% of the sampled forests and over 70% of trees, although in many other sites, the levels of damage fluctuate from low to moderate [5,8,9]. Nonetheless, IA and IP values may underestimate the damage associated with *C. undatus* [8]. The reason for this is that the damage observed when the cork is removed (i.e., the larval paths on the phellogenic region by which the cork plank was torn away) results from the galleries of the last brood of *C. undatus*, which are observable in the decorked stem. This does not mean that older galleries generated by previous broods (from the years preceding the cork extraction) may not be present and unseen because they were incorporated into the cork tissue. However, the mere presence of one gallery highly devaluates a cork plank.

Of the 97 observed trees, 43% were affected by *C. undatus*. However, the mean stand infestation index (IP) value was very low with an IP = 0.17 ± 0.09, thereby revealing low *C. undatus* population levels. This agrees with Branco et al. [38] who reported that the percentage of trees attacked by *C. undatus* in cork oak forests in Portugal varies between 0% and 50%.

3.2. Extractable Components in the "Inner Cork"

The extraction yields of the belly of cork planks ("inner cork") from cork oak trees with low and moderate damage from *C. undatus* are shown in Table 1. The total content was similar for the two groups of cork oaks, corresponding on average to 22% of the cork mass. The difference between the two groups of samples was not statistically significant ($t = 0.479$; $p = 0.637$).

Table 1. Extractives (% of the dry mass) of the inner cork layer of cork planks extracted from trees with low attack and with moderate attack from *Coraebus undatus* (mean and standard deviation).

	Low Attack	Moderate Attack
Total extractives	21.6 [a] ± 2.7	22.0 [a] ± 1.7
Dichloromethane extractives	2.3 [a] ± 1.2	1.7 [a] ± 1.5
Ethanol-water extractives	19.3 [a] ± 3.3	20.3 [a] ± 2.5

Different letters indicate significant differences at $p < 0.05$.

In both groups of cork oaks, a striking chemical feature of the inner cork extracts was the high proportion of polar compounds soluble in ethanol–water, which accounted for 91% of the total extractives (20% of the oven dry cork mass). The lipophilic compounds soluble in dichloromethane corresponded to only 9% of the total cork extractives (2% of the cork mass). The differences between the two groups of samples with different damage indexes were not statistically significant for lipophilic ($t = 0.955$; $p = 0.351$) and hydrophilic extractives ($t = -0.0443$; $p = 0.965$).

Overall, the content in extractives of the inner cork layer is well above the average value of 16.2% reported for *Q. suber* cork, although within the range of values for the species (8.6%–32.9%) [13]. There is also a difference in the chemical profile of the inner cork extractives in relation to that of the complete cork layer since the polar extractives represent only 42%–70% of the total extractives in the complete cork layer, while 91% is in the inner cork region [13,14,39]. The inner cork region close to the phellogen contains young cellular material, some of which is still in metabolic processing, and this explains the high content in polar compounds which comprise phenolic compounds, as reported further on, and also soluble carbohydrates.

Only very few studies have addressed the radial variation of cork chemical composition across the cork plank thickness. Jové et al. [40] analyzed the cork chemical composition in three radial positions near the back, mid-cork, and belly, and observed differences with the highest extractives content in the belly layer.

3.3. Lipophilic Extractives Composition

The identified compounds in lipophilic extracts of the inner cork layer from samples with low and moderate *C. undatus* attack are given in Table 2 in proportion to the total chromatogram area grouped by chemical family.

Table 2. Chemical composition (% of all chromatogram peak areas) of the lipophilic extract of the inner cork layer of cork planks extracted from trees with very low and moderate attack from *Coraebus undatus*.

	Low Attack	Moderate Attack
Dichloromethane Extractives, % Dry Mass	2.31 [a] ± 1.2	1.65 [a] ± 1.5
Alkanols	1.48 [a] ± 0.11	1.91 [a] ± 0.35
Hexadecan-1-ol	0.01 ± 0.01	0.04 ± 0.03
Octadecan-1-ol	0.04 ± 0.05	0.03 ± 0.05
Eicosan-1-ol	0.16 ± 0.11	0.23 ± 0.14
Docosan1-ol	0.67 ± 0.39	0.97 ± 1.07
Tetracosan-1-ol	0.53 ± 0.25	0.60 ± 0.74
Hexacosan-1-ol	0.07 ± 0.04	0.04 ± 0.06
Alkanoic acids	6.05 [a] ± 0.43	8.82 [b] ± 0.49
Saturated fatty acids	2.80 [a] ± 0.36	4.57 [b] ± 0.37
Hexadecanoic acid	1.45 ± 0.67	2.62 ± 0.87
Heptadecanoic acid	0.07 ± 0.03	0.08 ± 0.02
Octadecanoic acid	0.96 ± 0.85	1.47 ± 0.73
Eicosanoic acid	0.14 ± 0.16	0.19 ± 0.09
Docosanoic acid	0.18 ± 0.07	0.21 ± 0.05
Unsaturated fatty acids	3.30 [a] ± 0.60	4.35 [a] ± 0.64
9-cis-Hexadecenoic acid	0.53 ± 0.01	0.58 ± 0.25
9,12-Octadecadienoic acid	0.52 ± 0.51	1.34 ± 1.87
9,12,15-Octadecatrienoic acid	0.07 ± 0.09	0.37 ± 0.52
9-Octadecenoic acid	1.78 ± 2.03	1.89 ± 0.51
13-Octadecenoic acid	0.35 ± 0.38	0.09 ± 0.03
Substituted fatty acids	0.93 [a] ± 0.19	1.24 [a] ± 0.37
2-Hydroxy-2-methylpropanoic acid	0.76 ± 0.09	0.84 ± 0.57
2-Hydroxy-4-methylpentanoic acid	0.17 ± 0.29	0.39 ± 0.17
Dicarboxylic acids	0.27 [a] ± 0.13	0.3 [a] ± 0.20
Saturated dicarboxylic acid	0.17 [a] ± 0.20	0.21 [a] ± 0.19
Propanedioic acid	0.04 ± 0.06	0.03 ± 0.05
Nonanedioic acid	0.13 ± 0.25	0.18 ± 0.32
Substituted dicarboxylic acid	0.10 [a] ± 0.08	0.13 [a] ± 0.22
2-Hydroxydecanedioic acid	0.10 ± 0.08	0.13 ± 0.22
Glycerol derivatives	1.90 [a] ± 1.17	1.20 [a] ± 0.21
Glycerol	1.52 ± 0.93	0.91 ± 0.37
2,3-Dihydroxypropyl hexadecanoate	0.27 ± 0.19	0.14 ± 0.07
2,3-Dihydroxypropyl octadecanoate	0.11 ± 0.03	0.15 ± 0.20
Terpenes	77.95 [a] ± 3.48	76.12 [a] ± 2.29
Squalene	0.72 ± 0.95	0.62 ± 0.60
Lupeol	0.65 ± 0.15	0.74 ± 0.42
Friedelane-1-ene-3-one	17.56 ± 9.18	16.92 ± 10.06
Erythrodiol	0.28 ± 0.17	0.40 ± 0.34
Friedelin	37.56 ± 14.41	36.59 ± 5.03
Lup-20(29)-en-3-one	0.24 ± 0.19	0.19 ± 0.24
Betulin	0.97 ± 0.00	1.13 ± 0.55
Betulinic acid	15.71 ± 15.3	16.43 ± 10.21
Betulinaldehyde	0.32 ± 0.30	0.21 ± 0.15

Table 2. *Cont.*

	Low Attack	Moderate Attack
D:A-Friedooleanan-28-al, 3 oxo (Canophyllal)	0.63 ± 0.22	0.87 ± 0.27
D:A-Friedooleanan-3-one, 28-hydroxy-	1.06 ± 1.28	0.68 ± 0.63
D:A-Friedo-2,3-secooleanane-2,3-dioic acid, dimethyl ester, (4R)-	2.25 ± 2.53	1.33 ± 1.18
Sterols	3.19 [a] ± 0.37	4.32 [a] ± 0.22
β-Sitosterol	1.86 ± 1.00	3.05 ± 0.62
Lanosterol	0.42 ± 0.04	0.34 ± 0.09
Cycloartenol	0.30 ± 0.14	0.30 ± 0.25
Cycloeucalenol	0.39 ± 0.00	0.50 ± 0.10
Sitosteryl-3beta-D-Glucopiranoside	0.22 ± 0.28	0.13 ± 0.17
Aromatic compounds	1.90 [a] ± 0.17	1.54 [a] ± 0.13
Benzoic acid	0.19 ± 0.11	0.21 ± 0.11
Salicylic acid	0.08 ± 0.11	0.11 ± 0.14
Vanillin	0.49 ± 0.28	0.31 ± 0.20
Vanillin acid	0.18 ± 0.08	0.18 ± 0.07
Caffeic acid	0.01 ± 0.01	0.04 ± 0.06
Caffeic acids derivatives	0.72 [a] ± 0.14	0.50 [b] ± 0.12
Caffeic acid + Triacontanoic acid	0.42 ± 0.24	0.21 ± 0.11
Caffeic acid + Dotriacontanoic acid	0.22 ± 0.18	0.17 ± 0.25
Caffeic acid + Tetratiacontanoic acid	0.08 ± 0.00	0.12 ± 0.00
Others	6.84 ± 1.53	1.94 ± 0.29
Levoglucosan	1.11 ± 2.06	0.70 ± 0.61
Quinic acid	4.68 ± 3.98	-
Myo-inositol	0.51 ± 0.00	0.66 ± 0.00
(7E,11E,15E)-3-(methoxymethoxy)-3,7,16,20-tetramethylhenicosa-1,7,11,15,19-pentaene	0.43 ± 0.06	0.45 ± 0.19
Octacosahydro-9,9′-biphenanthrene	0.11 ± 0.01	0.13 ± 0.05
Identified	96.96 ± 7.90	92.04 ± 7.27

Different letters indicate significant differences at $p < 0.05$.

The lipophilic fraction in the inner cork layer contains essentially the same compounds and in a similar proportion in samples with *C. undatus* damage. Triterpenes represented the most abundant family, accounting for 77% of all the compounds (77.95 % and 76.12%), including predominantly friedelin (37.56 and 36.59%), friedelane-1-en-3-one (17.56%–16.92%), and betulinic acid (15.71%–16.43%) that together constitute about 93% of all the identified triterpenes and triterpenoids. Other triterpenoids such as betuline and lupeol, as well as β-sitosterol were also identified in small amounts.

The long-chain lipids represented only 6.05% and 8.82% of all the compounds, including mainly fatty acids with hexadecanoic (palmitic acid) and octadecanoic (stearic acid) acids as the most abundant saturated fatty acids, and octadec-9-enoic acid (oleic acid) as the most abundant unsaturated fatty acid. Long-chain aliphatic alcohols were present in the inner cork dichloromethane extract in small amounts (1.48% and 1.91%, respectively) with docosan-1-ol and tetracosan-1-ol as the major fatty alcohols. Glycerol and two monoglycerides (monopalmitin and monostearin) were also found, representing together 1.90 and 1.20 % of the total identified compounds, respectively.

Aromatic compounds were found in minor concentrations (1.3% of total compounds), mostly vanillin and vanillinic acid. Other compounds, such as quinic acid, were identified only in the inner cork layer from trees with low *C. undatus* galleries (4.68% of all compounds).

Overall, all of the identified lipophilic compounds have been previously reported in cork extractives with small differences in composition: lipophilic extracts of corks from *Q. suber* from Bulgaria and Turkey [38] showed the dominance of triterpenoids with betulinic acid and friedelan-3-one as the main components. The inner cork compared to the complete cork had a lower proportion of alkanoic acids (6.8% vs. 14.2% in reproduction cork) and a higher proportion of triterpenes (77% vs. 50% in reproduction cork) [23,25,26,28].

The insecticidal and phytotoxic potential against plant-pathogens of these lipophilic compounds has been reported, e.g., pentacyclic triterpenes have anti-insect properties [41] as well as friedelane triterpenes derived from the byproducts of cork processing [42]. Such compounds exist in the inner cork layer (Table 2) but in amounts without difference between trees with low and moderate *C. undatus* attacks. Therefore, the main terpenes in the inner cork layer (e.g., friedeli, friedelane-1-ene-3-one, and betulinic acid, Table 2) do not seem to have anti-food and anti-larvicidal properties against the larvae of *C. undatus*, at least in the present amounts since overall, the content of lipophilic compounds was very small (Table 1).

3.4. Composition of Hydrophilic Extractives

The composition of the hydrophilic extractives from the inner cork layer from samples with low and moderate *C. undatus* attack are presented in Table 3 regarding the content in phenolic compounds.

Table 3. Composition and antioxidant capacity of ethanol–water extracts of the inner cork layer of cork planks extracted from trees with very low and moderate attack from *Coraebus undatus* (mean and standard deviation).

	Low Attack	Moderate Attack
Total phenolics (mg GAE g^{-1} extract)	448.6 [a] $\pm$ 101.6	296.41 [b] $\pm$ 78.1
Total flavonoids (mg CE g^{-1} extract)	41.5 [a] $\pm$ 9.4	36.7 [a] $\pm$ 6.2
Proanthocyanidins (mg CE g^{-1} extract)	10.1 [a] $\pm$ 3.1	10.6 [a] $\pm$ 2.7
IC$_{50}$ (μg extract mL^{-1})	13.4 [a] $\pm$ 3.5	16.0 [a] $\pm$ 3.6

Different letters indicate significant differences at $p < 0.05$.

There was a clear difference in the levels of phenolic compounds with higher levels in extracts of inner cork from trees with low attack (448.6 mg GAE g^{-1} extract or 90.0 mg GAE g^{-1} of inner cork) as compared with samples with low attack intensity (296.4 mg GAE g^{-1} extract and 59.0 mg GAE g^{-1} of inner cork)($t = 3.773$; $p = 0.001$).

No differences were found between the two groups of samples (from trees with low and moderate *C. undatus* attack) regarding the content of flavonoids ($t = 1.333$; $p = 0.198$) or proanthocyanidins ($t = -0.345$; $p = 0.734$) nor regarding the antioxidant properties of the extract ($t = -1.035$; $p = 0.313$).

The comparison of the results obtained here for the inner-cork region with the literature data for the complete cork layer of reproduction cork shows some differences. The levels of phenolic compounds found in the inner cork layer were higher when compared to those found in extracts of reproduction cork (336.3 mg GAE g^{-1} of extract or 19.9 mg GAE g^{-1} dry cork) [43], or the mean values of 196.4 mg GAE g^{-1} of extract for ethanol–water extracts from cork [24]. The total flavonoid contents (36.7 mg and 41.5 mg CE g^{-1} extract in the inner cork layer) were lower compared with the values obtained for cork by Ferreira et al. [24]. As regards the antioxidant properties, the ethanol–water extracts of the inner cork revealed very low antioxidant activity with IC$_{50}$ values of 13.4 and 16.0 μg extract mL^{-1}, which is lower than previously reported for different cork extracts, namely 2.79 μg mL^{-1} in water extract, 3.58 μg mL^{-1} in methanol extract, and 5.84 μg mL^{-1} in methanol-water 50:50 extract [44] and for cork (3.2 μg mL^{-1}) [24]. The fact that inner cork extracts contain higher amounts of phenolics but have less antioxidant properties than complete cork extracts suggests that the antioxidant activity of the cork extracts is not directly related to the concentration of total phenolics and should result from specific phenolic compounds or from other compounds with antioxidant activity [26].

The main differences among the cork samples taken from trees with very low and moderate *C. undatus* attack are in the content of phenolic compounds (Table 3). Phenolics are frequently implicated in chemical defense mechanisms against pathogens in woody plants [45] that act as anti-nutritive agents or exert toxic effects on phytophagous insects [19,46,47]. Specifically for pine species, recent studies showed that phloem chemical composition, namely, the content in phenolic compounds was negatively related to pine nematode susceptibility [48,49]. The negative impact of specific phenolic compounds against some insect pests has been reported, for instance, of ellagic acid on the larvae of *Spodoptera litura* and its parasitoid *Bracon hebetor* [19] and on the southern armyworm *Spodoptera eridania* [20]. The predominant compounds detected in *Q. suber* reproduction cork extracts are ellagic, gallic, and protocatechuic acids [26–28,43], thus supporting their role in biotic defense.

Therefore, further studies targeted to analyze the phytochemical impact on cork borers' resistance should consider a detailed analysis of the phenolic and polyphenolic composition of the metabolites present in the inner cork extracts, including their bioactive properties that could be involved in protective functions, as well as a larger sampling of trees with differing *C. undatus* attacks.

Figure 2 shows the projection of variable loads (Figure 2A) and samples (Figure 2B) on the plane defined by the two first principal components (factors 1 and 2). This plane accounts for 69.5% of the variance of original data. PCA did not differentiate the two groups of samples with different *C. undatus* attack intensities since the projection of samples on this plane did not show any defined group. The PCA only showed that contents in dichloromethane extractives, condensed tannins, and flavonoids increase along the first PC while the content of ethanol–water extractives increase in the opposite sense (along the negative part of PC1) and IC50 increases along the second PC, opposite to the total phenol content. A CA was carried out to assess the presence of groups not detected by PCA (Figure 3). The groups formed are not related to the attack level. For instance, at a distance linkage of 50, we have three clusters where low-attacked samples and attacked samples are mixed.

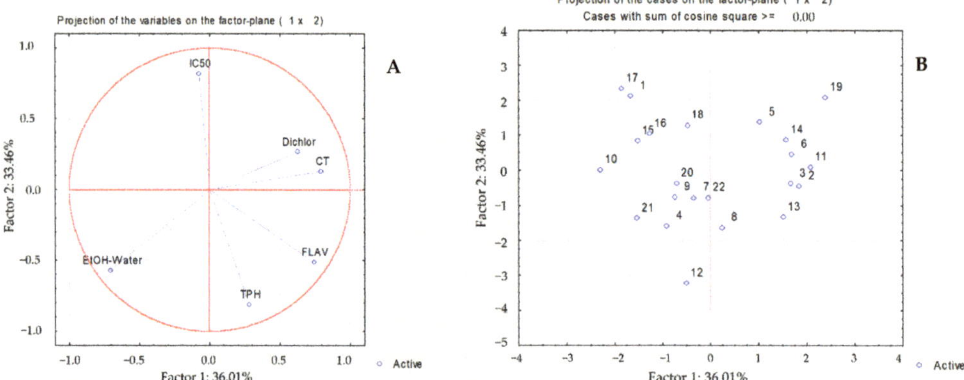

Figure 2. Principal components analysis of the chemical compounds present in the inner cork layer from tree samples with different *C. undatus* attack intensity levels: (**A**) score plots of the original variables; (**B**) samples (Samples 1 to 10 correspond to very low attack samples; sample codes from 11 to 22 correspond to moderate attack).

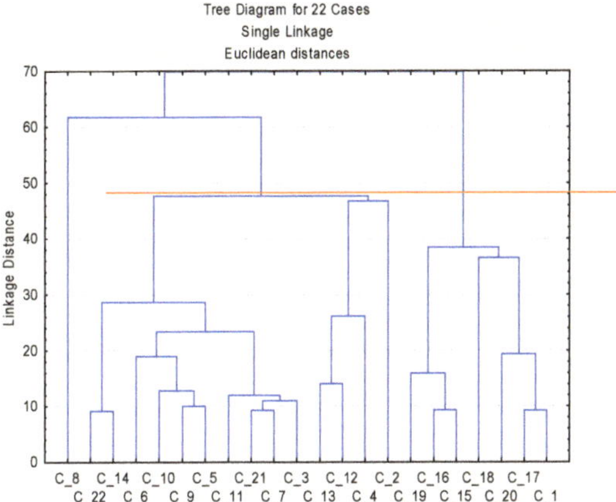

Figure 3. Dendrogram of the chemical compounds present in the inner cork layer from tree samples with different *C. undatus* attack intensity level.

4. Conclusions

This work reports for the first time the composition of extractives in the innermost cork layers (the belly) of cork planks from cork oak trees with very low and moderate *C. undatus* attack. No difference was found between the content of the total extractives, the proportion of apolar to polar compounds, and the composition of the lipophilic extractives in the inner cork layer from *Q. suber* trees with moderate or low *C. undatus* attack. However, a significant difference between the trees with moderate or very low attack was observed in the content of phenolics, which was 1.5 times higher in the trees with very low *C. undatus* attack. The potential toxic activity of phenolic compounds may have a role in decreasing larval feeding and survival. This result points out the need for targeted further research on a detailed analysis of the phytochemical composition of metabolites in the phellogenic and inner cork layers of *Q. suber* trees with differing *C. undatus* attacks. A relationship between phenolic contents and tree genome and physiological status should also be further investigated.

Author Contributions: Conceptualization, I.M., H.P. and M.B.; sampling and methodology, I.M., M.B., C.N., C.S.-S. and R.S.; experimental work, R.S. and C.C.; data curation, S.F.-D. and I.M.; writing—original draft preparation, I.M., H.P., S.F.-D. and R.S.; writing—review and editing, H.P., I.M., S.F.-D. and M.B. All authors have read and agreed to the published version of the manuscript.

Funding: This work was funded by PDR2020 through the project UNDERCORK-Gestão Integrada da cobrilha da cortiça and by FCT—Fundação para a Ciência e a Tecnologia, through funding of CEF-Forest Research Centre (UIDB/00239/2020).

Institutional Review Board Statement: Not applicable.

Conflicts of Interest: The authors declare no conflict of interest.

References

1. Bugalho, M.N.; Pinto-Correia, T.; Pulido, F. Human use of natural capital generates cultural and other ecosystem services in montado and dehesa oak woodlands. *Reconnecting Nat. Cult. Cap.* **2018**, *3*, 115–123.
2. Bugalho, M.N.; Caldeira, M.C.; Pereira, J.S.; Aronson, J.; Pausas, J.G. Mediterranean cork oak savannas require human use to sustain biodiversity and ecosystem services. *Front. Ecol. Environ.* **2011**, *9*, 278–286. [CrossRef]
3. Pereira, H. *Cork: Biology, Production and Uses*; Elsevier: Amsterdam, The Netherlands, 2007.
4. de Castro, A.; Avillez, F.; Rodrigues, V.; da Silva, F.G.; Santos, F.; Rebelo, F.; Jorge, M.N.; Aires, N. *The Cork Sector: From the Forest to the Consumer*; APCOR, Associação Portuguesa da Cortiça: Santa Maria de Lamas, Portugal, 2020.
5. Jiménez, A.; Gallardo, A.; Antonietty, C.A.; Villagrán, M.; Ocete, M.E.; Soria, F.J. Distribution of *Coraebus undatus* (Coleoptera: Buprestidae) in cork oak forests of southern Spain. *Int. J. Pest Manag.* **2012**, *58*, 281–288. [CrossRef]

6. Sallé, A.; Nageleisen, L.-M.; Lieutier, F. Bark and wood boring insects involved in oak declines in Europe: Current knowledge and future prospects in a context of climate change. *For. Ecol. Manag.* **2014**, *328*, 79–93. [CrossRef]

7. Martín, C. Notas sobre *Coraebus undatus* Fabr y Mars, plaga del alcornoque. *Bol. Serv. Plagas For.* **1964**, *7*, 95–101.

8. Gallardo, A.; Jiménez, A.; Antonietty, C.A.; Villagrán, M.; Ocete, M.E.; Soria, F.J. Forecasting infestation by *Coraebus undatus* (Coleoptera, Buprestidae) in cork oak forests. *Int. J. Pest Manag.* **2012**, *58*, 275–280. [CrossRef]

9. Cárdenas, A.M.; Toledo, D.; Gallardo, P.; Guerrero-Casado, J. Assessment and distribution of damages caused by the trunk-boring insects *Coraebus undatus* (Fabricius) (Coleoptera: Buprestidae) and *Reticulitermes grassei* Clément (Blattodea: Rhinotermitidae) in Mediterranean restored cork-oak forests. *Forests* **2021**, *12*, 1501. [CrossRef]

10. Gonzalez-Adrados, J.R.; Pereira, H. Classification of defects in cork planks using image analysis. *Wood Sci. Technol.* **1996**, *30*, 207–215. [CrossRef]

11. Oliveira, V.; Lopes, P.; Cabral, M.; Pereira, H. Influence of cork defects in the oxygen ingress through wine stoppers: Insights with X-ray tomography. *J. Food Eng.* **2015**, *165*, 66–73. [CrossRef]

12. Lauw, A.; Oliveira, V.; Lopes, F.; Pereira, H. Variation of cork quality for wine stoppers across the production regions in Portugal. *Eur. J. Wood Prod.* **2018**, *76*, 123–132. [CrossRef]

13. Pereira, H. Variability of the chemical composition of cork. *Bioresources* **2013**, *8*, 2246–2256. [CrossRef]

14. Pereira, H. Chemical composition and variability of cork from *Quercus suber* L. *Wood Sci. Technol.* **1988**, *22*, 211–218. [CrossRef]

15. Piasecka, A.; Jedrzejczak-Rey, N.; Bednarek, P. Secondary metabolites in plant innate immunity: Conserved function of divergent chemicals. *New Phytol.* **2015**, *206*, 948–964. [CrossRef]

16. López-Goldar, X.; Villari, C.; Bonello, P.; Borg-Karlson, A.K.; Grivet, D.; Zas, R.; Sampedro, L. Inducibility of plant secondary metabolites in the stem predicts genetic variation in resistance against a key insect herbivore in Maritime pine. *Front. Plant Sci.* **2018**, *9*, 1651. [CrossRef]

17. Chandramu, C.; Manohar, R.D.; Krupadanam, D.G.L.; Dashavantha, R. Isolation, characterization and biological activity of betulinic acid and ursolic acid from *Vitex negundo* L. *Phytother. Res.* **2003**, *17*, 129–134. [CrossRef]

18. Baskar, K.; Duraipandiyan, V.; Ignacimuthu, S. Bioefficacy of the triterpenoid friedelin against Helicoverpa armigera (Hub.) and Spodoptera litura (Fab.) (Lepidoptera: Noctuidae). *Pest Manag. Sci.* **2014**, *70*, 1877–1883. [CrossRef]

19. Punia, A.; Chauhan, N.S.; Kaur, S.; Sohal, S.K. Effect of ellagic acid on the larvae of *Spodoptera litura* (Lepidoptera: Noctuidae) and its parasitoid *Bracon hebetor* (Hymenoptera: Braconidae). *J. Asia Pac. Entomol.* **2020**, *23*, 660–665. [CrossRef]

20. Lindroth, R.L.; Peterson, S.S. Effects of plant phenols of performance of Southern armyworm larvae. *Oecologia* **1988**, *75*, 185–189. [CrossRef]

21. Fernandes, A.; Fernandes, I.; Cruz, L.; Mateus, N.; Cabral, M.; Freitas, V. Antioxidant and biological properties of bioactive phenolic compounds from *Quercus suber* L. *J. Agric. Food Chem.* **2009**, *57*, 11154–11160. [CrossRef]

22. Fernandes, A.; Sousa, A.; Mateus, N.; Cabral, M.; Freitas, V. Analysis of phenolic compounds in cork from *Quercus suber* L. by HPLC–DAD/ESI–MS. *Food Chem.* **2011**, *125*, 1398–1405. [CrossRef]

23. Castola, V.; Marongiu, B.; Bighelli, A.; Floris, C.; Lai, A.; Casanova, J. Extractives of cork (*Quercus suber* L.): Chemical composition of dichloromethane and supercritical CO_2 extracts. *Ind. Crops Prod.* **2005**, *21*, 65–69. [CrossRef]

24. Ferreira, F.; Santos, S.; Pereira, H. *In vitro* screening for acetylcholinesterase inhibition and antioxidant activity of *Quercus suber* cork and cork back extracts. *Evid. Based Complement. Alternat. Med.* **2020**, *2020*, 3825629. [CrossRef]

25. Şen, A.; Zhianski, M.; Glushkova, M.; Petkova, K.; Ferreira, J.; Pereira, H. Chemical composition and cellular structure of corks from *Quercus suber* trees planted in Bulgaria and Turkey. *Wood Sci. Technol.* **2016**, *50*, 1261–1276. [CrossRef]

26. Touati, R.; Santos, S.A.; Rocha, S.M.; Belhamel, K.; Silvestre, A.J. The potential of cork from *Quercus suber* L. grown in Algeria as a source of bioactive lipophilic and phenolic compounds. *Ind. Crops Prod.* **2015**, *76*, 936–945. [CrossRef]

27. Conde, E.; Cadahía, E.; García-Vallejo, M.C.; Fernández de Simóm, B. Polyphenolic composition of *Quercus suber* cork from different Spanish provenances. *J. Agric. Food Chem.* **1998**, *46*, 3166–3171. [CrossRef]

28. Conde, E.; García-Vallejo, M.; Pereira, C.; Cadahía, E. Waxes composition of reproduction cork from *Quercus suber* and its variability throughout the industrial processing. *J. Wood Sci. Technol.* **1999**, *33*, 229–244. [CrossRef]

29. Du Merle, P.D.; Attie, M. *Coroebus undatus* (Coleoptera Buprestidae) sur chêne-liége dans le sud-est de la France: Estimation des dégâts, relations entre ceux-ci et certains facteurs du milieu. *Ann Sci. For.* **1992**, *49*, 571–588. [CrossRef]

30. Eglinton, G.; Hunneman, D. Gas chromatographic-mass spectrometric studies of long-chain hydroxy acids-I. *Phytochemistry* **1968**, *7*, 313–322. [CrossRef]

31. Kolattukudy, P.; Agrawal, V. Structure and composition of aliphatic constituents of potato tuber skin. *Lipids* **1974**, *9*, 682–691. [CrossRef]

32. Marques, A.V.; Pereira, H. A methodological approach for the simultaneous quantification of glycerol and fatty acids from cork suberin in a single GC run. *Phytochem. Anal.* **2019**, *30*, 687–699. [CrossRef]

33. Singleton, V.L.; Rossi, J.A. Colorimetry of total phenolics with phosphomolybdic phosphotungstic acid reagents. *Am. J. Enol. Vitic.* **1965**, *16*, 144–158.

34. Chang, C.; Yang, M.; Wen, H.; Chern, J. Estimation of total flavonoid content in propolis by two complementary colorimetric methods. *J. Food Drug Anal.* **2002**, *10*, 178–182.

35. Abdalla, S.; Pizzi, A.; Ayed, N.; Bouthoury, F.C.; Charrier, B.; Bahabri, F.; Ganash, A. MALDI-TOF analysis of Aleppo pine (*Pinus halepensis*) bark tannin. *BioResource* **2014**, *9*, 3396–3406. [CrossRef]

36. Mirkin, B. Nonconvex Optimization and its Applications. In *Mathematical Classification and Clustering*; Springer Science & Business Media: New York, NY, USA, 1996.
37. Rencher, A.C. *Methods of Multivariate Analysis*; John Wiley and Sons: Hoboken, NJ, USA, 1995.
38. Branco, M.; Bragança, H.; Sousa, E.; Phillips, A.J. Pests and diseases in Portuguese forestry: Current and new threats. In *Forest context and policies in Portugal*; Reboredo, F., Ed.; Springer: Cham, Switzerland, 2014; pp. 117–154.
39. Cunha, M.; Lourenço, A.; Barreiros, S.; Paiva, A.; Simões, P. Valorization of cork using subcritical water. *Molecules* **2020**, *25*, 4695. [CrossRef] [PubMed]
40. Jové, P.; Olivella, M.À.; Cano, L. Study of the variability in chemical composition of bark layers of *Quercus suber* L. from different production areas. *BioResources* **2011**, *6*, 1806–1815.
41. González-Coloma, A.; López-Balboa, C.; Santana, O.; Reina, M.; Fraga, B.M. Triterpene-based plant defenses. *Phytochem. Rev.* **2011**, *10*, 245–260. [CrossRef]
42. Moiteiro, C.; Joao, M.; Curto, O.M.; Mohamed, N.; Bailen, M.; Martínez-Dıaz, R.; González-Coloma, A. Biovalorization of friedelane triterpenes derived from cork processing industry byproducts. *J. Agric. Food Chem.* **2006**, *54*, 3566–3571. [CrossRef]
43. Santos, S.A.O.; Villaverde, J.J.; Sousa, A.F.; Coelho, J.F.J.; Neto, C.P.; Silvestre, A.J.D. Phenolic composition and antioxidant activity of industrial cork by-products. *Ind. Crops Prod.* **2013**, *47*, 262–269. [CrossRef]
44. Santos, S.A.O.; Pinto, P.C.R.O.; Silvestre, A.J.D.; Neto, C.P. Chemical composition and antioxidant activity of phenolic extracts of cork from *Quercus suber* L. *Ind. Crops Prod.* **2010**, *31*, 521–526. [CrossRef]
45. Witzell, J.; Martin, J.A. Phenolic metabolites in the resistance of northern forest trees to pathogens—past experiences and future prospects. *Can. J. For. Res.* **2008**, *38*, 2711–2727. [CrossRef]
46. War, A.R.; Paulraj, M.G.; Ahmad, T.; Buhroo, A.A.; Hussain, B.; Ignacimuthu, S.; Sharma, H.C. Mechanisms of plant defense against insect herbivores. *Plant Signal Behav.* **2012**, *7*, 1306–1320. [CrossRef] [PubMed]
47. Lattanzio, V.; Lattanzio, V.M.T.; Cardinali, A. Role of phenolics in the resistance mechanisms of plants against fungal pathogens and insects. *Phytochem. Adv. Research.* **2006**, *661*, 23–67.
48. Pimentel, C.S.; Firmino, P.N.; Calvão, T.; Ayres, M.P.; Miranda, I.; Pereira, H. Pinewood nematode population growth in relation to pinephloem chemical composition. *Plant Pathol.* **2016**, *5*, 856–865.
49. Simões, R.; Pimentel, C.; Ferreira-Dias, S.; Miranda, I.; Pereira, H. Phytochemical characterization of phloem in maritime pine and stone pine in three sites in Portugal. *Heliyon* **2021**, *7*, e06718. [CrossRef] [PubMed]

Article

An Integrated Similarity Analysis of Anatomical and Physical Wood Properties of Tropical Species from India, Mozambique, and East Timor

Fernanda Bessa [1], Vicelina Sousa [1,2,*], Teresa Quilhó [1,2] and Helena Pereira [1,2]

1   Centro de Estudos Florestais, Instituto Superior de Agronomia, Universidade de Lisboa, Tapada da Ajuda, 1349-017 Lisboa, Portugal
2   Laboratório para a Sustentabilidade do Uso da Terra e dos Serviços dos Ecossistemas, Tapada da Ajuda, 1349-017 Lisboa, Portugal
*   Correspondence: vsousa@isa.ulisboa.pt

Abstract: Tropical species are highly valued timber sources showing a large diversity of wood characteristics. Since there are major concerns regarding the sustainability of these tropical species in many tropical regions, knowledge of the variability in wood properties is therefore a valuable tool to design targeted exploitation and to enlarge the wood resources base, namely by identifying alternatives for CITES-listed species. In this study, 98 tropical wood species belonging to 73 genera from India, Mozambique, and East Timor were investigated regarding wood anatomy and physical properties. Numerical taxonomy, by means of cluster analysis and principal component analysis grouped species with anatomical and physical similarities from different geographical origins. In addition to wood density, ray and vessel characteristics as well as wood moisture and wood shrinkage properties explained the main variability of these species. The contribution of wood color patterns was highlighted as consistently separating the Mozambique woods. A distinct geographical pattern was not observed, reinforcing that species from India, Mozambique, and East Timor show similar anatomical and physical wood properties, which could be useful to increase timber trade diversity. The multivariate analysis showed that species from Mozambique, such as *Morus mesozygia*, and *Millettia stuhlmannii* and *Swartzia madagascariensis*, could be alternatives for the CITES-listed species *Cedrela odorata* and *Dalbergia melanoxylon*, respectively.

Keywords: tropical species; wood anatomy; wood density; wood color; multivariate analysis; species diversity

Citation: Bessa, F.; Sousa, V.; Quilhó, T.; Pereira, H. An Integrated Similarity Analysis of Anatomical and Physical Wood Properties of Tropical Species from India, Mozambique, and East Timor. *Forests* **2022**, *13*, 1675 . https://doi.org/10.3390/f13101675

Academic Editor: Giai Petit

Received: 2 September 2022
Accepted: 6 October 2022
Published: 12 October 2022

Publisher's Note: MDPI stays neutral with regard to jurisdictional claims in published maps and institutional affiliations.

1. Introduction

Tropical species play an important role in world forest diversity as well as in timber trade. The sustainability of tropical forests is a matter of global concern, and international conventions have been crucial in controlling illegal wood trade. Wood identification and the characterization of tropical species are increasingly considered tools for trade monitoring (e.g., [1–4]). Most commercial tropical woods are mainly appreciated for their high wood density and aesthetic properties, such as wood color, although a high natural species variability is found. In fact, wood density varies within species and/or genera, ranging, for example, from 100 kg/m^3 in *Ochroma* sp. (balsa) and 380 kg/m^3 in *Triplochiton scleroxylon* K. Schum. (obeche) to over 1000 kg/m^3 in *Diospyros* sp. (ebony), *Tamarindus indica* L. (tamarind), and *Lophira alata* Banks ex Gaertn. (azobé) [5,6]. Wood color variability ranges from yellowish white in *Triplochiton scleroxylon* and *Turraenthus africana* Harms (avodire) to reddish brown in *Pterocarpus soyauxii* Taub. (padauk) and black in *Diospyrus crassiflora* Hiern and *Dalbergia melanoxylon* Guill. and Perr. (ebony) [7]. Even if there is this species diversity and wood properties variability, there are global concerns regarding the sustainability of the tropical forests for which the contribution of international conventions has been crucial by increasing the interest in wood identification and wood characterization of the tropical species as a tool to control illegal trade.

The anatomical wood characteristics and their interactions across different taxa (genera, families, and species) and climatic and ecological gradients have been extensively analyzed and are of great interest for phylogenetic, taxonomic, and wood identification purposes (e.g., [8–14]). However, tropical species represent important challenges due to their large diversity and the similarity of anatomical wood patterns in many genera and species (e.g., [15,16]. Therefore, genetic and chemical data are also being successfully studied to discriminate between tropical species (e.g., [4,17,18]).

Wood density is of high interest for taxonomic and phylogenetic studies since it is greatly controlled by genetic factors [19,20]. Thus, the genetic and/or geographical provenance effects on wood density are often studied for the most valued woods due to their practical implications for tree breeding and conservation programs (e.g., [21–25]).

The species characterization of the anatomical (e.g., tissue composition, cellular dimension and arrangement) and physical wood properties (e.g., wood density) is essential to analyze the specific wood quality and product suitability [26]. Given the within-species variability regarding wood properties, interactions across different geographical locations and ecological gradients are often studied to analyze the potential for wood production and/or species conservation [27–30].

Compared to temperate forests, less studies are found on the variability in the anatomical and physical wood properties from different tropical species across different geographical regions. Therefore, the systematic wood characterization of different tropical wood species is challenging and may allow their better sustainable use, i.e., by highlighting the importance of valuable secondary species that may increase the number of available timber species, thereby mitigating the tropical deforestation and species over-exploitation that are expressed by the endangered timber species listed in the Convention on International Trade of Endangered Species in Wild Fauna and Flora (CITES). About 25% of the traded timber in industrial countries in the northern hemisphere comes from tropical forests, namely from the Asia–Pacific region, followed by South America and Africa [31]. Some of the most valuable tropical woods, such as *Spirostachys africana* Sond. (tamboti, African sandalwood) and *Dalbergia melanoxylon*, are found in Mozambique; *Santalum album* L. (sandalwood) is found in East Timor; and *Tectona grandis* L. (teak) is found in India and East Timor. However, in these regions, other lesser-known wood species, such as *Pseudolachnostylis maprounaefolia* Pax, *Pericopsis angolensis* Meeuwen, *Sterculia appendiculata* K. Schum., and *Sterculia quinqueloba* K. Schum. or lesser-known provenances, for instance, of *Tectona grandis*, may reveal promising end uses [21,32,33].

In this study, the anatomical and physical wood characteristics from 98 tropical species from India, Mozambique, and East Timor were analyzed and compared by means of a multivariate analysis. The overarching goal of this research was to contribute to the better use of tropical wood resources by obtaining and adding knowledge on these tropical wood species based on the anatomical and physical wood characterization. The specific aims were to obtain the species classifications and analyze the similarities in the anatomical and physical properties of wood between species from different origins.

2. Materials and Methods

2.1. Collections

The data included in this paper are based on original observations of wood samples and published results [34–38]. The studied samples were selected from wood collections of the Forest Research Center of the School of Agriculture (Portugal LISFLw xylarium) and the Tropical Botanical Garden (Portugal LISJCw xylarium), both from the University of Lisbon (UL) in Portugal and from the University of Eduardo Mondlane (UEM) in Maputo, Mozambique. The wood samples were obtained from mature trees and correspond to mature wood. Biometric data and the ages of the trees were often not recorded. A total of 98 wood samples from tropical species were studied: 17 from India (I), corresponding to the former "Portuguese India" (Goa); 33 from East Timor (T); and 48 from Mozambique (M—from the UL or N—from the UEM), as described in Table 1.

Table 1. List of the 98 studied tropical species by code according to each geographical provenance: India (I), Mozambique (M or N), and East Timor (T). The species with more than one geographical provenance and the genera with more than one species are marked in bold. Species marked with an asterisk are listed in the Convention on International Trade in Endangered Species of Wild Fauna and Flora (CITES) Appendix II.

Code	Species	Code	Species
I1	*Acacia catechu* Willd.	T1	*Albizia lebbeckoides* (DC) Benth.
I2	*Aegle marmelos* Corrêa	T2	*Aleurites moluccana* Willd.
I3	*Albizia lebbeck* Benth.	T3	*Alstonia scholaris* (L.) R. Br.
I4	*Artocarpus integrifolia***L.**	T4	*Artocarpus integrifolia* L.
I5	*Bombax malabaricum* DC.	T5	*Bischofia javanica* Blume
I6	*Careya arborea* Roxb.	T6	*Calophyllum inophyllum* L.
I7 *	*Dalbergia sissoo* Roxb.	T7	*Canarium commune* L.
I8	*Eugenia jambolana* Lam.	T8	*Cassia fistula* L.
I9	*Ficus indica* Roxb.	T9	*Casuarina junghuhniana* Miq.
I10	*Lagerstroemia parviflora* Roxb.	T10 *	*Cedrela toona* var *australis* Roxb. C. DC.
I11	*Mangifera indica* L.	T11	*Decaspermum paniculatum* Kurz
I12	*Polyalthia fragans* Benth. and Hook	T12	*Elaeocarpus sphaericus* K. Schum.
I13	*Tectona grandis***L.**	T13	*Ficus macrophylla* Roxb.
I14	*Terminalia bellirica* Roxb.	T14	*Ganophyllum falcatum* Blume
I15	*Terminalia paniculata* Roth	T15	*Hibiscus tiliaceus* L.
I16	*Terminalia tomentosa* W. et Arn.	T16	*Homalium tomentosum* Benth.
I17	*Xylia dolabriformis* Benth.	T17	*Intsia bijuga* O. K.
M1	*Adina microcephala* (del.) Hiern	T18	*Macaranga tanarius* Muell.
M2	*Afrormosia angolensis* (Bak.) Harms	T19	*Melaleuca leucadendron* **L.**
M3	*Afzelia quanzensis* Welw.	T20	*Pometia pinnata* Forst.
M4	*Albizia adianthifolia* W. F. Wight	T21	*Pterocarpus indicus* Willd.
M5	*Albizia versicolor* Welw. ex Oliv.als	T22	*Pterospermum acerifolium* Will.
M6	*Amblygonocarpus obtusangulus* Harms	T23	*Pygeum* sp.
M7	*Androstachys johnsonii* Prain	T24	*Santalum album* L.
M8	*Bombax rhodognaphalon* K. Schum. ex. Engl.	T25	*Sarcocephalus cordatus* Miq.
M9	*Burkea africana* Hook.	T26	*Schleichera oleosa* Merr.
M10	*Celtis durandii* Engl.	T27	*Sterculia foetida* L.
M11	*Celtis kraussiana* Bernh.	T28	*Tamarindus indica* L.
M12	*Chlorophora excelsa* (*Milicia excelsa*) (Welw.) Benth. Hook	T29	*Tectona grandis* **L.**
M13	*Colophospermum mopane* Kirk.	T30	*Terminalia catappa* **L.**
M14	*Combretum imberbe* Wawra	T31	*Thespesia populnea* Soland, ex Corrêa
M15	*Cordyla africana* Lour.	T32	*Timonius rumphii* DC.
M16 *	*Dalbergia melanoxylon* Guill. and Perr	T33	*Vitex pubescens* Vahl
M17	*Dialium schlechteri* Harms	N1	*Acacia robusta* Burch

Table 1. *Cont.*

Code	Species	Code	Species
M18	*Diospyros mespiliformis* Hochst. ex A. DC.	N2	***Amblygonocarpus*** *andongensis* (Welw. ex Oliv.) Excell and Torre
M19	***Erythrophleum*** *africanum* (Benth.) Harms	N3	*Berchemia discolor* (Klotzsch) Hemsl.
M20	***Erythrophleum*** *guineense* Don	N4 *	***Cedrela*** *odorata* L.
M21	*Khaya* sp.	N5	*Cleistanthus schlechteri* (Pax) Hutch.
M22	*Khaya* sp.	N6	***Combretum*** *zeyheri* Sond.
M23	*Millettia stuhlmannii* Taub.	N7	*Diplorhynchus condylocarpon* (Mull. Arg.) Pichon
M24	***Morus*** *lactea* Mildbr. (*Celtis lactea* Sim.)	N8	***Melaleuca leucadendron*(L.) L.**
M25	*Ostryoderris stuhlmannii* Dunn ex Baker f.	N9	***Morus*** *mesozygia* Stapf
M26	*Piliostigma thonningii* (Schumach.) Milne-Redhead	N10	***Pterocarpus*** *antunesii* (Tab.) Harms
M27	*Piptadenia buchananii* Bak. (*Newtonia buchananii*)	N11	*Rhodognaphalon schumannianum* A. Robyns
M28	*Pteleopsis myrtifolia* (Lawson) Engl. and Diels	N12	*Schrebera trichoclada* Welw.
M29	***Pterocarpus*** *angolensis* DC.	N13	*Syncarpia glomulifera* (Sm.) Wield.
M30	*Ricinodendron rautanenii* (Schinz) Radcl-Sm	N14	*Syringa vulgaris* L.
M31	*Spirostachys africana* Sond.	N15	***Xylia*** *torreana* Brenan
M32	***Sterculia*** *quinqueloba* (Garcke) K. Schum.		
M33	*Swartzia madagascariensis* Desv.		

* According Palacio et al. [14], since 2017, all species of the genus *Dalbergia* have been listed in CITES Appendix II, except *D. nigra*, which has been listed in Appendix I since 1992; *Cedrela* P. Browne comprises 18 species, all of them listed in CITES Appendix II.

2.2. Wood Characterization

2.2.1. Anatomical Characterization

Microscopic slides were prepared from wood samples following the usual methods of softening, sectioning, staining, and mounting. The wood samples were softened in boiling water or in a 50/50 glycerin solution. Sections of 17–20 μm thickness were cut with a sliding microtome, stained with safranin, and mounted with Euparal. Macerations were prepared using Jeffrey's solution and stained with 1% gentian violet. Macro- and microphotographs were taken for each wood section of each species. Samples were observed by light microscopy (Leitz Dialux 20EB), and 40 measurements, at least, per feature were performed using image analysis software (Leica Qwin Standard). The fiber dimensions were measured in individualized cells (maceration); the vessel diameters were measured in transverse sections as well as the vessel frequency and the vessel wall thickness; the vessel element lengths and vessel pits were measured in tangential sections; the ray widths and ray lengths were measured in tangential sections; the ray frequency was measured in the tangential; and uniseriate, biseriate, and multiseriate rays were counted as one ray each. The anatomical descriptions followed the recommendations of the IAWA Committee [39].

2.2.2. Physical Properties Determination

The wood density, moisture content (MC), and wood shrinkage values were determined according to the following methods. Cubic samples with 30 mm of edge and faces corresponding to the transverse, tangential, and radial sections were prepared. Volumes and weights were measured for each sample. Samples were saturated, air-dried, and oven-dried, first at 60 °C and then at 103 ± 2 °C to a constant weight for the determination of wood basic density, referred to here as wood density, calculated on the basis of oven dry weight. Moisture content and wood shrinkage determinations were made according to the Portuguese standards (NP-614 and NP-615), i.e., samples were at 60% relative humidity and 20 °C until they reached a constant mass, and wood density was also assessed.

2.2.3. Color Measurements

Wood color was measured directly on the tangential surface of the standard wood samples after a fine sanding of the surface to represent the natural wood color. The CIELab chromatic system from the Commission Internationale d'Eclairage (CIE) was used to measure lightness (L*), ranging from 0 (black) to 100 (white); redness (a*), from +a*(red) to −a* (green); and yellowness degree (b*), from +b* (yellow) to –b* (blue), from 0 to 60. Four measurements were performed per each tangential face using a Minolta CM–3630 Spectrophotometer, and data were analyzed by Papercontrol v. 2. A color intensity qualitative character ranging from 1 (light) to 27 (black), was conceived using a scale of 33 different wood species and wood textures with an expert-based classification taking into account the visual observation and comparative analysis.

2.3. Character Selection

A total of 29 variables related to 13 anatomical and 16 physical (including 4 colorimetric) characteristics were selected and are compiled in Table 2. The mean value for each wood species was used in the different analyses.

Table 2. Descriptions and codes of the anatomical and physical wood variables.

Code	Character Description (Units)	Code	Character Description (Units)
V1	Mean number (Nr) of vessels/mm^2	FIS1	Wood density, 12% MC (g/cm^3)
V4	Mean vessel pit diameter (μm)	FIS2	Oven-dry wood density (g/cm^3)
V5	Mean vessel wall thickness (μm)	FIS3	MC, dry basis (%)
V6	Mean vessel element length (μm)	FIS4	MC, wet basis (%)
V7	Mean vessel diameter (μm)	FIS5	Volumetric shrinkage (%)
R1	Mean Nr of rays/mm	FIS6	Tangential shrinkage (%)
R3	Ray height in cell numbers (#)	FIS7	Radial shrinkage (%)
R5	Mean ray height (μm)	FIS8	Axial shrinkage (%)
R7	Ray width in cell numbers (#)	FIS9	Coefficient of volumetric shrinkage (%)
R8	Mean ray width (μm)	FIS10	Coefficient of tangential shrinkage (%)
F1	Mean fiber wall thickness (μm)	FIS11	Coefficient of radial shrinkage (%)
F2	Mean fiber length (μm)	FIS12	Coefficient of axial shrinkage (%)
F4	Mean fiber width (μm)	C4	L* polished sample/natural wood
		C5	a* polished sample/natural wood
		C6	b* polished sample/natural wood
		C7	Color intensity

2.4. Data Analysis

The similarity of the anatomical and physical wood characteristics of the different tropical wood species was quantified by means of cluster analysis (CA) and principal component analysis (PCA), generating classification systems accordingly [40,41]. The CA was performed using the UPGMA method (unweighted pair-group average) to express the relationships existing between taxa based on their pairwise similarities. The distortion degree of the generated dendrogram was given by the cophenetic correlation coefficient (c) to measure how faithfully the pairwise distances between the original unmodeled data points were preserved.

The PCA was conducted to identify directions along which the variance in the data was maximal. Eigenvalues, eigenvectors, and principal component (PC) axis scores were produced for the datasets. PCs were plotted against one another to reveal clustering or structure in the dataset due to similarities between the wood species. The overlapping of PCA projections with MST (minimum spanning tree), also known as the SCN (shortest connection network) method, resulted in a network of connections between the different tropical woods, detecting pairs that, even if they were close within the PC1 and PC2 plan, were in fact distant, taking into consideration the plan of PC1 and PC3. The principal component axis scores of PCs 1 vs. 2 and PCs 1 vs. 3 were plotted to identify clustering patterns.

The number of tropical wood species and the variables included in each analysis are shown in Table 3. Some species and variables were excluded from the analysis to avoid biased results, namely species with missing values or highly correlated variables (e.g., ratios between variables and standard deviation values), while the variables were set as equally important, as required by the multivariate analysis criteria.

Table 3. Number and list of the tropical wood species (codes as in Table 1) and variables (codes as in Table 2) included in each analysis.

Species (#)	Tropical Species Code	Variable Code
81	I1 to I17, M1 to M33, T1 to T17, T20 to T33	V1, V4, V5, V6, V7, R1, R3, R5, R7, R8, F1, F2, F4, C4, C5, C6, C7, FIS1, FIS3, FIS5
87	I1 to I17, M1 to M6, M8 to M21, M23 to M29, M31, M33, T2 to T23, T25, T26, T28 to T33, N2 to N9, N11 to N13	V1, V4, V5, V6, V7, R1, R3, R5, R7, R8, F1, F2, F4
54	I1 to I17, M1 to M6, M8 to M30, M33, T9, T10, T17, T20, T21, T25, T28	FIS1, FIS4 to FIS12, C4, C5, C6, C7

The statistical analysis was performed using the NTSYSpc (v.2.1, New York, NY, USA) software program.

3. Results

3.1. Species Variability Based on Anatomical and Physical Wood Characteristics

A high variability was found for the anatomical and physical wood properties, as observed in the range of values shown in Table 4. For the main anatomical variables, such as vessel frequency, ray and fiber tissues, wood density, and wood color, the following were found: the highest vessel frequency was 193 vessels per mm^2 in *Androstachys jobnsonii* (M7), and the lowest was in *Bombax malabaricum* (I5); the tallest and largest rays were 1500 μm × 215 μm in *Sterculia quinqueloba* (M32), and the shortest rays were 101 μm in *Dalbergia sissoo* (I7); the longest fibers measured 3780 μm in *Syringa vulgaris*, and the widest were 46 μm in *Ricinodendron rautanenii* (M30); and the highest value of wood density was 1400 kg/m^3 for *Tamarindus indica* (T28), while the lowest was 230 kg/m^3 for *Ricinodendron rautanenii* (M30). The wood color varied from the lightest-colored wood in *Aleurites moluccana* (T2) to the reddest in *Pterocarpus indicus* (T21), the yellowest in *Morus lactea* (M24), and the darkest wood in *Dalbergia melanoxylon* (M16). The mean values are summarized in Table S1.

Table 4. Minimum and maximum values found for each of the studied anatomical or physical variables and the scientific names and codes of the corresponding species. For code characters and abbreviations see Table 2.

Character (Units)	Min–Max Values/Species Name/Species Code
V1 (Nr vessels/mm^2)	1–193/*Bombax malabaricum–Androstachys jobnsonii*/I5–M7
V4 (μm)	1.16–15.85/*Cleistanthus schlechteri–Ricinodendron rautanenii*/N5–M30
V5 (μm)	3.1–15.6/*Khaya–Xylia torreana*/M21–N15
V6 (μm)	150–850/*Dalbergia sissoo–Aleurites moluccana*/I7–T2
V7 (μm)	45–285/*A. jobnsonii–R. rautanenii*/M7–M30
V8 (μm)	5–85/*Schrebera trichoclata–Sterculia quinqueloba*/N12–M32
R1 (Nr of rays/mm)	2–23/*Acacia robusta–Pterocarpus antunesii*/N1–N10
R3 (#)	5–67/*Calophyllum inophyllum–A. robusta*/T6–N1

Table 4. *Cont.*

Character (Units)	Min–Max Values/Species Name/Species Code
R5 (μm)	101–1500/*Dalbergia sissoo–S. quinqueloba*/I7–M32
R7 (#)	1–11/*Lagerstromia parviflora–S. quinqueloba*/I10–M32
R8 (μm)	13–215/*Ganophyllum falcatum–Albizia lebbeckoides*/T14–T1
F1 (μm)	2.4–7.2/*Elaeocarpus sphaericus–O. stuhlmannii*/T12–M25
F2 (μm)	700–3780/*Dalbergia melanoxylon–Syringa vulgaris*/M16–N14
F4 (μm)	12–46/*Colophospermum mopane–R. rautanenii*/M13–M30
C4	25.91–85.11/*D. melanoxylon–A. moluccana*/M16–T2
C5	1.86–20.29/*D. melanoxylon–P. indicus*/M16–T21
C6	0.97–34.89/*D. melanoxylon–Morus lactea*/M16–M24
C7	1–27/*Diospyros mespiliformis–D. melanoxylon*/M18–T2
FIS1 (g/cm^3)	0.23–1.37/*R. rautanenii–Tamarindus indica*/M30–T28
FIS2 (g/cm^3)	0.21–1.31/*R. rautanenii–T. indica*/M30–T28
FIS3 (%)	10.0–16.9/*P. indicus–Bischofia javanica*/T21–T5
FIS4 (%)	9.1–31.0/*P. indicus–Ficus indica*/T21–I9
FIS5 (%)	3.50–14.33/*Cordyla africana–Terminalia tomentosa*/M15–I16
FIS6 (%)	1.6–9.17/*Cordyla africana–Aegle marmelos*/M15–I2
FIS7 (%)	1.20–5.17/*Albizia adianthifolia–Terminalia tomentosa*/M4–I16
FIS8 (%)	0.01–0.65/*Aegle marmelos–Terminalia bellirica*/I2–I14
FIS9 (%)	0.26–0.77/*R. rautanenii–Casuarina junghuniana*/M30–T9
FIS10 (%)	0.14–0.50/*R. rautanenii–C. junghuniana*/M30–T9
FIS11(%)	0.09–0.30/*R. rautanenii–T. indica*/M30–T28
FIS12 (%)	0.00–0.04/*Eugenia jambolana–P. indicus*/I8–T21

The dendrogram based on 13 anatomical and 7 physical wood characteristics of 81 tropical species (listed in Tables 1 and 3) originated several groups and showed that *Androstachys johnsonii* (M7) and three groups of species (T5 and T1; M30, T2, and T12; and I5, M32, and T27) were clearly different from all other sets (Figure 1).

Cumulatively, PCs 1–3 accounted for 49.6% of the total amount of variance in the overall dataset of 81 species (Tables 5 and S2). From PC axis score plots, the clustering of wood species was apparent, showing its relative importance (Figure 2a). PC1 explained 23.0% of the variation, and the highest loadings were found for eight anatomical (vessel element length, V6; vessel diameter, V7; number of rays/mm, R1; ray height, R5 and R7; ray width, R8; fiber length, F2; and fiber width, F4) and two physical (b* parameter, C6, and wood density, FIS1) variables. Along PC1, the species showing the highest wood density (FIS1), shortest vessel elements (V6), thinner fibers (F4), and lowest values of the b* parameter (C6) were grouped in the right-hand region: *Casuarina junghuhniana* (T9), *Colophospermum mopane* (M13), *Dialium schlechteri* (M17), *Dalbergia sissoo* (I7), *Tamarindus indica* (T28), and *Dalbergia melanoxylon* (M16) (Figure 2b). In the left-hand region, the wood species from India and Mozambique and some from East Timor were grouped, showing the taller and wider rays (R5, R7, and R8), the yellower colored wood (high values for the b* parameter) (C6), the longer (F2) and larger (F4) fibers, and the longer (V6) and wider (V7) vessels. PC2 explained 15.4% of the residual variation, and three color (L* parameter, C4; a* parameter, C5; and qualitative designation, C7) and two anatomical (ray height, R3, and fiber wall thickness, F1) variables showed the highest loadings. The lighter woods (with the highest values for the L* parameter) (C4) were grouped in the upper region, composed of *Aleurites moluccana* (T2), *Alstonia scholaris* (T3), *Elaeocarpus sphaericus* (T12), and *Ficus macrophylla* (T13) from East Timor as well as *Ricinodendron rautanenii* (M30) from Mozambique. In the opposite region, the woods correlated with high values for the cell number of ray height (R3), fiber wall thickness (F1), and color intensity (C7) were grouped, corresponding to species from Mozambique (darker woods). The remaining variation (11.2%) was explained by PC3, grouping the species showing high vessel frequency (V1), in opposition to species showing high values for mean vessel pit diameter (V4), vessel wall thickness (V5), high moisture content (FIS3), and volumetric shrinkage (FIS5), corresponding to a high number of species from East Timor (Figure S1).

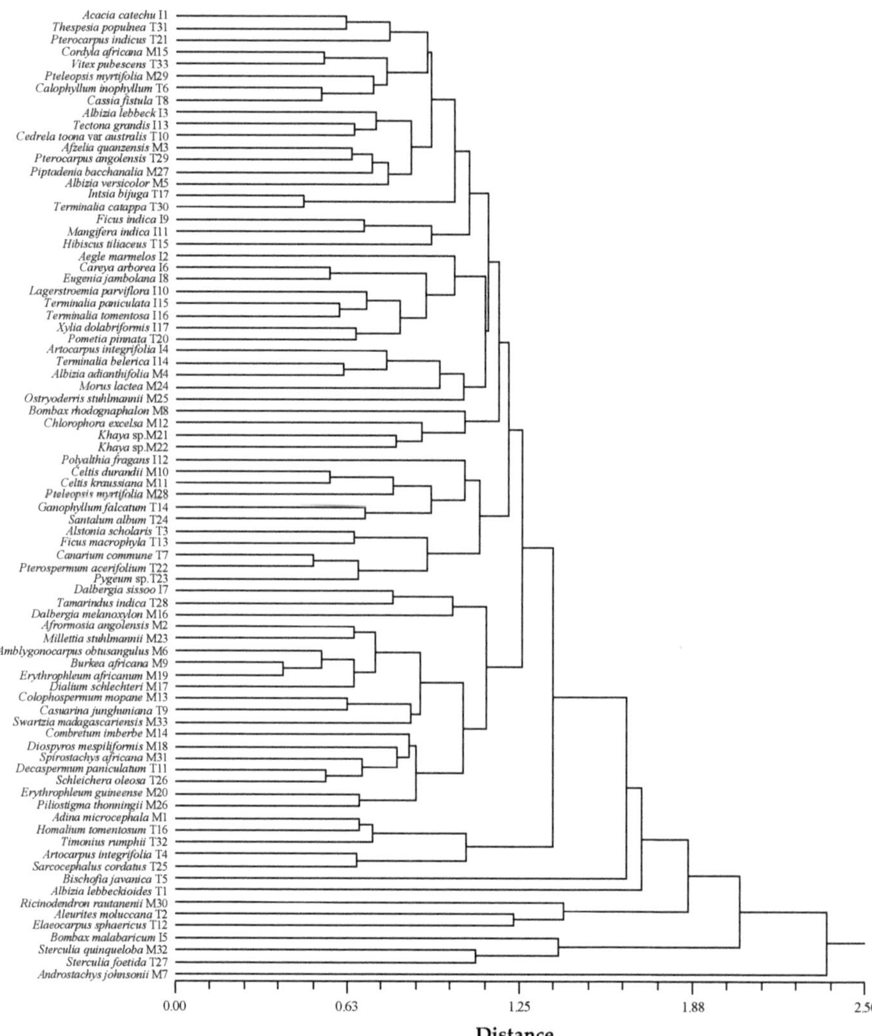

Figure 1. Classification of 81 tropical species (listed in Tables 1 and 3) based on 13 anatomical and 7 physical wood characteristics (coded in Table 2), obtained by the UPGMA clustering method ($c = 0.798$).

Table 5. Statistics of the principal component analyses (PCA) for the anatomical and physical wood characteristics (coded in Table 2) of the tropical species sets (as described in Tables 1 and 3). The contributions of components 1–3 to the wood species variation (eigenvalue (E.V.), proportion explained (P.E.), and cumulative proportion (C.P.)).

	81 Species			87 Species			54 Species		
Variable	PC1	PC2	PC3	PC1	PC2	PC3	PC1	PC2	PC3
E.V.	4.606	3.079	2.240	3.433	2.217	2.065	4.213	3.925	1.992
P.E. (%)	23.0	15.4	11.2	26.7	17.1	15.9	30.1	28.0	14.2
C.P. (%)	23.0	38.4	49.6	26.7	43.8	59.7	30.1	58.1	72.3

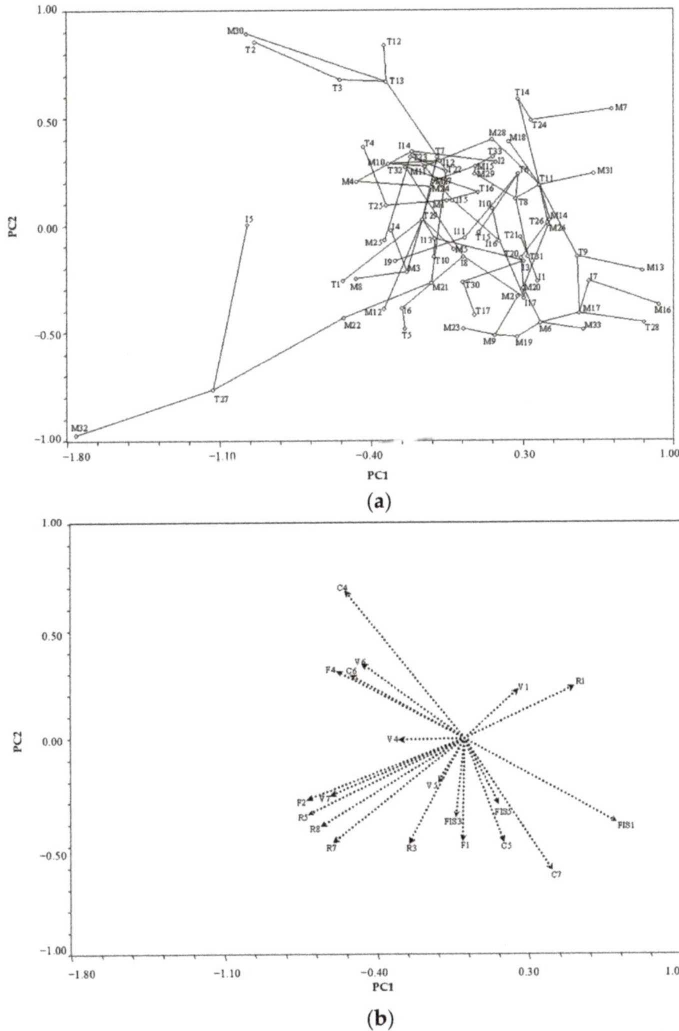

Figure 2. (**a**) PC1 and PC2 projection plots of the 81 tropical species (as listed in Tables 1 and 3) based on 20 anatomical and physical wood characteristics (coded in Table 2), overlapped by the MST method (similarity coefficient of 0.87). (**b**) The indication of the correlation degree between the variables based on the position of the vector.

As shown in Figure 2a, the most different wood species formed one group composed of M30, T2, T3, T12, and T13 (at the top) and a second group of M22, M32, T1, T27, and I5 (on the bottom of the left side), while *Androstachys johnsonii* (M7) was kept separate from all sets in accordance with Figure 1.

3.2. Species Variability Based on Anatomical Characteristics

The cluster analysis of the correlation matrix of 87 species (as listed in Tables 1 and 3) based on 13 anatomical quantitative characteristics showed a low similarity coefficient (r = 0.695), i.e., the dendrogram was not a good representation of the original distances (Figure S2).

The PCA revealed that PCs 1–3 accounted for 59.7% of the total variation of 87 wood species (Table 5). PC1 explained 26.7% of the variation, and the variables with the highest loadings were ray frequency (R1), ray height (R3), ray width (R7 and R8), and fiber length (F2). The variables with the highest loadings for PC2 were the vessel frequency (V1), ray height (R5), and fiber wall thickness (F1), in opposition to the vessel wall thickness (V5) and vessel diameter (V7). Along PC2, the wood species with high vessel frequency (V1), taller rays (R5), and thicker fibers (F1) are found in the upper region, in opposition to wood species with vessels with thicker walls (V5) and wider vessels (V7) (Figure 3). A group of wood species composed of *Bombax malabaricum* (I5), *Bombax rhodognaphalon* (M8), *Chlorophora excelsa* (*Milicia excelsa*) (M12), *Afzelia quanzensis* (M3), *Careya arborea* (I6), and *Ficus indica* (I9) is found in the left-hand region and is quite different from all other sets, in opposition to a group composed of *Decaspermum paniculatum* (T11), *Homalium tomentosum* (T16), *Casuarina junghuhniana* (T9), *Pteleopsis myrtifolia* (M28), *Spirostachys africana* (M31) *Ganophyllum falcatum* (T14), *Schleichera oleosa* (T26), and *Melaleuca leucadendron* (T19). The variables with the highest loadings for PC3 were vessel pit diameter (V4), vessel element length (V6), and fiber width (F4), forming two groups of wood species of East Timor, one comprising *Aleurites moluccana* (T2), *Bischofia javanica* (T5), *Artocarpus integrifolia* (T4), *Sarcocephalus cordatus* (T25), and *Timonius rumphii* (T32) and the other comprising *Macaranga tanarius* (T18), *Elaeocarpus sphaericus* (T12), and *Alstonia scholaris* (T3), which were considered different from all other sets (Figure S3).

A tendency to group tropical wood species from either East Timor or Mozambique was observed, while wood species from India were found to be more dispersed.

3.3. Species Variability Based on Physical Characteristics

The dendrogram based on 14 physical characteristics of 54 species (listed in Tables 1 and 3) showed two groups composed of 29 species mostly from Mozambique and a third group of 13 species from India, all of them forming subgroups (Figure 4).

Cumulatively, PCs 1–3 explained 72.3% of the total variance of 54 wood species (Table 5). PC1 explained 30.1% of the variation, and the variable with the highest loadings was the axial shrinkage coefficient (FIS12), in opposition to the other shrinkage characters (volumetric shrinkage, FIS5; tangential shrinkage, FIS6; radial shrinkage, FIS7; the volumetric shrinkage coefficient, FIS9; the tangential shrinkage coefficient, FIS10; and the axial shrinkage coefficient, FIS11). Along PC1, in the top left-hand quadrant, a group composed mainly of the species from India (I17, I16, I6, I8, I10, I13, I11, I15, I9, I4, I5, I12, and I2) is shown to be clearly different from all other sets, showing high values of moisture content (FIS4) and volumetric, tangential, and axial shrinkage (FIS5, FIS6, and FIS7) while showing lower axial shrinkage coefficients (FIS12) (Figure 5). *Albizia versicolor* (M5), *Pterocarpus angolensis* (M29), *Cordyla africana* (M15), and *Ricinodendron rautanenii* (M30) are found on the right-hand side, showing lower values for those characteristics. The second component explained 28.0% of the variation, and the variables of the lightness of the wood (L*, C4) and the moisture content (FIS4) showed high loadings, in opposition to the wood redness (C5) and wood density (FIS1). Along PC2, one group of species from India (I8, I10, I11, I15, I9, I4, I12, I5, and I2) showing high values for those characteristics was formed, while the redder (higher value of a*, C5) species from Mozambique (M1, M6, M16, M33, M13, M9, M19, M20, and M2) and some from East Timor (T28, T9, and T21) showing high wood density (FIS1) were grouped on the bottom side. The species from India showed a mean wood density of 650 kg/m^3, 24.9 % MC, and 10.7 % wood (volumetric) shrinkage, in opposition to 770 kg/m^3, 11.3% MC, and 5.9 % mean values obtained for the species from Mozambique. In the third component, 14.2% of the remaining variation was explained by the yellowness (C6) and axial shrinkage (FIS8), in opposition to the color intensity (C7). Along PC3, the wood species *Cedrela toona* var *australis* (T10), *Ostryoderris stuhlmannii* (M25), and *Terminalia bellirica* (I14), with yellower woods (high values of b*, C6) and high axial shrinkage (FIS8) are found in the upper region, in opposition to the darker woods of *Dalbergia melanoxylon*

(M16), *Ficus indica* (I9), and *Mangifera indica* (I11), which showed the highest color intensity values (C7) (Figure S4).

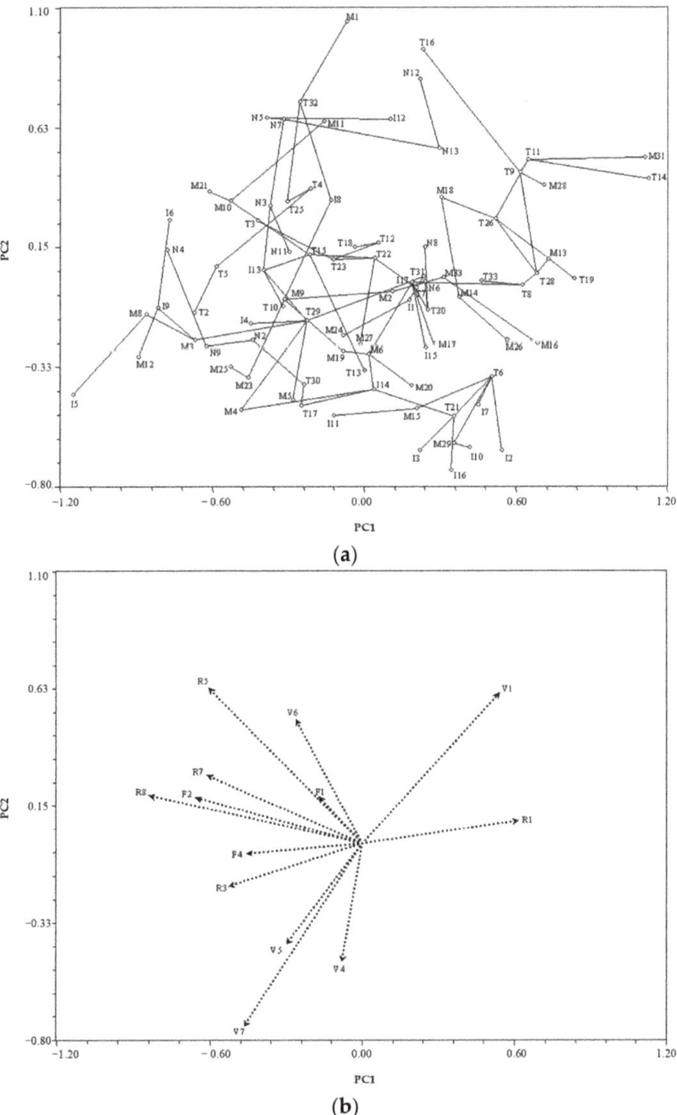

Figure 3. (**a**) PC 1 and PC 2 projection plots of the 87 tropical species (listed in Tables 1 and 3) based on anatomical wood characteristics (coded in Table 2), overlapped with the MST method (c = 0.87). (**b**) The correlation degree between the variables based on the position of the vector.

3.4. Similarity within Species and Genus

3.4.1. Similarity Based on Both Anatomical and Physical Wood Characteristics

Similarity based on both anatomical and physical properties within species (specimens showing more than one geographical provenance) was only observed for *Tectona grandis* from India (I13) and East Timor (T29). Four other weaker connections were found for *Tectona grandis* from East Timor (T29) with *Albizia lebbeckoides* (T1), *Chlorophora excelsea*

(M12), *Albizia versicolor* (M5), and *Afzelia quanzensis* (M3), while *Tectona grandis* from India (I13) was only connected to one more species, *Terminalia catappa* (T30) (Figures 2a and S5).

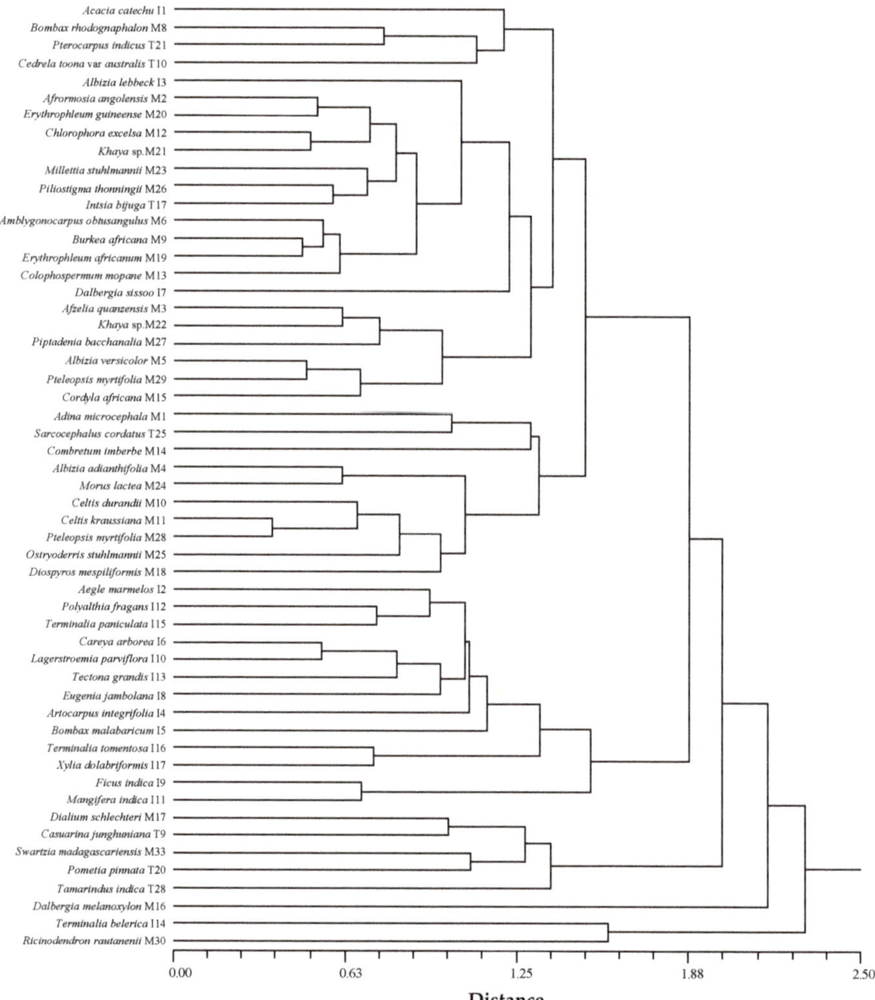

Figure 4. Classification of 54 tropical species (listed in Tables 1 and 3) based on 14 physical wood characteristics (coded in Table 2), obtained by the UPGMA clustering method ($c = 0.720$).

The *Artocarpus integrifolia* from India (I4) and East Timor (T4) were found to be connected to species from different genera: *Tectona grandis* (T29) and *Sarcocephalus cordatus* (T25), respectively.

Within the genus, i.e., from the genera showing more than one species (*Acacia* spp., *Albizia* spp., *Ambylognocarpus* spp., *Bombax* spp., *Cedrela* spp., *Celtis* spp., *Combretum* spp., *Dalbergia* spp., *Erythroplhleum* spp., *Ficus* spp., *Morus* spp., *Pterocarpus* spp., *Sterculia* spp., *Terminalia* spp., and *Xylia* spp.), four connections were found: *Dalbergia melanoxylon* (M16) with *Dalbergia sissoo* (I7) and *Celtis durandii* (M10) with *Celtis kraussiana* (M11), both from Mozambique, *Sterculia quinqueloba* (M32) with *Sterculia foetida* (T27), *Terminalia bellirica* (I14) with *Terminalia paniculate* (I15), and *Terminalia tomentosa* (I16) with *Terminalia paniculate* (I15), even if a long Euclidean distance was observed that highlighted their differences.

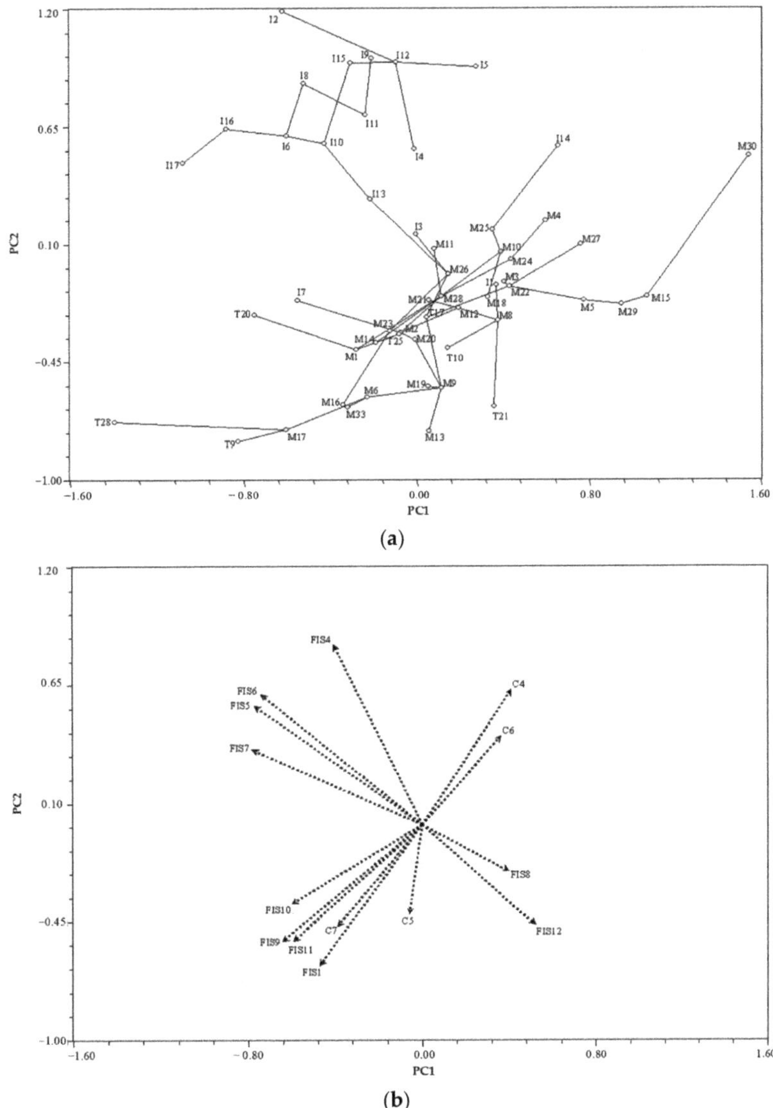

Figure 5. (**a**) PC 1 and PC 2 projection plots of the 54 tropical species (listed in Tables 1 and 3) based on physical wood characteristics (coded in Table 2), overlapped by the MST method (*c* = 0.92). (**b**) The correlation degree between the variables based on the position of the vector.

Stronger similarities were found between different genera compared to the previously mentioned cases, such as *Tectona grandis* (I13) and *Cedrela toona* (T10), *Careya arborea* (I6) and *Bischofia javanica* (T5), *Elaeocarpus sphaericus* (T12) and *Ficus macrophylla* (T13), *Millettia stuhlmannii* (M23) and *Burkea africana* (M9), *Erythrophleum africanum* (M19) and *Burkea africana* (M9), *Amblygonocarpus obtusangulus* (M6) and *Erythrophleum africanum* (M19), *Dalbergia sissoo* (I7) and *Dialium schlechteri* (M17), *Ganophyllum falcatum* (T14) and *Santalum album* (T24), *Cordyla africana* (M15) and *Pterocarpus angolensis* (M29), and *Pterospermum acerifolium* (T22) and *Canarium commune* (T7).

The CITES-listed species *Dalbergia melanoxylon* (M16) from Mozambique showed both anatomical and physical similarity to the CITES-listed species *Dalbergia sissoo* from India (I7).

3.4.2. Similarity Based on Anatomical Characteristics

Within the three cases of species showing more than one geographical origin, only *Tectona grandis* from India (I13) and from East Timor (T29) were connected, while *Artocarpus integrifolia* from East Timor (I4) was connected to *T. grandis* (T29), *Melaleuca leucadendron* from East Timor (T19) was connected to *Schleichera oleosa* (T26), and *M. leucadendron* from Mozambique (N8) was connected to *Pometia pinnata* (T20) (Figures 3a and S6).

Within the genus (a total of 15 different genera), i.e., from *Acacia* spp., *Albizia* spp., *Ambylognocarpus* spp., *Bombax* spp., *Cedrela* spp., *Celtis* spp., *Combretum* spp., *Dalbergia* spp., *Erythroplhleum* spp., *Ficus* spp., *Morus* spp., *Pterocarpus* spp., *Sterculia* spp., *Terminalia* spp., and *Xylia* spp.), only *Celtis durandii* (M10) and *Celtis kraussiana* (M11), both from Mozambique, were shown to be connected, even if a long Euclidean distance was observed, thereby highlighting their differences.

In fact, stronger similarities compared to the previously mentioned cases were found, such as *Tectona grandis* (I13) and *Hibiscus tiliaceus* (T15), *Tectona grandis* (I13) and *Cedrela toona* (T10), *Tectona grandis* (T29) and *Artocarpus integrifolia* (I4), *Melaleuca leucadendron* (T19) and *Schleichera oleosa* (T26), *Terminalia catappa* (T30) and *Intsia bijuga* (T17), and *Combretum imberbe* (M14) and *Piliostigma thonningii* (M26).

Anatomical similarity was also observed for other species from the set, such as *Acacia catechu* (I1) and *Thespesia populnea* (T31), *Amblygonocarpus obtusangulus* (M6) and *Terminalia bellirica* (I14) as well as *Xylia dolabriformis* (I17), *Xylia dolabriformis* (I17) and *Pometia pinnata* (T20), *Melaleuca leucadendron* (N8) and *Pometia pinnata* (T20), *Canarium commune* (T7) and *Pterospermum acerifolium* (T22), *Afrormosia angolensis* (M2) and *Sterculia quinqueloba* (M32) as well as *Burkea africana* (M9), *Ficus indica* (I9) and *Afzelia quanzensis* (M3), and *Ostryoderris stuhlmannii* (M25) and *Millettia stuhlmannii* (M23).

The CITES-listed species *Cedrela odorata* (N4) and *Dalbergia melanoxylon* (M16) showed similarity to non-CITES species *Morus mesozygia* (N9) and *Swartzia madagascariensis* (M33), respectively.

3.4.3. Similarity Based on Physical Characteristics

No similarity was observed within the genus, i.e., none of *Albizia* spp. (M4, M5, and I3), *Bombax* spp. (I5 and M8), *Celtis* spp. (M10 and M11), *Dalbergia* spp. (M16 and I7), *Erythrophleum* spp. (M19 and M20), *Pterocarpus* spp. (M29 and T21), or *Terminalia* spp. (I14, I15, and I16) were connected between themselves (other species were not analyzed due to missing physical data) (Figures 5a and S7).

The species with stronger similarities were: *Pteleopsis myrtifolia* (M28) and *Celtis kraussiana* (M11), *Burkea africana* (M9) and *Erythrophleum africanum* (M19), and *Albizia versicolor* (M5) and *Pterocarpus angolensis* (M29). More distant connections were found between the species: *Burkea africana* (M9) and *Colophospermum mopane* (M13), *Erythrophleum guineense* (M20) and *Burkea africana* (M9), *Burkea Africana* (M9) and *Amblygonocarpus obtusangulus* (M6), and *Afrormosia angolensis* (M2) and *Erythrophleum guineense* (M20).

Concerning the CITES-listed species found in the set, physical similarity was observed for *D. melanoxylon* (M16) and the non-CITES species *Millettia stuhlmannii* (M23).

4. Discussion

4.1. Species Clustering

The grouping of species by their geographical location (i.e., India, East Timor, and Mozambique) based on both anatomical and physical characteristics was more evident for the 81 species set, with a stronger tendency to group species from Mozambique. The most dispersed species were *Ricinodendron rautanenii* (M30), *Aleurites moluccana* (T2), *Alstonia scholaris* (T3), *Elaeocarpus sphaericus* (T12), *Ficus macrophylla* (T13), *Khaya* sp. (M22), *Sterculia quinqueloba* (M32), *Albizia lebbeckoides* (T1), *Sterculia foetida* (T27), *Bombax malabaricum* (I5), and *Androstachys johnsonii* (M7). In fact, different geographical distribution trends were found when separately analyzing the anatomical and physical characteristics, i.e., anatomical similarity grouped species from East Timor and Mozambique, and dispersed species from India, while the physical similarity aggregated species from India. The analyses showed large clusters of species from India, Mozambique, and East Timor, which reflects their wood properties similarity. The CA clustering was, in general, confirmed by the PCA results. Thus, wood properties were, in general, of limited value in the delimitation of the geographical origin of wood species, and specific features should be considered according to specific species and each region. In fact, numerical taxonomy methods have been very useful in many comparative and phylogenetic studies within different taxa and/or across different ecological gradients, even if they have been mostly based on qualitative anatomical characters [11,42–46].

4.2. Variability and Interaction of Wood Properties

4.2.1. Both Anatomical and Physical Properties

The diagnostic value of the different wood characters was confirmed by PCA, showing that the variability of the tropical species, based on both anatomical and physical characteristics, was mostly explained by wood density (FIS1), ray (height, R5, and width, R7 and R8), and fiber (length, F2, and width, F4) (Figure 2). Color characteristics (lightness, C4; redness, C5; and intensity, C7) accounted for the species variability. The diagnostic value of vessel-related features was less consistent, with the highest values for the vessel element length (V6) and vessel diameter (V7). Wood density (FIS1) was crucial to grouping wood species. Those showing high wood density were grouped in opposition to wood species showing narrow fibers (F4) and longer vessel elements (V6). Other studies highlighted the relationship between high wood density and longer and thinner fibers in *Eucalyptus* spp. in Brazil [47]. Moreover, the wood (twigs) density variation of different tropical species was more controlled by the wood (fiber and parenchyma) tissues surrounding vessels rather than the vessel lumen proportion, although ambiguous ecological patterns for fiber–parenchyma tissues were found [29,48]. However, studies have usually focused on specific anatomical features, and wood density variability, explained as a function of multiple anatomical traits, is still not well-understood.

Another relatively consistent distribution pattern was found based on wood color (C4, C5, and C7), and it was interesting to note the importance of the lightness of the wood (C4) to forming geographical groups, i.e., the darker woods were mainly from Mozambique (*Afrormosia angolensis* (M2), *Amblygonocarpus obtusangulus* (M6), *Burkea africana* (M9), *Erythrophleum africanum* (M19), *Millettia stuhlmannii* (M23), and *Swartzia madagascariensis* (M33)), while the lighter woods were mainly from East Timor (*Aleurites moluccana* (T2), *Alstonia scholaris* (T3), *Elaeocarpus sphaericus* (T12), and *Ficus macrophylla* (T13)). The formation of wood color is associated with several factors, among others, wood density, wood structure (e.g., growth rings), and anatomy as well as chemical features, namely the content and composition of extractives [19]. Since edaphoclimatic conditions may influence all these wood features [19], it is possible to find a wood color range within the same species. This is certainly a matter where more studies are needed.

The distribution pattern related to vessel features was also observed in wood species showing wider vessels (V7) in opposition to high vessel frequency (V1). Within the present study, the explained variance between the two sets was different, and the lowest vessel frequency (V1) was found for *Bombax malabaricum* (I5) from India (one vessel per mm^2), showing a vessel diameter of 264 µm. This type of distribution is more studied due to the importance of the safety and efficiency of the conductive system that is being reported as habitat and geographically related. The values reported here agree with the literature values. Species from tropical Africa and America, Southeast Asia, and India show 5 to 20 vessels per mm^2 and diameters over 100 µm, in contrast with species from Europe, North America, and Temperate Asia that show narrow vessels and more than 40 vessels per mm^2 [10,49–52]. In fact, tropical species showing a low frequency of vessels and wider vessels are considered safer and more efficient [49,53]. For example, vessel frequency and vessel diameter as well as wood density were effective for the identification and delimiting *Pterocarpus santalinus* L.f. within the genus, including species from different geographical origins [46]. Moreover, *Dalbergia cearensis* Ducke could be separated from *D. nigra* on the basis of vessel frequency, showing the highest vessel frequency (over 10 vessels per mm^2) [11]. Vessel diameter was found to be relevant to differentiating the wood from the species of the Araucaria Forest in Brazil [54]. Other anatomical characters, such as the number of rays per mm^2 (R1), vessel pit diameter (V4), and fiber wall thickness (F1), contributed less to the set variability (PC3).

4.2.2. Quantitative Wood Anatomy

The variability exclusively explained by the quantitative wood anatomy was mainly due to the ray frequency (R1), height (R3), and width (R7 and R8) and the fiber length (F2). The vessel frequency (V1), vessel wall thickness (V5), vessel diameter (V7), and fiber wall thickness (F1) were also important, while the vessel pit diameter (V4), vessel element length (V6), and fiber width (F4) accounted less for the species set variation. The contributions of vessel frequency, diameter, and element length are consistent with the analysis discussed above. Species showing longer vessel elements (V6) and high values for vessel pit diameter (V4) only explained part of the residual variation (PC3) but made it possible to distinguish some species from East Timor: *Aleurites moluccana* (T2), *Bischofia javanica* (T5), *Artocarpus integrifolia* (T4), *Sarcocephalus cordatus* (T25), *Timonius rumphii* (T32), *Macaranga tanarius* (T18), *Elaeocarpus sphaericus* (T12), and *Alstonia scholaris* (T3). The vessel pitting of Lauraceae species allowed their distinction between other families in Brazilian forests [54]. In fact, the variability in vessel pits is also gaining more interest to explain species distribution [12,50,51,55,56]. With regards to the fiber-related geographical pattern, it has been reported that half of the species from tropical and South Africa show fibers with extremely thick walls [10], while very thin-walled fibers are more prevalent at higher latitudes and in more humid environments and milder temperatures [8]. In this study, *Elaeocarpus sphaericus* (T12) from East Timor showed the thinnest fibres (2.4 μm), in opposition to *Ostryoderris stuhlmannii* (M25) (7.2 μm) from Mozambique and Southeast Africa.

Theoretically there is a compromise between mechanical resistance (fiber thickness) and conductive efficiency (wide vessels) and safety (narrow vessels) [57]. However, most genera present thinner to thicker fibers as well as all the vessel diameter ranges [10]. The fiber–parenchyma variation of several species from Australia was poorly correlated with vessel features and with the climate at different sites [48]. In the present analysis, the fiber thickness (V4) was less relevant compared to the vessel diameter (V6) for explaining the species variability, as it was observed in previous analyses, including physical properties. Ray characters accounted for most of the anatomical variability (PC1) but are considered to be ecologically or geographically related [8,10]. For instance, uniseriate rays were correlated with lower latitudes, rays two cells wide were found more often in warmer environments, and rays three cells wide tended to be found in higher latitudes, but they did not indicate ecological trends related to ray composition [10]. In contrast, quantitative ray features such as the number of rays per mm and ray height are established variables in systematic wood anatomy [39], which might suggest their diagnostic function overlap. For example, ray frequency was distinct between *Dalbergia miscolobium* Benth. and *Dalbergia nigra* [11]. The long and wide rays of *Citronella paniculata* (Mart.) R. A. Howard and *Myrsine coriacea* (Sw.) R.Br. ex Roem. and Schult. allowed their anatomical wood discrimination [54]. With regards to wood potential, for example, quantitative ray characters were also important to select *Dalbergia nigra* as a tropical wood species showing good quality for musical instruments [58].

The lack of distinct anatomical geographical patterns, as found here, might be related to the fact that external (e.g., climate and edaphic) and internal (e.g., age and stem sampling location) factors that account for wood variability within species were not included in the analyses. Those influences may overlap the taxonomic value of some anatomical quantitative features of wood [13,43,59]. In addition, the qualitative anatomical features, such as the type of growth ring porosity, vessel grouping and arrangement, vessel perforation plates, and parenchyma were also not included and vary by geographic region [8,10,11].

4.2.3. Physical Properties

Despite the wood density (FIS1) importance, the wood shrinkage and wood color, contributions were highlighted for the species distribution based on physical properties. In fact, wood density is greatly controlled by genetic factors [19], allowing the study of the effects of geographical provenance on wood density for a higher number of species from temperate to tropical regions due to their practical implications for tree improvement programs (e.g., [22–25,60]). In neotropical species, for example, wood density was found to be highly phylogenetically conserved [20]. Opposite results showing a small wood density variation (500 kg/m^3 to 800 kg/m^3) with no distinct geographical pattern were found for tropical species from Asia (Malaysia, Sri Lanka, and the east of India), America (Brazil, Venezuela, Guyana, and Surinam), and Africa (Cameroon and Gabon) [61]. In fact, factors other than the climate of origin, such as soil fertility, should also be determinant for the genetic patterns [30]. For example, the wood density variation of *T. grandis* was high among different provenances [23], and the wood density of tropical species from Costa Rica, Panama, and Peru was found to be inversely correlated with site fertility [62].

Due to the stronger genetic control of wood density, wood shrinkage is generally estimated based on wood density, and a positive and strong correlation is generally accepted [19]. However, shrinkage genetic control in *Calycophyllum spruceanum* Benth. has suggested its potential to improve wood quality [63]. Other wood properties are also of interest to explain wood shrinkage, such as the high extractives content in *Bagassa guianensis* Aubl. (from Africa and French Guinea), even if it is strongly controlled by the wood density in heartwood [64] and the high resistance of vessel walls in *Eucalyptus resinifera* Sm. [65].

The geographical wood color pattern found here, i.e., showing Mozambican species with darker woods and Asian species with lighter woods, could also be explained by the genetic effects on the lightness of the wood (L*). For instance, the stronger genetic control on wood lightness (L*) compared to the redness (a*) and yellowness (b*) was reported for African and Peruvian tropical species [66,67]. Moreover, the climatic conditions explain more of the color variation compared to the edaphic conditions for *T. grandis* in Costa Rica [68] and in African species, for instance, in Mali [67]. The ranges of values were in agreement with other species from Africa (30.5–72.3) and Asia (48.1–55.1) that are expected to be lower when compared to the temperate woods (e.g., Europe and North America), which are known to be lighter (51.1–84.5) [69]. In fact, this color variation is also explained due to extractives formation during the heartwood development, which is higher in tropical woods when compared to temperate species [19].

However, there are few multispecies datasets, and analyses and comparisons are complex due to different wood density determination methods and unaccounted internal (genetic, phylogenetic, or ontogenetic) or external (climatic) effects [20,70]. Moreover, internal and external factors, as discussed above, were not included, and their influence was not accounted for in the species distribution based on the physical characteristics.

4.3. Similarity within Species and Genus

Tectona grandis from India (I13) and from East Timor (T29) showed anatomical and physical similarity, which is in accordance with previous wood anatomy (macroscopic and microscopic) and physical descriptions mentioning similar vessel, ray, and fiber characteristics as well as wood density [36,37]. However, tree growth (measured as ring width) shows considerable variation between regions, even if found comparable between teak from India and East Timor [71,72]. Moreover, the ring structure, namely the proportion of fibers and parenchyma cells have been related to have more effects on the wood quality of teak in India, being influenced by geographical conditions [71]. These aspects were not studied here, and a careful interpretation is needed. Moreover, the wood density of teak is considered to show strong genetic control [23], and it could not be compared here due to missing values.

The species *Albizia adianthifolia* (M4), *Albizia versicolor* (M5), *Albizia lebbeck* (I3), and *Albizia lebbeckoides* (T1), from the same geographical origin, did not show any similarity, while the species from the genus *Cedrela* (N4 and T10), *Sterculia* (T27 and M32), *Dalbergia* (M16 and I7), and *Terminalia* (I14, I15, and I16) showed similarity on 13 anatomical and 7 physical characteristics. In fact, for the *Albizia* genus, several attempts have been made to distinguish species based in their anatomical characteristics, such as the seriation of rays and the presence of septa in fibers, for species in India [73]. Even quantitative characteristics such as fiber and vessel diameter have shown taxonomic significance in Nigeria species [74]. This high anatomical wood variability might be related to the wide natural distribution in different continents, such as in the *Albizia* genus, or due to wide world plantation for other species, such as *Cedrela*. Thus, besides the diagnostic value, within the genus, of the different anatomical characteristics, the physical characteristics such as the color parameters, wood density, moisture content, and volumetric shrinkage could be also considered, for example, for wood identification and discrimination analysis.

The species *Pterocarpus antunesii* (N10), *Terminalia paniculata* (I15), and *Sterculia quinqueloba* (M32) were found to be clearly different from all studied wood species. Few anatomical comparative studies are found for these species. In fact, due to the general morphological character variability and overlap in some species such as in *Terminalia* spp., other methods have also been applied to species identification [75]. Moreover, the superficial similarity between species may generate confusion, as seen in the example of *S. quinqueloba* and *Sterculia appendiculata* K. Schum, with the former known essentially for bark uses and the latter for its soft wood. Nevertheless, species classification based on anatomical characteristics was found to be reasonable for the CITES-listed species such as *P. santalinus* [46]. Descriptions of specific anatomical characters such as chambered prismatic crystals have also shown value for distinguishing *Sterculia comorensis* Baill. and *Sterculia appendiculata* K.Schum. in Mozambique [33].

The similarity of *Cedrela odorata* (N4) and *Dalbergia melanoxylon* (M16), as CITES-listed species, to non-CITES species is of particular interest. The commercial substitution of high-value CITES-listed species with alternative species certainly requires a full wood characterization, including mechanical properties, chemical features, and durability performance. In the present study, we were restricted to the available information and to nondestructive measurements of the xylarium samples. Wood density and wood shrinkage were chosen in this study to be indicative of wood quality. Therefore, the proposal made for the commercial introduction of alternative species into the timber market should be considered as indicative and as a first step requiring further characterization, including postharvest processing, namely drying and sawing performance. The similarity found for *Cedrela odorata* (N4) and non-CITES *Morus mesozygia* (N9) was interesting, even if *M. mesozygia* (N9) showed longer fibers (F2) and narrower and shorter vessels (V6 and V7) compared to *Cedrela odorata* [76,77]. Nevertheless, *M. mesozygia* (N9) is also recommended for sliced veneer, furniture, flooring, decorative artefacts, and toys [76–79]. For *Dalbergia melanoxylon* (M16), the anatomical similarity was found to be discreet for non-CITES *Swartzia madagascariensis* (M33), while the physical similarity was more evident for non-CITES *Millettia stuhlmannii* (M23), even if the physical similarity with *Millettia stuhlmannii* (M23) was not observed in the dendrogram, which showed a weaker representation of the original distances (0.720 vs. 0.920). However, the anatomical similarity between the *Dalbergia* and *Swartzia* genera is also considered for other CITES-listed species such as *Dalbergia nigra* and both *Swartzia leiocalycina* Benth. (Guiana and Suriname) and *Swartzia benthamiana* Miq. (Brasil and Colombia), with the latter two *Swartzia* species frequently being mistaken in commerce [80]. This genus similarity is confirmed here. Specifically, *Swartzia madagascariensis* (M33) showed similarity to *Dalbergia melanoxylon* regarding its wood color, wood density, and wood moisture content, as non-CITES *Swartzia madagascariensis* (M33) is also endemic to Mozambique, which could increase species diversity. The physical similarity of *Dalbergia melanoxylon* and *Millettia stuhlmannii* (M23) was also interesting, even if *Dalbergia melanoxylon* is denser (1250 kg/cm^3 vs. 868 kg/m^3), being used for wood carving, while *Millettia stuhlmannii* is used for construction, ship and boat building, railway sleepers, flooring, and furniture [7,32,34,38].

Different results were obtained with regards to species similarity, suggesting that anatomical or physical characteristics may be chosen to be analyzed depending on the purpose of the study or diagnosis. Thus, the identification of characters with the maximum variability between different species by PCA might be very useful for wood selection. However, the cumulative variance found here was not high, and similarities within species and genera might have been compromised by the constraints discussed above for the lack of distinct geographical patterns.

5. Conclusions

This was the first insight into an integrated analysis of anatomical and physical wood characteristics and the potential geographical patterns of tropical species from India, Mozambique, and East Timor. The anatomical and physical wood characteristics revealed to be independent, generating different species distributions. No geographical distribution was found, i.e., species from different origins showed similar wood properties, and therefore different wood origins may be considered to increase the diversity in the tropical timber trade. The wood potential of species, such as *Millettia stuhlmannii* and *Swartzia madagascariensis* from Mozambique, as alternatives for the CITES-listed species *Dalbergia melanoxylon* were highlighted as well as *Morus mesozygia* as an alternative for the CITES-listed *Cedrela odorata*. The multivariate analyses by PCA may be applied for wood identification if combined with comparative wood anatomy as well as for the control and management of sustainable wood production and legal timber trade.

However, this study is a general overview, and information should be taken carefully, considering the response of a single species to different climatic conditions and its use.

Supplementary Materials: The following supporting information can be downloaded at: https://www.mdpi.com/article/10.3390/f13101675/s1, Table S1: Mean values for each of the studied anatomical or physical wood variables and scientific names and codes of the corresponding species (as listed in Table 1). For code characters and abbreviations, see Table 2; Table S2: Statistics of the principal component analyses (PCA) for the anatomical and physical wood characteristics (coded in Table 2) of the tropical species sets (as described in Tables 1 and 3). Loadings of the anatomical and physical variables to principal components (PC) 1–3 of variation; Figure S1: (a) PC1 and PC3 projection plots of the 81 tropical species (as listed in Tables 1 and 3) based on both anatomical and physical wood characteristics (coded in Table 2), overlapped by the MST method. (b) The indication of the correlation degree between the variables based on the position of the vector; Figure S2: Classification of 87 tropical species (listed in Tables 1 and 3) based on 13 anatomical wood characteristics (coded in Table 2), obtained by the UPGMA clustering method (c = 0.695); Figure S3: (a) PC1 and PC3 projection plots of the 87 tropical species (as listed in Tables 1 and 3) based on anatomical wood characteristics (coded in Table 2), overlapped by the MST method. (b) The indication of the correlation degree between the variables based on the position of the vector; Figure S4: (a) PC1 and PC3 projection plots of the 54 tropical species (as listed in Tables 1 and 3) based on physical wood characteristics (coded in Table 2), overlapped by the MST method. (b) The indication of the correlation degree between the variables based on the position of the vector; Figure S5: The minimum spanning tree (MST), also known as SCN (shortest connection network), showing the connections of the 81 tropical species set (distances are not to scale); Figure S6: The minimum spanning tree (MST), also known as SCN (shortest connection network), showing the connections of the 87 tropical species set (distances are not to scale); Figure S7: The minimum spanning tree (MST), also known as SCN (shortest connection network), showing the connections of the 54 tropical species set (distances are not to scale).

Author Contributions: Conceptualization, F.B., T.Q. and H.P.; methodology, F.B., T.Q. and H.P.; validation, F.B. and V.S.; formal analysis, F.B. and V.S.; investigation, F.B., T.Q. and H.P.; resources, F.B., T.Q. and H.P.; data curation, F.B.; writing—original draft preparation, F.B. and V.S.; writing—review and editing V.S., T.Q. and H.P.; visualization, V.S.; supervision, T.Q. and H.P. All authors have read and agreed to the published version of the manuscript.

Funding: This research was funded by the Forest Research Centre, a research unit funded by Fundação para a Ciência e a Tecnologia I.P. (FCT), Portugal (UIDB/00239/2020). The first author acknowledges a Science and Technology Management Grant (SFRH/BGCT/15380/2005), and the second author acknowledges a research contract (DL57/2016/CP1382/CT0004), both by FCT.

Informed Consent Statement: Not applicable.

Data Availability Statement: The data are provided in the Supplementary Materials.

Acknowledgments: We are grateful to Luís Cruz Carneiro for support and guidance in the statistical analysis and to Cristiana Alves (Instituto Superior de Agronomia) and António Pereira da Silva (Instituto de Pesquisas Tecnológicas de São Paulo) for laboratory assistance.

Conflicts of Interest: The authors declare no conflict of interest. The funders had no role in the design of the study; in the collection, analyses, or interpretation of data; in the writing of the manuscript; or in the decision to publish the results.

References

1. Gasson, P.; Baas, P.; Wheeler, E. Wood Anatomy of Cites-Listed Tree Species. *IAWA J.* **2011**, *32*, 155–198. [CrossRef]
2. Perdigão, C.R.V.; Júnior, M.M.B.; Gonçalves, T.A.P.; de Araujo, C.S.; Mori, F.A.; Barbosa, A.C.M.C.; Souza, F.I.B.d.; Motta, J.P.; Melo, L.E.d.L. Forestry Control in the Brazilian Amazon I: Wood and Charcoal Anatomy of Three Endangered Species. *IAWA J.* **2020**, *41*, 490–509. [CrossRef]
3. Koch, G.; Haag, V. Control of Internationally Traded Timber—The Role of Macroscopic and Microscopic Wood Identification against Illegal Logging. *J. Forensic Res.* **2015**, *6*, 317. [CrossRef]
4. Gasson, P.E.; Lancaster, C.A.; Young, R.; Redstone, S.; Miles-Bunch, I.A.; Rees, G.; Guillery, R.P.; Parker-Forney, M.; Lebow, E.T. WorldForestID: Addressing the Need for Standardized Wood Reference Collections to Support Authentication Analysis Technologies; a Way Forward for Checking the Origin and Identity of Traded Timber. *Plants People Planet* **2021**, *3*, 130–141. [CrossRef]
5. Dulbecco, P.; Luco, D. *L' Essentiel Sur Le Bois*; Centre Technique du Bois et de L'ameublement: Paris, France, 2001; ISBN 978-2-85684-040-5.
6. CITAB. *Guide Pour Le Choix Des Bois En Menuiserie*; FCBA, CIRAD: Montpellier, France, 2002; ISBN 978-2-85689-025-7.
7. ATIBT. *Tropical Timber Atlas—Volume One—Africa*; ATIBT: Nogent-sur-Marne, France, 1986.
8. Alves, E.S.; Angyalossy-Alfonso, V. Ecological trends in the wood anatomy of some brazilian species. 2. Axial parenchyma, rays and fibres. *IAWA J.* **2002**, *23*, 391–418. [CrossRef]
9. Jansen, S.; Robbrecht, E.; Beeckman, H.; Smets, E. A Survey of the Systematic Wood Anatomy of the Rubiaceae. *IAWA J.* **2002**, *23*, 1–67. [CrossRef]
10. Wheeler, E.; Baas, P.; Rodgers, S. Variations In Dieot Wood Anatomy: A Global Analysis Based on the Insidewood Database. *IAWA J.* **2007**, *28*, 229–258. [CrossRef]
11. Gasson, P.; Miller, R.; Stekel, D.J.; Whinder, F.; Zieminska, K. Wood Identification of Dalbergia Nigra (CITES Appendix I) Using Quantitative Wood Anatomy, Principal Components Analysis and Naive Bayes Classification. *Ann. Bot.* **2010**, *105*, 45–56. [CrossRef]
12. Fichtler, E.; Worbes, M. Wood Anatomical Variables in Tropical Trees and Their Relation to Site Conditions and Individual Tree Morphology. *IAWA J.* **2012**, *33*, 119–140. [CrossRef]
13. Tarelkin, Y.; Delvaux, C.; Ridder, M.D.; Berkani, T.E.; Cannière, C.D.; Beeckman, H. Growth-ring distinctness and boundary anatomy variability in tropical trees. *IAWA J.* **2016**, *37*, 275–294. [CrossRef]
14. De Palacios, P.; Esteban, L.G.; Gasson, P.; García-Fernández, F.; de Marco, A.; García-Iruela, A.; García-Esteban, L.; González-de-Vega, D. Using Lenses Attached to a Smartphone as a Macroscopic Early Warning Tool in the Illegal Timber Trade, in Particular for CITES-Listed Species. *Forests* **2020**, *11*, 1147. [CrossRef]
15. Bodin, S.C.; Scheel-Ybert, R.; Beauchêne, J.; Molino, J.-F.; Bremond, L. CharKey: An Electronic Identification Key for Wood Charcoals of French Guiana. *IAWA J.* **2019**, *40*, 75-S20. [CrossRef]
16. Kitin, P.; Espinoza, E.; Beeckman, H.; Abe, H.; McClure, P.J. Direct Analysis in Real-Time (DART) Time-of-Flight Mass Spectrometry (TOFMS) of Wood Reveals Distinct Chemical Signatures of Two Species of Afzelia. *Ann. For. Sci.* **2021**, *78*, 31. [CrossRef]
17. Scholz, G.; Windeisen, E.; Liebner, F.; Bäucker, E.; Bues, C.-T. Wood Anatomical Features and Chemical Composition of Prosopis Kuntzei from the Paraguayan Chaco. *IAWA J.* **2010**, *31*, 39–52. [CrossRef]
18. Beeckman, H.; Blanc-Jolivet, C.; Boeschoten, L.; Braga, J.; Cabezas, J.A.; Chaix, G.; Crameri, S.; Degen, B.; Deklerck, V.; Dormontt, E.; et al. *Overview of Current Practices in Data Analysis for Wood Identification. A Guide for the Different Timber Tracking Methods*; HAL Open Science: Lyon, France, 2020.
19. Zobel, B.J.; Van Buijtenen, J. *Wood Variation. Its Causes and Control*; Springer: Berlin/Heidelberg, Germany, 1989.
20. Chave, J.; Muller-Landau, H.C.; Baker, T.R.; Easdale, T.A.; ter Steege, H.; Webb, C.O. Regional and Phylogenetic Variation of Wood Density across 2456 Neotropical Tree Species. *Ecol. Appl.* **2006**, *16*, 2356–2367. [CrossRef]

21. Miranda, I.; Sousa, V.; Pereira, H. Wood Properties of Teak (*Tectona grandis*) from a Mature Unmanaged Stand in East Timor. *J. Wood Sci.* **2011**, *57*, 171–178. [CrossRef]
22. Weber, J.C.; Sotelo Montes, C. Geographic Variation in Tree Growth and Wood Density of *Guazuma crinita* Mart. in the Peruvian Amazon. *New For.* **2008**, *36*, 29. [CrossRef]
23. Nocetti, M.; Rozenberg, P.; Chaix, G.; Macchioni, N. Provenance Effect on the Ring Structure of Teak (*Tectona grandis* L.f.) Wood by X-Ray Microdensitometry. *Ann. For. Sci.* **2011**, *68*, 1375–1383. [CrossRef]
24. Nazari, N.; Bahmani, M.; Kahyani, S.; Humar, M.; Koch, G. Geographic Variations of the Wood Density and Fiber Dimensions of the Persian Oak Wood. *Forests* **2020**, *11*, 1003. [CrossRef]
25. Sousa, V.; Silva, M.E.; Louzada, J.L.; Pereira, H. Wood Density and Ring Width in *Quercus rotundifolia* Trees in Southern Portugal. *Forests* **2021**, *12*, 1499. [CrossRef]
26. Moore, J.; Cown, D. Wood Quality Variability—What Is It, What Are the Consequences and What Can We Do about It? *N. Z. J. For.* **2015**, *59*, 3–9.
27. Preston, K.; Cornwell, W.; Denoyer, J. Wood Density and Vessel Traits as Distinct Correlates of Ecological Strategy in 51 California Coast Range Angiosperms. *New Phytol.* **2006**, *170*, 807–818. [CrossRef] [PubMed]
28. Martínez-Cabrera, H.I.; Jones, C.S.; Espino, S.; Schenk, H.J. Wood Anatomy and Wood Density in Shrubs: Responses to Varying Aridity along Transcontinental Transects. *Am. J. Bot.* **2009**, *96*, 1388–1398. [CrossRef]
29. Ziemińska, K.; Butler, D.W.; Gleason, S.M.; Wright, I.J.; Westoby, M. Fibre Wall and Lumen Fractions Drive Wood Density Variation across 24 Australian Angiosperms. *AoB Plants* **2013**, *5*, plt046. [CrossRef]
30. Nabais, C.; Hansen, J.K.; David-Schwartz, R.; Klisz, M.; López, R.; Rozenberg, P. The Effect of Climate on Wood Density: What Provenance Trials Tell Us? *For. Ecol. Manag.* **2018**, *408*, 148–156. [CrossRef]
31. Sedjo, R.A. Wood Use and Trade. In *ternational Trade in Wood Products. In Encyclopedia of Forest Sciences*; Burley, J., Ed.; Elsevier: Oxford, UK, 2004; pp. 1857–1863. ISBN 978-0-12-145160-8.
32. Ali, A.C.; Uetimane, E.; Lhate, I.A.; Terziev, N. Anatomical Characteristics, Properties and Use of Traditionally Used and Lesser-Known Wood Species from Mozambique: A Literature Review. *Wood Sci. Technol.* **2008**, *42*, 453–472. [CrossRef]
33. Junior, E.U.; Terziev, N.; Daniel, G. Wood Anatomy of Three Lesser Known Species from Mozambique. *IAWA J.* **2009**, *30*, 277–291. [CrossRef]
34. Ferreirinha, M.P. *Catálogo da Madeiras de Moçambique—Parte I*; Memórias, Série Botânica II; Ministério do Ultramar, Junta de Investigações do Ultramar: Lisbon, Portugal, 1955.
35. Freitas, M.C. *Estudo das Madeiras de Timor—I Contribuição*; Memórias, Série Botânica III; Ministério do Ultramar, Junta de Investigações do Ultramar: Lisbon, Portugal, 1955.
36. Freitas, M.C. *Estudo das Madeiras de Timor. II Contribuição*; Memórias da Junta de Investigações do Ultramar; Ministério do Ultramar: Lisbon, Portugal, 1958.
37. Freitas, M.C. *Madeiras da Índia Portuguesa*; Memórias da Junta de Investigações do Ultramar; Ministério do Ultramar: Lisbon, Portugal, 1963; Volume 47.
38. Freitas, M.C. *Madeiras de Moçambique. Características Anatómicas Físicas e Mecânicas*; Instituto de Investigação Científica Tropical: Lisbon, Portugal, 1986.
39. IAWA Committee. IAWA List of Microscopic Features for Hardwood Identification. *IAWA Bull.* **1989**, *10*, 219–232.
40. Sneath, P.H.A.; Sokal, R.R. *Numerical Taxonomy: The Principles and Practice of Numerical Classification*; A Series of Books in Biology; W. H. Freeman: San Francisco, CA, USA, 1973; ISBN 978-0-7167-0697-7.
41. Sokal, R.R.; Rohlf, F.J. The Comparison of Dendrograms by Objective Methods. Taxon 11: 33–40. *Taxon* **1962**, *11*, 33–40. [CrossRef]
42. Martínez-Cabrera, H.I.; Cevallos-Ferriz, S.R.S. New Species of Tapirira (Anacardiaceae) from Early Miocene Sediments of the El Cien Formation, Baja California Sur, Mexico. *IAWA J.* **2004**, *25*, 103–117. [CrossRef]
43. Barros, C.F.; Marcon-Ferreira, M.L.; Callado, C.H.; Lima, H.R.P.; da Cunha, M.; Marquete, O.; Costa, C.G. Tendências Ecológicas Na Anatomia Da Madeira de Espécies Da Comunidade Arbórea Da Reserva Biológica de Poço Das Antas, Rio de Janeiro, Brasil. *Rodriguésia* **2006**, *57*, 443–460. [CrossRef]
44. Wickremasinghe, B.; Herat, T. A Comparative Wood Anatomical Study of the *Genus diospyros* L. (Ebenaceae) in Sri Lanka. *Ceylon J. Sci.* **2006**, *35*, 115–136.
45. Pande, P.K.; Negi, K.; Singh, M. Wood Anatomy of Shorea of White Meranti (Meranti Pa'ang) Group of the Malay Peninsula. *Curr. Sci.* **2007**, *92*, 1748–1754.
46. MacLachlan, I.R.; Gasson, P. PCA of Cites Listed *Pterocarpus santalinus* (Leguminosae) Wood. *IAWA J.* **2010**, *31*, 121–138. [CrossRef]
47. Trugilho, P.; Mori, F.; Lima, J. Correlação Canônica Das Características Químicas e Físicas Da Madeira de Clones de *Eucalyptus grandis* e *Eucalyptus saligna*. *CERNE* **2003**, *9*, 66–80.
48. Ziemińska, K.; Westoby, M.; Wright, I.J. Broad Anatomical Variation within a Narrow Wood Density Range—A Study of Twig Wood across 69 Australian Angiosperms. *PLoS ONE* **2015**, *10*, e0124892. [CrossRef]
49. Zimmermann, M. *Xylem Structure and the Ascent of Sap*; Springer: Berlin/Heidelberg, Germany, 1983.
50. Noshiro, S.; Baas, P. Systematic Wood Anatomy of Cornaceae and Allies. *IAWA J.* **1998**, *19*, 43–97. [CrossRef]
51. Noshiro, S.; Baas, P. Latitudinal Trends in Wood Anatomy within Species and Genera: Case Study in Cornus S.L. (Cornaceae). *Am. J. Bot.* **2000**, *87*, 1495–1506. [CrossRef]

52. Carlquist, S. *Comparative Wood Anatomy Systematic, Ecological, and Evolutionary Aspects of Dicotyledon Wood*, 2nd ed.; Springer: Berlin/Heidelberg, Germany, 2001.
53. Baas, P.; Ewers, F.; Davis, S.; Wheeler, E. Evolution of Xylem Physiology. In *The Evolution of Plant Physiology, Linnean Society symposium series Number 21*; Academic Press: Oxford, UK, 2004; pp. 273–295. ISBN 978-0-12-339552-8.
54. Vieira, H.; Rios, A.; Quilhó, T.; Cunha, A.; Brand, M.; Danielli, D.; Florez, J.; Stange, R.; Buss, R.; Higuchi, P. Agrupamento e Caracterização Anatômica Da Madeira de Espécies Nativas Da Floresta Ombrófila Mista. *Rodriguésia* **2019**, *70*, 1–16. [CrossRef]
55. Hacke, U.G.; Sperry, J.S.; Pockman, W.T.; Davis, S.D.; McCulloh, K.A. Trends in Wood Density and Structure Are Linked to Prevention of Xylem Implosion by Negative Pressure. *Oecologia* **2001**, *126*, 457–461. [CrossRef]
56. Hacke, U.G.; Sperry, J.S.; Wheeler, J.K.; Castro, L. Scaling of Angiosperm Xylem Structure with Safety and Efficiency. *Tree Physiol* **2006**, *26*, 689–701. [CrossRef]
57. Ewers, F.W. Xylem' Structure and Water Conduction in Conifer Trees, Dicot Trees, and Llanas. *IAWA J.* **1985**, *6*, 309–317. [CrossRef]
58. Bessa, F. Caracterização anatómica, física, química e acústica de madeiras de várias espécies para a construção de instrumentos musicais—uma aplicação à viola dedilhada. Master's Thesis, Instituto Superior de Agronomia, Universidade Técnica de Lisboa, Lisboa, Portugal, 2000.
59. Aguilar-Rodríguez, S.; Terrazas, T.; López-Mata, L. Anatomical Wood Variation of Buddleja Cordata (Buddlejaceae) along Its Natural Range in Mexico. *Trees* **2005**, *20*, 253. [CrossRef]
60. Miranda, I.; Almeida, M.H.; Pereira, H. Provenance and Site Variation of Wood Density in *Eucalyptus globulus* Labill. at Harvest Age and Its Relation to a Non-Destructive Early Assessment. *For. Ecol. Manag.* **2001**, *149*, 235–240. [CrossRef]
61. Reyes, G.; Brown, S.; Chapman, J.; Lugo, A. *Wood Densities of Tropical Tree Species*; US Department of Agriculture: New Orleans, LA, USA, 1992.
62. Muller-Landau, H.C. Interspecific and Inter-Site Variation in Wood Specific Gravity of Tropical Trees. *Biotropica* **2004**, *36*, 20–32. [CrossRef]
63. Sotelo Montes, C.; Beaulieu, J.; Hernández, R.E. Genetic Variation in Wood Shrinkage and Its Correlations with Tree Growth and Wood Density of *Calycophyllum spruceanum* at an Early Age in the Peruvian Amazon. *Can. J. For. Res.* **2007**, *37*, 966–976. [CrossRef]
64. Bossu, J.; Beauchêne, J.; Estevez, Y.; Duplais, C.; Clair, B. New Insights on Wood Dimensional Stability Influenced by Secondary Metabolites: The Case of a Fast-Growing Tropical Species Bagassa Guianensis Aubl. *PLoS ONE* **2016**, *11*, e0150777. [CrossRef]
65. Lima, I.; Longui, E.; Freitas, M.L.; Zanatto, A.; Zanata, M.; Florsheim, S.; Jr, G. Physical-Mechanical and Anatomical Characterization in 26-Year-Old *Eucalyptus resinifera* Wood. *Floresta E Ambiente* **2013**, *21*, 91–98. [CrossRef]
66. Sotelo Montes, C.; Hernández, R.E.; Beaulieu, J.; Weber, J.C. Genetic Variation in Wood Color and Its Correlations with Tree Growth and Wood Density of *Calycophyllum spruceanum* at an Early Age in the Peruvian Amazon. *New For.* **2008**, *35*, 57–73. [CrossRef]
67. Sotelo Montes, C.; Weber, J.C.; Garcia, R.A.; Silva, D.A.; Muñiz, G.I.B. Variation in Wood Color among Natural Populations of Five Tree and Shrub Species in the Sahelian and Sudanian Ecozones of Mali. *Can. J. For. Res.* **2013**, *43*, 552–562. [CrossRef]
68. Moya, R.; Calvo-Alvarado, J. Variation of Wood Color Parameters of Tectona Grandis and Its Relationship with Physical Environmental Factors. *Ann. For. Sci.* **2012**, *69*, 947–959. [CrossRef]
69. Bakali, I.; Yagi, S.; Merlin, A.; Deglise, X. A Screening Study of Natural Colour of Wood from Different Geographical Regions. *Res. J. For.* **2011**, *5*, 162–168. [CrossRef]
70. Rungwattana, K.; Hietz, P. Radial Variation of Wood Functional Traits Reflect Size-Related Adaptations of Tree Mechanics and Hydraulics. *Funct. Ecol.* **2018**, *32*, 260–272. [CrossRef]
71. Bhat, K.M.; Priya, P.B. Influence of provenance variation on wood properties of teak from the Western Ghat region in India. *IAWA J.* **2004**, *25*, 273–282. [CrossRef]
72. Sousa, V.B.; Cardoso, S.; Quilhó, T.; Pereira, H. Growth Rate and Ring Width Variability of Teak, *Tectona grandis* (Verbenaceae) in an Unmanaged Forest in East Timor. *Rev. Biol. Trop.* **2012**, *60*, 483–494. [CrossRef] [PubMed]
73. Chauhan, L.; Dayal, R. Wood Anatomy of Indian Albizias. *IAWA J.* **1985**, *6*, 213–218. [CrossRef]
74. Odiye, M.; Owolabi, S.; Akinloye, A.; Folorunso, A.; Ayodele, A. Comparative Wood Anatomical Studies in the *Genus albizia* Durazz in Nigeria and Their Potential for Papermaking. *Plants Environ.* **2019**, *1*, 70–82.
75. Nithaniyal, S.; Parani, M. Evaluation of Chloroplast and Nuclear DNA Barcodes for Species Identification in *Terminalia* L. *Biochem. Syst. Ecol.* **2016**, *68*, 223–229. [CrossRef]
76. InsideWood. Published on the Internet 2004. Available online: http://insidewood.lib.ncsu.edu/search (accessed on 1 May 2020).
77. Wheeler, E.A. Inside Wood—A Web Resource for Hardwood Anatomy. *IAWA J.* **2011**, *32*, 199–211. [CrossRef]
78. Bolza, E.; Keating, W.G. *African Timbers—The Properties, Uses and Characteristics of 700 Species*; Division of Building Research CSIRO: Melbourne, Australia, 1972.
79. Keating, W.G.; Bolza, E. *Characteristics, Properties and Uses of Timber Southeast Asia, Northern Australia and the Pacific*; Inkata Press: Melbourne, Australia, 1982; Volume 1.
80. Affre, A.; Kathe, W.; Raymakers, C. *Looking Under the Veneer. Implementation Manual on EU Timber Trade Control: Focus on CITES-Listed Trees*; Report to the European Commission; TRAFFIC Europe: Brussels, Belgium, 2004.

forests

Article

Quantitative Anatomical Characteristics of Virgin Cork in *Quercus variabilis* Grown in Korea

Denni Prasetia [1], Byantara Darsan Purusatama [2], Jong-Ho Kim [1], Go-Un Yang [1], Jae-Hyuk Jang [3], Se-Yeong Park [1], Seung-Hwan Lee [1] and Nam-Hun Kim [1,*]

[1] Department of Forest Biomaterials Engineering, College of Forest and Environmental Sciences, Kangwon National University, Chuncheon 24341, Korea
[2] Institute of Forest Science, Kangwon National University, Chuncheon 24341, Korea
[3] FC Korea Land Co., Ltd., Seoul 07271, Korea
* Correspondence: kimnh@kangwon.ac.kr

Abstract: The quantitative anatomical characteristics of *Quercus variabilis* virgin cork grown in Korea were observed by scanning electron microscopy and compared with *Quercus suber* reproduction cork from Portugal to obtain basic data for further utilization of domestic cork resources in Korean cork industries. *Q. variabilis* virgin cork showed a smaller growth ring width and higher latecork percentage than *Q. suber* reproduction cork. *Q. variabilis* showed a smaller proportion of cork cells and a higher proportion of lenticular channels than *Q. suber*, whereas sclereid and dark-brown zones were found only in *Q. variabilis*. The frequency of pentagonal cork cells in the transverse and radial sections was higher in the cork of *Q. suber* than in *Q. variabilis*. In the tangential section, *Q. variabilis* displayed a lower frequency of heptagonal cells and a higher frequency of pentagonal cells than *Q. suber*. *Q. variabilis* cork had a smaller cell width, lumen diameter, cell wall thickness, prism base edge and area, total cell volume, and solid volume of the cell wall than *Q. suber* cork. The fractional solid volume and number of cells per cm^3 were higher in *Q. variabilis* than *Q. suber*.

Keywords: dark-brown zone; lenticular filling tissue; quantitative anatomical characteristics; *Quercus suber*; *Quercus variabilis*; reproduction cork; scanning electron microscopy; sclereid; virgin cork

Citation: Prasetia, D.; Purusatama, B.D.; Kim, J.-H.; Yang, G.-U.; Jang, J.-H.; Park, S.-Y.; Lee, S.-H.; Kim, N.-H. Quantitative Anatomical Characteristics of Virgin Cork in *Quercus variabilis* Grown in Korea. *Forests* **2022**, *13*, 1711. https://doi.org/10.3390/f13101711

Academic Editors: Vicelina Sousa, Helena Pereira, Teresa Quilhó and Isabel Miranda

Received: 13 September 2022
Accepted: 15 October 2022
Published: 17 October 2022

Publisher's Note: MDPI stays neutral with regard to jurisdictional claims in published maps and institutional affiliations.

1. Introduction

Cork is a non-timber forest product and a part of the periderm in the bark system that surrounds the stems, branches, and roots of dicotyledonous plants [1]. Cork has high economic value owing to its remarkable properties, including its impermeability to liquids and gases, its excellent thermal and sound insulation, shock absorption [2,3], and a high coefficient of friction and resistance to microbial activity [4]. Therefore, cork is widely used as a renewable and sustainable raw material in industry, including in wine stoppers [1], floors and ceilings [5], food product packaging [6], insulation for energy absorption [7], and the surfacing of walking areas [8].

There are three types of cork in the process of cork production: virgin cork, second cork, and reproduction cork. Virgin cork is found in the first periderm [9], while second cork is produced by the regenerated phellogen after removing virgin cork. Successive cork layers are called reproduction cork and are harvested from trees at nine-year intervals [4]. Virgin and second corks are generally used as triturations for agglomerate production because of the uneven structure of the cork tissue. In contrast, reproduction cork is the most essential material in the cork industry owing to its structure for producing solid cork products, such as wine stoppers [1].

The primary resource of cork in the global industry is that obtained from *Q. suber*, a species of plant that is widely distributed in the Western Mediterranean; Southwestern Europe, including Portugal, Spain, southern France, and Italy; and in North Africa, such as Morocco, Algeria, and Tunisia [10]. *Q. variabilis* is another species that contains a large amount of cork in its bark periderm, widely found in Eastern Asia, including China, Korea, and Japan [11]. The cork of *Q. variabilis* has been cultivated and exploited in China for a limited scale of cork production [12].

The anatomical characteristics of cork, including growth ring characteristics, cork element composition, and cork cell structure, are the basis of many properties of the materials, such as very low permeability, hydrophobic behavior, biological inertia, high elasticity in compression, and dimensional recovery [1,13–15]. Many studies have reported on the quantitative aspects of the anatomical characteristics of cork in *Q. variabilis* grown in China. According to Yafang et al. [16], the three-dimensional structure of virgin cork cells in *Q. variabilis* was a prism, with the cork frequently showing a hexagonal shape in the transverse, radial, and tangential sections (53.6%, 52.5%, and 52.3% of cells, respectively). These authors also mentioned that cork cells showed a pentagonal shape in the transverse, radial, and tangential sections (22.6%, 23.4%, and 25.90%, respectively), as well as a heptagonal shape in the transverse, radial, and tangential sections (19.5%, 14.9%, and 14.3%, respectively). Additionally, the prism height was found to decrease from earlycork to latecork cells, whereas the radial cell wall thickness increased from earlycork to latecork cells. The prism base edge of the earlycork cells was 8.0–14.0 μm, with an average base area of 200–500 μm^2. Miranda et al. [12] reported that *Q. variabilis* reproduction cork frequently showed a hexagonal shape in the tangential section, which accounted for approximately 60.0% of cork cells. In the non-tangential sections, 39.5% of the cork cells were pentagonal, while 37.1% were hexagonal. The earlycork cells of *Q. variabilis* showed a higher prism height and total cell volume than latecork cells, while the radial cell wall thickness and solid volume fraction increased from earlycork to latecork cells. Ferreira et al. [13] reported that 40.5% and 41.2% of the cells in the tangential section of the virgin and reproduction corks in *Q. variabilis* were hexagonal, while 31.3% and 32.1% were pentagonal. In the non-tangential sections, virgin cork dominantly showed a pentagonal shape (44.4% of cells), while reproduction cork commonly showed a hexagonal shape (46.9% of cells). The prism height, total cell volume, and solid volume fraction in virgin cork were smaller than in reproduction cork; however, the prism base edge and average base area in virgin cork were larger than in reproduction cork. Li et al. [17] reported that the virgin and reproduction corks of *Q. variabilis* displayed lenticular channels and sclereids surrounded by dark and hard layers. In addition, lenticular channels and sclereids surrounded by dark and hard layers of the reproduction cork were less frequent than those of the virgin cork.

At present, Kim [18] investigated the quantitative anatomical characteristics of *Q. variabilis* reproduction cork grown in Korea and reported that *Q. variabilis* reproduction cork consisted of 87.2% cork cells, 9.0% lenticels, 0.8% sclereids, and 3.0% dark-brown zones. The growth ring width of *Q. variabilis* was found to be narrower than *Q. suber* reproduction cork (0.82 mm and 2.06 mm, respectively). Earlycork cells showed a higher prism height than latecork cells, whereas the prism edge length and cell wall thickness were greater in latecork than in earlycork cells.

In Korea, *Q. variabilis* wood has been used historically as a raw material for timber, firewood, charcoal [19–21], musical instruments, and fuel [22]. *Q. variabilis* also has a large amount of cork in its bark, which can be used as a sustainable resource in the cork industry [13]; nevertheless, no reproduction cork from *Q. variabilis* is available for industrial use in Korea. Therefore, *Q. suber* reproduction cork is largely imported from Portugal.

In Korean cork industry, the main raw material for various products (e.g., wine stoppers, insulation boards, and surfacing products for pavements and sidewalls) is *Q. suber* reproduction cork imported from Portugal; however, with an increasing demand for cork products, there is a need to identify alternative cork resources from domestic oak species to ensure the success of the Korean cork industry. In Korea, *Q. variabilis* is widely distributed, and virgin cork can be easily obtained from the trees; however, there remains a lack of information regarding the quality of cork resources produced by this species. Thus, to gain insights into cork quality and effectively utilize the cork of *Q. variabilis* as sustainable raw material for various products, the quantitative anatomical characteristics of *Q. variabilis* virgin cork grown in Korea were investigated and compared with those of commercial cork (*Q. suber* reproduction cork from Portugal).

2. Materials and Methods

2.1. Materials

Quercus variabilis virgin cork was collected from three trees at breast height in the research forest of Kangwon National University, Chuncheon, Korea (37°77′ N, 127°81′ E). Two planks of *Q. suber* reproduction cork from Castelo Branco cork forest of Amorim Group (Mozelos, Portugal) were provided by FC Korea Land Co., Ltd. (Seoul, Korea). Photographs of *Q. variabilis* virgin cork and *Q. suber* reproduction cork are presented in Figure 1, and basic information on the cork samples is presented in Table 1.

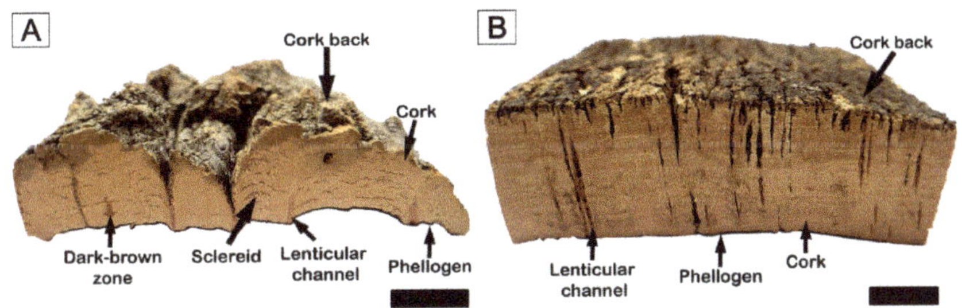

Figure 1. *Q. variabilis* virgin cork (**A**) and *Q. suber* reproduction cork (**B**). Black scale bar represents 20 mm.

Table 1. Basic information of the sample cork.

Species	Cork Type	Location	Cork Thickness (mm)
Q. variabilis	Virgin cork	Research forest of Kangwon National University, Chuncheon, Korea (37°77′ N, 127°81′ E)	10–30
Q. suber	Reproduction cork	Castelo Branco cork forest of Amorim Group, Mozelos, Portugal	30–40

2.2. Measurement of Growth Ring Characteristics

Samples of both species with dimensions of 20–40 (radial) × 20 (tangential) × 20 (longitudinal) mm^3 were prepared, and the transverse section was trimmed using a sliding microtome (MSL-H; Nippon Optical Works, Nagano, Japan). The growth ring characteristics, such as quantity, ring width, and percentage of latecork, in both species were measured in the transverse section of three samples from each species. The growth ring number and width were observed using a measuring microscope (MM-40; Nikon, Tokyo, Japan) connected to an image analysis system (IMT i-solution lite, version 9.1; Burnaby, British Columbia, Canada). For latecork percentage, the cork samples with dimensions of 10 (radial) × 10 (tangential) × 10 (longitudinal) mm^3 were coated with gold using a sputter coater (Cressington sputter coater 108; Watford, UK) and observed using a scanning electron microscope (SEM) (JSM-5510, 15 kV; Tokyo, Japan).

2.3. Cellular Structure Observations

Samples with dimensions of 20 (radial) × 20 (tangential) × 20 (longitudinal) mm^3 were prepared, and the transverse, radial, and tangential sections were trimmed using a sliding microtome (MSL-H; Nippon Optical Works, Nagano, Japan). The cork tissue, such as cork cells, lenticular channel, sclereid, and dark-brown zone, were determined as the ratio of the area of each element to the total area of 400 mm^2, and the measurements were performed using 20 samples from each species. The images of the three sections were captured using a mobile phone and recorded in digital format of 3024 × 3024 pixels with a resolution of 72 dpi and 24-bits depth (Samsung Note 20, 12MP with F1.8; Suwon-si, Gyeonggi-do, Korea). The images were analyzed using ImageJ (version Java 1.80_172, 64 bits; Bethesda, MD, USA).

2.4. Cork cell Dimension

Cork samples with dimensions of 10 (radial) × 10 (tangential) × 10 (longitudinal) mm^3 were used to observe cork cell dimensions. The cork samples were coated with gold using a sputter coater (Cressington sputter coater 108; Watford, UK) and observed using SEM (JSM-5510, 15 kV; Tokyo, Japan).

The number of edges and the two- and three-dimensional characteristics were detected from 400 cork cells in both species.

The frequency of the edge number in a cork cell (fi) was measured in the transverse, radial, and tangential sections and calculated using Equation (1). Then, the dispersion (μ_2) of the function in relation to the mean of cell shape (i_m) was calculated using Equation (2) [4]. The dispersion value presented homogeneity of the cell shape in each section of both species. If all cells show one type of cell shape, the dispersion value would be 0 [12].

$$fi = \left(Ni / \sum Ni \right) \times 100 \ (\%) \tag{1}$$

$$\mu_2 = \sum (i - i_m)^2 \times fi \ (\%) \tag{2}$$

where Ni represents the number of cells with i is edges, and $\sum Ni$ is the total number of cells.

The two-dimensional characteristics of earlycork and latecork cells, such as radial width or prism height, radial lumen diameter, radial wall thickness, tangential width, tangential lumen diameter, and tangential wall thickness, were measured in the transverse section. In addition, the radial and tangential wall thicknesses of the earlycork and latecork cells were calculated using Equation (3) [12], as follows:

$$Cell\ wall\ thickness = (\text{cell dimension} - \text{lumen dimension})/2 \ (\mu m) \tag{3}$$

A three-dimensional illustration of a cork cell as a hexagonal prism is shown in Figure 2. The three-dimensional characteristics of earlycork and latecork cells, including radial width or prism height, total cell volume, solid volume, fractional solid volume of cell walls, and number of cork cells per cm^3, were measured in the transverse section. Additionally, the prism base edge and base area of earlycork cells were measured in the tangential section.

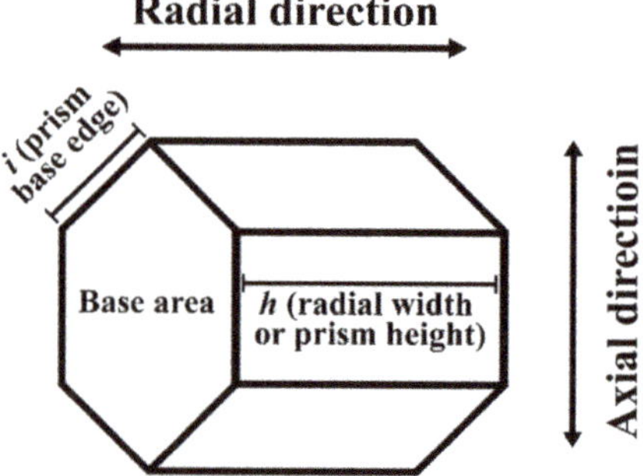

Figure 2. Three-dimensional illustration of a cork cell as a hexagonal prism (modified from Miranda et al. [12]).

The total volume (V), solid volume (V_s), and lumen volume (V_o) of earlycork and latecork cells were calculated according to Pereira [1], as shown in Equation (4). The fractional solid volume (V_{sf}) was calculated using Equation (5):

$$V = 3\sqrt{\frac{3}{2}}i^2 \times h \ (\mu m^3) \tag{4}$$

$$
\begin{aligned}
V_0 &= 3\sqrt{\frac{3}{2}}\left(i - \frac{e}{\sqrt{3}}\right)^2 \times (h - e) \ (\mu m^3) \\
V_s &= 3\sqrt{\frac{3}{2}}i^2 \times h - 3\sqrt{\frac{3}{2}}\left(i - \frac{e}{\sqrt{3}}\right)^2 \times (h - e) \ (\mu m^3) \\
V_{sf} &= \frac{V_s}{V} \times 100 \ (\%)
\end{aligned}
\tag{5}
$$

where i is the base edge, h is the prism height, and e is the wall thickness.

The number of cork cells per cm^3 (N) was calculated using Equation (6).

$$N = \frac{\text{Cork volume of 1 } cm^3}{\text{Average cork cell volume in earlycork and latecork cells}} \tag{6}$$

2.5. Statistical Analysis

Statistical differences in the quantitative anatomical characteristics between species were analyzed using one-way analysis of variance (ANOVA), followed by post hoc Tukey's honestly significant difference (HSD) test ($p \leq 0.05$) (SPSS, version 24; IBM Corp., New York, NY, USA).

3. Results and Discussion

3.1. Growth Ring Characteristics

The growth ring widths in *Q. variabilis* virgin cork and *Q. suber* reproduction cork are shown in Table 2. The growth ring width was 0.54 mm for *Q. variabilis* cork and 3.07 mm for *Q. suber* cork. A significant difference was observed in terms of the width of the growth ring between the two species.

Table 2. Growth ring widths of *Q. variabilis* virgin cork and *Q. suber* reproduction cork.

Q. variabilis (mm)				*Q. suber* (mm)		
Tree 1	Tree 2	Tree 3	Average	Plank 1	Plank 2	Average
0.58 (0.24)	0.61 (0.2)	0.43 (0.22)	0.54 [a] (0.10)	3.08 (0.54)	3.05 (0.26)	3.07 [b] (0.02)

Note: Numbers in parentheses are standard deviations (SD). The listing of the same superscript lowercase letters beside the mean values in the same row denotes nonsignificant outcomes at the 5% significance level for comparisons between species.

The radial variation in the growth ring width from the cork back to the phellogen is shown in Figure 3. *Q. variabilis* virgin cork had a 48–52 growth ring number, whereas *Q. suber* reproduction cork showed 10 to 11 growth rings. The growth ring width of *Q. variabilis* virgin cork rapidly decreased from the cork back to the 25th growth ring, becoming constant after the 25th growth ring. By contrast, the growth ring width of *Q. suber* reproduction cork gradually decreased from the cork back to the phellogen. Kim [18,23] reported that the average growth ring width in *Q. variabilis* and *Q. suber* reproduction cork was 0.82 mm and 2.06 mm, respectively. The growth ring width of *Q. variabilis* reproduction cork in previous studies was higher than that of virgin cork in the present study.

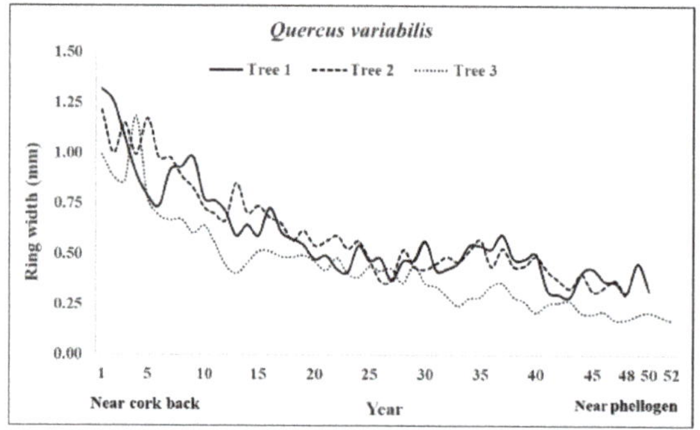
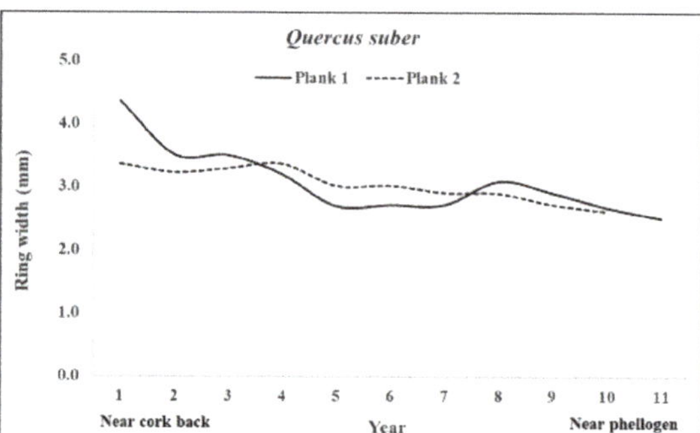

Figure 3. Variation in growth ring width from the cork back to the phellogen in *Q. variabilis* virgin cork and *Q. suber* reproduction cork.

The latecork percentage and radial variation in the cork of both species are presented in Table 3 and Figure 4, respectively. The latecork percentages in *Q. variabilis* virgin cork and *Q. suber* reproduction cork were 7.41% and 0.97%, respectively. Both species showed an increase in the latecork cell percentage from the near cork back to the near phellogen stage. Kim [18,23] mentioned that the latecork percentage of *Q. variabilis* was higher than that of *Q. suber*, which was 14.00% in *Q. variabilis* and 10.60% in *Q. suber*. The latecork percentage in the previous study was higher than that in the present study. Pereira et al. [4] also reported that, within a growth ring, *Q. suber* virgin cork showed a higher proportion of latecork cells than *Q. suber* reproduction cork.

Table 3. Latecork percentage in *Q. variabilis* virgin cork and *Q. suber* reproduction cork.

Q. variabilis (%)				Q. suber (%)		
Tree 1	Tree 2	Tree 3	Average	Plank 1	Plank 2	Average
7.56 (4.17)	5.55 (2.8)	9.13 (5.3)	7.41 [a] (1.79)	0.96 (0.25)	0.98 (0.14)	0.97 [b] (0.01)

Note: Numbers in parentheses are the SD. The listing of the same superscript lowercase letters beside the mean values in the same row denotes nonsignificant outcomes at the 5% significance level for comparisons between species.

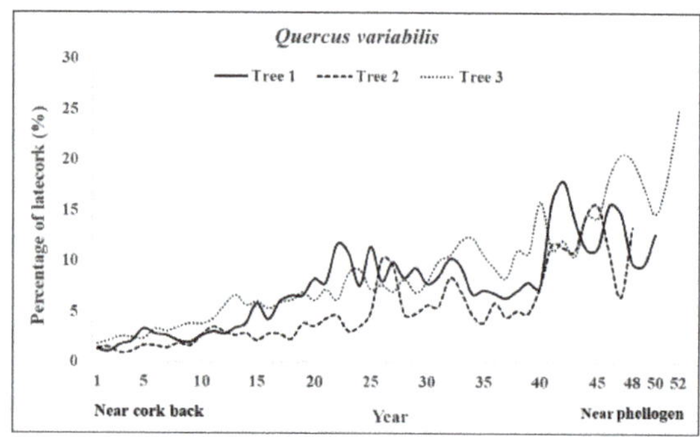
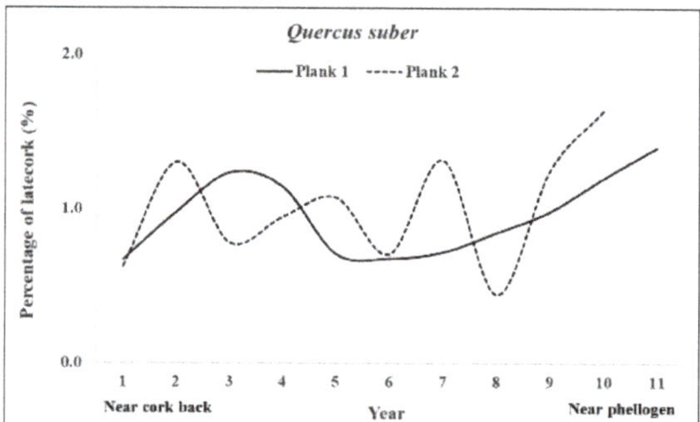

Figure 4. Variation of latecork percentage from the cork back to the phellogen in *Q. variabilis* virgin cork and *Q. suber* reproduction cork.

3.2. Cellular Structure Observations

SEM images in the transverse, radial, and tangential sections of *Q. variabilis* virgin cork and *Q. suber* reproduction cork are displayed in Figure 5. In the transverse and radial sections, the earlycork and latecork cells of both species showed a brick wall-type structure.

In the tangential section, the cork cells of both species showed a honeycomb structure with various shapes, such as rectangular, pentagonal, hexagonal, heptagonal, octagonal, and nonagonal.

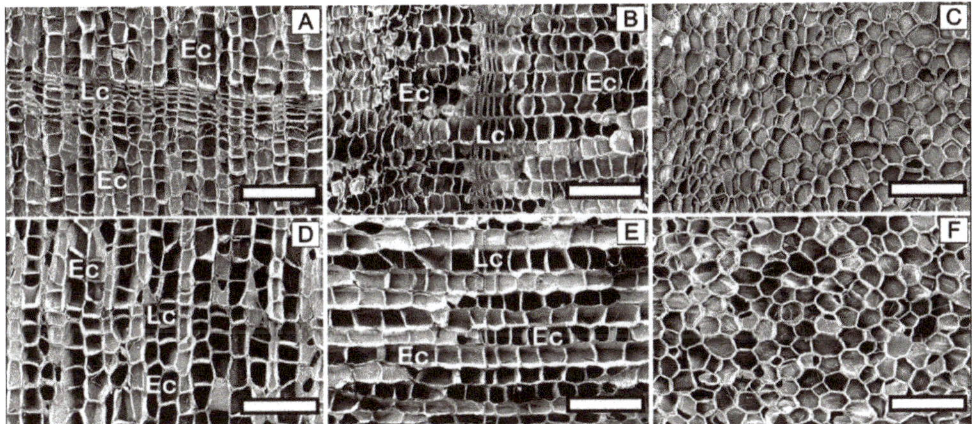

Figure 5. Transverse (**A,D**), radial (**B,E**), and tangential (**C,F**) sections of *Q. variabilis* virgin cork (**A–C**) and *Q. suber* reproduction cork (**D–F**). Earlycork (Ec) and latecork (Lc). White scale bar represents 100 µm.

The results of previous studies are in line with those reported in the present study. *Q. variabilis* and *Q. suber* corks showed a brick wall-type structure in the transverse and radial sections and a honeycomb structure in the tangential section [4,12,13,16].

The percentage of cork tissue of *Q. variabilis* virgin cork and *Q. suber* reproduction cork in the three sections are shown in Table 4. *Q. variabilis* virgin cork consisted of cork cells, lenticular channels, dark-brown zones, and sclereids, whereas *Q. suber* reproduction cork consisted of cork cells and lenticular channels. The proportions of cork cells in the transverse, radial, and tangential sections of *Q. variabilis* were 86.85%, 84.06%, and 84.68%, respectively, whereas those of *Q. suber* were 95.45%, 91.43%, and 93.03%, respectively. Both species had a higher proportion of cork cells in the transverse section than in the radial and tangential sections; however, the proportion of cork cells in all the sections of *Q. suber* was significantly higher than that of *Q. variabilis*.

Table 4. Cork tissue percentages in *Q. variabilis* virgin cork and *Q. suber* reproduction cork.

Elements	*Q. variabilis* (%)			*Q. suber* (%)		
	Transverse	Radial	Tangential	Transverse	Radial	Tangential
Cork cell	86.85 [abc] (4.39)	84.06 [a] (8.92)	84.68 [ab] (2.84)	95.45 [d] (0.96)	91.43 [bcd] (1.09)	93.03 [cd] (1.02)
Lenticular channel	10.68 [ab] (4.10)	13.00 [b] (9.48)	12.01 [b] (2.40)	4.55 [a] (0.96)	8.57 [ab] (1.09)	6.97 [ab] (1.02)
Dark-brown zone	2.33 [a] (0.30)	2.77 [a] (0.79)	2.89 [a] (0.33)	-	-	-
Sclereid	0.15 [a] (0.03)	0.17 [a] (0.06)	0.41 [b] (0.10)	-	-	-

Note: Numbers in parentheses are the SD. The listing of the same superscript lowercase letters beside the mean values in the same row denotes nonsignificant outcomes at the 5% significance level for comparisons between sections in both species.

The lenticular channel proportions in the transverse, radial, and tangential sections of *Q. variabilis* cork were 10.68%, 13.00%, and 12.01%, respectively, whereas those of *Q. suber* were 4.55%, 8.57%, and 6.97%, respectively. The radial section of both species had the highest proportion of lenticular channels among the three sections. Additionally, there was a significant difference in the proportion of lenticular channels in the three sections between the two species.

The proportions of the dark-brown zones in the transverse, radial, and tangential sections were 2.33%, 2.77%, and 2.89%, respectively, whereas those of sclereid were 0.15%, 0.17%, and 0.41%, respectively. The highest proportion of dark-brown zones and sclereids in *Q. variabilis* virgin cork was observed in the tangential section. Additionally, the proportion of dark-brown zones and sclereids was the lowest in the transverse section. There was no significant difference in the proportion of dark-brown zone between sections in *Q. variabilis* cork, whereas the proportion of sclereids in the tangential section showed a significant difference than in the transverse and radial sections.

The results of previous studies are in line with those reported in the present study. Kim [18,23] reported that *Q. variabilis* and *Q. suber* reproduction corks consisted of 87.20% and 93.00% cork cells, 9.00% and 4.60% lenticels, 0.80% and 0.60% sclereids, and 3.00% and 1.80% dark-brown zones. The author also reported a significant difference in the proportion of cork cells and lenticels between the two species. Notwithstanding, no significant difference was observed in terms of the proportion of sclereids and dark-brown zones between the two species. Li et al. [17] reported that the virgin and reproduction corks of *Q. variabilis* showed lenticular channels and sclereids encircled by dark and hard layers. Additionally, lenticular channels and sclereids surrounded by the dark and hard layers were less frequent in reproduction cork compared to virgin cork.

3.3. Cork Cell Dimensions

3.3.1. Number of Edges in a Cork Cell

The frequency of edge numbers in *Q. variabilis* virgin cork and *Q. suber* reproduction cork are presented in Table 5. In the transverse and radial sections, the cork cells in *Q. variabilis* virgin cork frequently exhibited a hexagonal shape (59.75% and 56.50%, respectively), while pentagonal shape was observed in 19.00% and 19.75% of cells in the transverse and radial sections, respectively. Moreover, 16.75% and 18.50% of cork cells showed a heptagonal shape in the transverse and radial sections, respectively. A small amount of octagonal and rectangular shapes was observed in the transverse section (2.00% and 2.50%, respectively) and radial section (3.75% and 1.50%, respectively). In the tangential section, a hexagonal shape was observed in 50.00% of the cork cells in *Q. variabilis* virgin cork. Pentagonal and heptagonal shapes were also observed in the tangential section (34.00% and 10.75%, respectively). Cork cells also displayed a small percentage of octagonal and rectangular shapes (1.00% and 4.00%, respectively). In addition, the dispersion of the edge number in the tangential section was 67.00%, which was higher than that in the transverse and radial sections (53.75% and 59.25%, respectively).

In *Q. suber*, cork cells frequently showed a hexagonal shape of 57.00% in the transverse section and 55.00% in the radial section. Cork cells in the transverse and radial sections also exhibited pentagonal shape (22.75% and 26.00%, respectively). Moreover, cork cells were heptagonal in shape in the transverse and radial sections (16.25% and 15.75%, respectively). In addition, octagonal and rectangular cork cells were observed in the transverse section (1.75% and 2.25%, respectively) and radial section (1.50% and 1.75%, respectively). In the tangential section, 48.00% of cork cells were hexagonal cell, 26.75% were pentagonal, 19.00% were heptagonal, 4.25% were rectangular, and 1.75% were octagonal. Furthermore, both species showed a small proportion of nonagonal shapes in the tangential section (0.25%). In addition, the dispersion of the edge number in the tangential section was higher than those in the transverse and radial sections (72.00%, 55.00%, and 54.75%, respectively).

Table 5. Frequency of the edge number in *Q. variabilis* virgin cork and *Q. suber* reproduction cork.

Shape	*Q. variabilis* (%)			*Q. suber* (%)		
	Transverse	Radial	Tangential	Transverse	Radial	Tangential
Triangular	0.00	0.00	0.00	0.00	0.00	0.00
Rectangular	2.50	1.50	4.00	2.25	1.75	4.25
Pentagonal	19.00	19.75	34.00	22.75	26.00	26.75
Hexagonal	59.75	56.50	50.00	57.00	55.00	48.00
Heptagonal	16.75	18.50	10.75	16.25	15.75	19.00
Octagonal	2.00	3.75	1.00	1.75	1.50	1.75
Nonagonal	0.00	0.00	0.25	0.00	0.00	0.25
μ_2 (dispersion)	53.75	59.25	67.00	55.00	54.75	72.00

In the transverse and radial sections, a smaller number of cork cells with a pentagonal shape was observed in *Q. variabilis* virgin cork compared to *Q. suber*. In the tangential section, *Q. variabilis* had a higher proportion of pentagonal cells than *Q. suber*. In addition, *Q. variabilis* cork cells also showed a smaller proportion of heptagonal shapes compared to *Q. suber*. The dispersion of the edge number in the transverse and tangential sections in *Q. variabilis* was smaller than that in *Q. suber*, whereas that in the radial section was higher in *Q. variabilis* than in *Q. suber*. Overall, both cork species showed a comparable distribution of cell shapes in the three sections, in line with the results of previous studies.

Pereira et al. [4] reported that 52.60%, 56.20%, and 47.80% of the cork cells in *Q. suber* virgin and reproduction corks had a hexagonal shape in the transverse, radial, and tangential sections, respectively. These authors also reported that pentagonal and heptagonal cork cells accounted for 22.60% and 17.80% of cork cells in the transverse section, 20.30% and 17.20% in the radial section, and 24.90% and 21.60% in the tangential section, respectively. The dispersion of the number of sides in the tangential section was 71.00%, which was higher compared to those in the transverse and radial sections (70.00%, and 62.00%, respectively). Yafang et al. [16] reported that 53.60%, 52.50%, and 52.30% of virgin cork cells in *Q. variabilis* were hexagonal in shape in the transverse, radial, and tangential sections, respectively. The cork cells in the transverse, radial, and tangential sections also exhibited a pentagonal shape (23.40%, 25.90%, and 22.60%, respectively). The heptagonal shape of the transverse, radial, and tangential sections accounted for 14.90%, 14.30%, and 19.50% of cork cells, respectively. The authors also reported that the dispersion of the number of sides of cork cells in the transverse, radial, and tangential sections were 72.00%, 69.00%, and 73.00%, respectively, which were higher in the tangential section than in the transverse and radial sections. Miranda et al. [12] reported that the number of edges in *Q. variabilis* reproduction cork frequently showed a hexagonal shape in the tangential section, accounting for 60.00% of cork cells. Furthermore, 19.40% and 17.60% of cells exhibited pentagonal and heptagonal shapes in the tangential section, respectively, while 39.50%, 37.10%, 18.50%, and 4.80% exhibited pentagonal, hexagonal, rectangular, and heptagonal shapes in the non-tangential sections, respectively. Additionally, the dispersion of the number of edges in the non-tangential sections was higher than that in the tangential section (74.80% vs. 49.10%, respectively). Ferreira et al. [13] reported that there were 40.50% and 41.20% hexagonal cells, 31.30% and 32.10% pentagonal cells, and 18.20% and 17.70% heptagonal cells in the tangential section of the virgin and reproduction corks of *Q. variabilis*, respectively. The dispersion of the number of cell edges in the tangential section of virgin cork was higher than that of reproduction cork (91.30% vs. 84.60%, respectively). The authors also reported that in the non-tangential sections, virgin cork commonly showed a pentagonal shape (44.40%), whereas reproduction cork generally showed a hexagonal shape (46.90%). The percentages of hexagonal and rectangular shapes were 32.10% and 21.20%, respectively, in virgin cork. In reproduction cork, pentagonal and rectangular shapes were observed in

26.30% and 22.10% of the cork cells, respectively. Virgin cork showed a smaller dispersion of the number of edges of cells in the non-tangential sections than reproduction cork, which was 60.20% and 75.90%, respectively.

3.3.2. Two-Dimensional Characteristics of Cork Cells

The two-dimensional characteristics of the earlycork and latecork cells in *Q. variabilis* virgin cork and *Q. suber* reproduction cork are summarized in Table 6.

Table 6. Two-dimensional characteristics of the cork cells in *Q. variabilis* virgin cork and *Q. suber* reproduction cork.

	Q. variabilis (μm)		*Q. suber* (μm)	
	Earlycork	Latecork	Earlycork	Latecork
Radial width	15.81 [c] (3.11)	7.10 [a] (2.36)	37.61 [d] (5.06)	14.33 [b] (3.35)
Radial lumen diameter	14.59 [c] (3.10)	5.66 [a] (2.40)	35.90 [d] (5.21)	12.17 [b] (3.27)
Radial cell wall thickness	1.21 [a] (0.24)	1.45 [b] (0.32)	1.50 [b] (0.27)	2.00 [c] (0.43)
Tangential width	22.78 [a] (2.78)	22.62 [a] (3.20)	27.16 [c] (3.68)	26.16 [b] (3.83)
Tangential lumen diameter	21.84 [b] (2.76)	21.10 [a] (3.13)	25.36 [d] (3.76)	23.81 [c] (3.81)
Tangential cell wall thickness	1.14 [a] (0.22)	1.53 [b] (0.35)	1.45 [b] (0.29)	2.31 [c] (0.50)

Note: Numbers in parentheses are the SD. The listing of the same superscript lowercase letters beside the mean values in the same row denotes nonsignificant outcomes at the 5% significance level for comparisons between earlycork and latecork in both species.

In *Q. variabilis*, the radial and tangential widths of earlycork cells were higher than those of latecork cells as 15.81 μm and 22.78 μm for earlycork cells and 7.10 μm and 22.62 μm for latecork cells, respectively. There were significant differences in radial width between earlycork and latecork cells, whereas, in tangential width, there was no significant difference between earlycork and latecork cells. For the radial and tangential cell walls thickness: 1.21 μm and 1.14 μm for the earlycork cells and 1.45 μm and 1.53 μm for the latecork cells. The radial and tangential cell walls were significantly thicker in the latter than in the earlycork cells. Earlycork and latecork cells showed tangential lumen diameter of 21.84 μm and 21.10 μm, respectively. The radial lumen diameter was greatly higher in the earlycork cells than in the latecork cells showing 14.59 μm and 5.66 μm, respectively. The difference between earlycork and latecork cells in radial and tangential lumen diameters was statistically confirmed.

In *Q. suber*, the radial width of earlycork and latecork cells was 37.61 μm and 14.33 μm, respectively, while the tangential width of earlycork and latecork cells was 27.16 μm and 26.16 μm, respectively. The radial width of the earlycork cells was greater than that of the latecork cells, while the tangential width of earlycork cells was slightly larger than that of latecork cells. The difference between earlycork and latecork cells in radial and tangential widths was statistically significant. The thickness of radial and tangential walls was 1.50 μm and 1.45 μm for earlycork cells and 2.00 μm and 2.31 μm for latecork cells, respectively. Both cell walls were significantly thicker in the latecork cells than in the earlycork cells. The tangential lumen diameter of earlycork cells was slightly larger than that of latecork cells (25.36 μm and 23.81 μm, respectively), whereas the radial lumen diameter of earlycork cells was larger than that of latecork cells (35.90 μm and 12.17 μm, respectively). The difference between earlycork and latecork cells in radial and tangential lumen diameters was statistically confirmed. Additionally, the earlycork and latecork cells in *Q. suber* reproduction cork were significantly larger in cell width and lumen diameter and thicker in cell wall thickness than those in *Q. variabilis* virgin cork.

In this study, differences were observed in the two-dimensional characteristics between species and between the earlycork and latecork cells in each species, in line with previous studies. Pereira et al. [4] reported that the radial width of the cork cells of *Q. suber* was

30.00–40.00 µm in the earlycork cells and 10.00–15.00 µm in the latter. The authors also mentioned that the radial cell wall thickness in the earlycork cells was 1.00–1.50 µm and approximately twice as large in the latecork cells (2.00–3.00 µm). Kim [18] found that the radial width in the reproduction cork cells of *Q. variabilis* from Korea and *Q. suber* from Portugal were 26.30 µm and 44.20 µm in the earlycork cells and 6.90 µm and 10.60 µm in the latecork cells, accordingly. The author also found that the cork cell walls in *Q. variabilis* and *Q. suber* were 0.60 µm and 1.00 µm in the earlycork cells and 1.00 µm and 1.80 µm in the latecork cells. Yafang et al. [16] reported that the radial width of *Q. variabilis* virgin cork was 15.00–35.00 µm in the earlycork cells and 10.00 µm in the latter. These authors also mentioned that the earlycork cells showed smaller radial cell walls thickness than the latecork cells (1.00–1.50 µm vs. 3.00 µm). Miranda et al. [12] reported that earlycork cells in *Q. variabilis* reproduction cork from China showed a larger radial width than latecork cells (21.40 µm vs. 10.40 µm). The authors also reported that the earlycork cells showed smaller radial cell walls thickness than the latecork cells (1.20 µm vs. 2.80 µm). Ferreira et al. [13] also reported that the radial width of earlycork cells in *Q. variabilis* virgin cork was smaller than that in reproduction cork (19.20 µm vs. 22.70 µm, respectively).

3.3.3. Three-Dimensional Characteristics of Cork Cells

The three-dimensional characteristics of the cork cells in *Q. variabilis* virgin cork and *Q. suber* reproduction cork are presented in Table 7. The prism base edge of the cork cells in *Q. variabilis* virgin cork was smaller than that in *Q. suber* reproduction cork (14.81 µm vs. 15.88 µm). The aspect ratio of prism height to prism base edge in *Q. variabilis* virgin cork was 1.07 for earlycork cells and 0.48 for latecork cells, while that in *Q. suber* reproduction cork was about 2.37 for earlycork cells and 0.90 for latecork cells. The prism base area in the earlycork cells of *Q. variabilis* was smaller than that of *Q. suber* (623 µm² vs. 830 µm²). Significant differences were observed in the prism base edge and area of earlycork cells between the two species.

Table 7. Three-dimensional characteristics of the cork cells in *Q. variabilis* virgin cork and *Q. suber* reproduction cork.

	Q. variabilis		Q. suber	
	Earlycork	**Latecork**	**Earlycork**	**Latecork**
Prism height, µm	15.81 [c] (3.11)	7.10 [a] (2.36)	37.61 [d] (5.06)	14.33 [b] (3.35)
Prism base edge, µm	14.81 [a] (2.91)	14.81 [a] (2.91)	15.88 [b] (4.12)	15.88 [b] (4.12)
Prism base area, cm²	6.23×10^{-6} [a] (1.64×10^{-6})	6.23×10^{-6} [a] (1.64×10^{-6})	8.30×10^{-6} [b] (2.65×10^{-6})	8.30×10^{-6} [b] (2.65×10^{-6})
Cork cell volume, cm³	1.27×10^{-8} [b] (0.25×10^{-8})	0.57×10^{-8} [a] (0.18×10^{-8})	3.47×10^{-8} [c] (0.47×10^{-8})	1.32×10^{-8} [b] (0.31×10^{-8})
Solid volume, cm³	0.21×10^{-8} [b] (0.04×10^{-8})	0.17×10^{-8} [a] (0.04×10^{-8})	0.49×10^{-8} [d] (0.08×10^{-8})	0.34×10^{-8} [c] (0.08×10^{-8})
Fractional solid volume, %	16.40 [b] (3.23)	30.75 [d] (7.67)	14.29 [a] (2.51)	26.39 [c] (4.77)
Number of cells per cm³	7.86×10^7	17.50×10^7	2.88×10^7	7.60×10^7

Note: Numbers in parentheses are the SD. The listing of the same superscript lowercase letters beside the mean values in the same row denotes nonsignificant outcomes at the 5% significance level for comparisons between earlycork and latecork in both species.

The cork cell volume in *Q. variabilis* was 1.27×10^{-8} cm³ for earlycork cells and 0.57×10^{-8} cm³ for latecork cells, while that in *Q. suber* was 3.47×10^{-8} cm³ for earlycork cells and 1.32×10^{-8} cm³ for latecork cells. In both species, the cork cell volume of earlycork cells was larger than that of latecork cells. Additionally, significant differences were observed in the volume of earlycork and latecork cells between species.

The solid volume in *Q. variabilis* was 0.21×10^{-8} cm^3 for earlycork cells and 0.17×10^{-8} cm^3 for latecork cells, whereas that in *Q. suber* was 0.49×10^{-8} cm^3 for earlycork cells and 0.34×10^{-8} cm^3 for latecork cells, which was significantly bigger than those in *Q. variabilis*. The solid volume of latecork cells in both species was smaller than that of earlycork cells, which was statistically confirmed.

The fractional solid volume of the cork cell in *Q. variabilis* was 30.75% for latecork cells and 16.40% for earlycork cells, while that in *Q. suber* was 26.39% for latecork cells and 14.29% for earlycork cells. The fractional solid volume in *Q. suber* was significantly smaller than that in *Q. variabilis*. The volume of latecork cells in both species was nearly double that of earlycork cells, showing significant difference between species.

Q. variabilis virgin cork had 7.86×10^7 earlycork cells per cm^3 and 17.50×10^7 latecork cells per cm^3, whereas *Q. suber* reproduction cork *had* 2.88×10^7 earlycork cells per cm^3 and 7.60×10^7 latecork cells per cm^3, respectively. *Q. variabilis* virgin cork had more cork cells per cm^3 than *Q. suber* reproduction cork.

Kim [18] reported that *Q. variabilis* and *Q. suber* reproduction corks showed no difference in terms of the prism edge length between earlycork and latecork cells. The prism edge length of *Q. variabilis* was smaller than that of *Q. suber*. These results are in line with those reported in the present study. In a previous study, Yafang et al. [16] reported that the prism base edge and the average base area of *Q. variabilis* virgin cork were 8.00–14.00 µm and 200–500 µm^2 in the earlycork cells. Miranda et al. [12] reported that the prism base edge and the average base area of *Q. variabilis* reproduction cork were 17.20 µm and 764 µm^2, respectively, while the aspect ratio between prism height and prism base edge in earlycork cells was approximately 1:10. The authors also reported that the total cell volume of the earlycork cells was 1.60×10^{-8} cm^3, which was higher than that of the latecork cells (0.80×10^{-8} cm^3). Earlycork cells showed a much smaller solid volume fraction than latecork cells (13.10% vs. 40.00%). Ferreira et al. [13] examined earlycork cells in the virgin and reproduction corks of *Q. variabilis* and found that the prism base edge in virgin cork was higher than that in reproduction cork (14.30 µm vs. 13.60 µm). Virgin cork also had a higher average base area than reproduction cork (532 µm^2 vs. 477 µm^2). In contrast, reproduction cork had a higher total cell volume than virgin cork (1.08×10^{-8} cm^3 vs. 1.02×10^{-8} cm^3). The solid volume fraction of reproduction cork was 14.70%, while that of virgin cork was 11.50%. The three-dimensional characteristics of *Q. variabilis* virgin cork reported in this study were comparable to those of *Q. variabilis* virgin cork reported by Yafang et al. [16] and Ferreira et al. [13]. In addition, the three-dimensional characteristics of reproduction cork reported by Miranda et al. [12] were higher than those of virgin cork in the present study.

4. Conclusions

Q. variabilis virgin cork exhibited a narrower growth ring than *Q. suber* reproduction cork. The latecork percentage of *Q. variabilis* was significantly higher than that of *Q. suber*. *Q. variabilis* cork had a smaller proportion of cork cells and a higher proportion of lenticular channels in the three sections than *Q. suber* cork. Sclereid and dark-brown zones were only observed in *Q. variabilis* cork.

In both species, hexagonal cell shape was commonly observed in all three sections (50–60%). The frequency of pentagonal and heptagonal cell shapes was also higher than that of rectangular, octagonal, and nonagonal shapes in both species.

The cork cells of *Q. variabilis* had smaller cell width, lumen diameter, and cell wall thicknesses than those of *Q. suber*. The earlycork cells of both species had larger radial cells widths and lumen diameters than the latter, while, in both species, the tangential cells width and lumen diameter of earlycork cells were slightly higher than those of the latecork cells.

The prism base edge and area, cork cell volume, and solid volume of the cell wall in *Q. suber* cork were significantly greater than those of *Q. variabilis* cork, whereas the fractional solid volume and number of cells per cm^3 in *Q. variabilis* cork were significantly greater than those of *Q. suber* cork. In both species, earlycork cells showed higher solid and total cell volumes than latecork cells; nevertheless, the fractional solid volume and number of cells per cm^3 in the latecork cells were greater than those in the earlycork cells.

In conclusion, *Q. variabilis* virgin cork grown in Korea shows distinctive quantitative anatomical characteristics compared to *Q. suber* reproduction cork grown in Portugal. Due to the structural characteristics of *Q. variabilis* virgin cork, applications require trituration to cork granules and agglomeration to produce cork composite products, while its cellular features allow considering it for insulation, surfacing, and sealant products. The results of this study may be used to evaluate the quality and identify *Q. variabilis* virgin cork grown in Korea for further utilization.

Author Contributions: Conceptualization, D.P. and N.-H.K.; methodology, D.P., B.D.P., J.-H.K., J.-H.J., G.-U.Y., S.-Y.P., S.-H.L. and N.-H.K.; software, D.P.; validation, N.-H.K.; formal analysis, D.P.; investigation, D.P.; resources, N.-H.K.; data curation, D.P.; writing—original draft preparation, D.P.; writing—review and editing, D.P., B.D.P., S.-H.L. and N.-H.K.; visualization, D.P.; supervision, N.-H.K.; project administration, J.-H.K.; and funding acquisition, N.-H.K., S.-H.L. and J.-H.J. All authors have read and agreed to the published version of the manuscript.

Funding: This research was supported by the Science and Technology Support Program through the National Research Foundation of Korea (NRF) funded by the Ministry of Science and ICT (MSIT) (NRF-2019K1A3A9A01000018 and No. 2022R1A2C1006470), Basic Science Research Program through the NRF funded by the Ministry of Education (NRF-2016R1D1A1B01008339 and No. 2018R1A6A1A03025582), and R&D Program for Forest Science Technology (Project No. 2021350C10-2223-AC03) provided by the Korea Forest Service (Korea Forestry Promotion Institute).

Data Availability Statement: The datasets generated and analyzed during the current study are not publicly available but are available from the corresponding author upon reasonable request.

Conflicts of Interest: The authors declare no conflict of interest.

References

1. Pereira, H. *Cork: Biology, Production, and Uses*; Elsevier Publications: Amsterdam, The Netherlands, 2007.
2. Aronson, J.; Pereira, J.S.; Pausas, J.G. *Cork Oak Woodlands on Edge: Ecology, Adaptive Management, and Restoration*; Island Press Publication: Washington, DC, USA, 2009.
3. Gibson, L.J.; Easterling, K.E.; Ashby, M.F. The structure and mechanics of cork. *Proc. R. Soc. Lond. A Math. Phys. Sci.* **1981**, *377*, 99–117. [CrossRef]
4. Pereira, H.; Rosa, M.E.; Fortes, M.A. The cellular structure of cork from *Quercus suber*. *IAWA Bull.* **1987**, *8*, 213–218. [CrossRef]
5. Knapic, S.; Oliveira, V.; Machado, J.S.; Pereira, H. Cork as a building material: A review. *Eur. J. Wood Wood Prod.* **2016**, *74*, 775–791. [CrossRef]
6. Gibson, L.J.; Ashby, M.F. *Cellular Solids: Structure and Properties*, 2nd ed.; Cambridge University Press: Cambridge, UK, 1997.
7. Pereira, H. The rationale behind cork properties: A review of structure and chemistry. *Bioresources* **2015**, *10*, 6207–6229. [CrossRef]
8. Pereira, H.; Tome, M. Cork Oak. In *Encyclopedia of Forest Sciences*; Burley, J., Ed.; Elsevier Ltd.: Oxford, UK, 2004; pp. 613–620. [CrossRef]
9. Leite, C.; Pereira, H. Cork-containing barks–A review. *Front. Mater.* **2017**, *3*, 1–19. [CrossRef]
10. Sousa, V.B.; Leal, S.; Quilhó, T.; Pereira, H. Characterization of cork oak (*Quercus suber*) wood anatomy. *IAWA Bull.* **2009**, *30*, 149–161. [CrossRef]
11. Chen, D.; Zhang, X.; Kang, H.; Sun, X.; Yin, S.; Du, H.; Yamanaka, N.; Gapare, W.; Wu, H.X.; Liu, C. Phylogeography of *Quercus variabilis* based on chloroplast DNA sequence in East Asia: Multiple glacial refugia and mainland-migrated island populations. *PLoS ONE* **2012**, *7*, e47268. [CrossRef]
12. Miranda, I.; Gominho, J.; Pereira, H. Cellular structure and chemical composition of cork from the Chinese cork oak (*Quercus variabilis*). *J. Wood Sci.* **2013**, *59*, 1–9. [CrossRef]
13. Ferreira, J.; Miranda, I.; Sen, U.; Pereira, H. Chemical and cellular features of chemical and cellular features of virgin and reproduction cork from *Q. variabilis*. *Ind. Crops Prod.* **2016**, *94*, 638–648. [CrossRef]
14. Anjos, O.; Pereira, H.; Rosa, M.E. Effect of quality, porosity and density on the compression properties of cork. *Holz Roh-Und Werkst.* **2008**, *66*, 295–301. [CrossRef]
15. Anjos, O.; Pereira, H.; Rosa, M.E. Tensile properties of cork in axial stress and influence of porosity, density, quality and radial position in the plank. *Eur. J. Wood Prod.* **2011**, *69*, 85–91. [CrossRef]
16. Yafang, L.; Yanzhen, L.; Wei, Z.; Jingfeng, Z. The microstructure of cork from *Quercus variabilis*. *Sci. Silvae Sin.* **2009**, *45*, 167–172. (In Chinese)

17. Li, J.; Bi, J.; Song, X.; Qu, W.; Liu, D. Surface and dynamic viscoelastic properties of cork from *Quercus variabilis*. *Bioresources* **2019**, *14*, 607–618.

18. Kim, B.R. Studies on the physical and mechanical properties of imported and domestic corks. *J. Korean Wood Sci. Technol.* **1993**, *21*, 45–54. (In Korean)

19. Kang, H.Y.; Kim, S.W. Air-klin drying the boards and disks of *Quercus variabilis*. *J. Korean Wood Sci. Technol.* **2004**, *32*, 52–58. [CrossRef]

20. Kim, N.H.; Hanna, R.B. Morphological characteristics of *Quercus variabilis* charcoal prepared at different temperatures. *Wood Sci. Technol.* **2006**, *40*, 392–401. [CrossRef]

21. Kwon, S.M.; Kim, N.H.; Cha, D.S. An investigation on the transition characteristics of the wood cell walls during carbonization. *Wood Sci. Technol.* **2009**, *43*, 487–498. [CrossRef]

22. Jeon, W.S.; Lee, H.M.; Park, J.H. Comparison of anatomical characteristics for wood damaged by oak wilt and sound wood from *Quercus mongolica*. *J. Korean Wood Sci. Technol.* **2020**, *48*, 807–819. [CrossRef]

23. Kim, B.R. The physical properties of the bark of *Quercus suber* L. *J. Agric. Sci.* **1991**, *9*, 86–95. (In Korean)

 *forests*

Article

Chemical Composition and Optimization of Liquefaction Parameters of *Cytisus scoparius* (Broom)

Luísa Cruz-Lopes [1,2,*], Daniela Almeida [1], Yuliya Dulyanska [2], Idalina Domingos [2,3], José Ferreira [2,3], Anabela Fragata [4] and Bruno Esteves [2,3]

[1] Department of Environmental Engineering, Polytechnic Institute of Viseu, Av. Cor. José Maria Vale de Andrade, 3504-510 Viseu, Portugal

[2] Centre for Natural Resources, Environment and Society-CERNAS-IPV Research Centre, Av. Cor. José Maria Vale de Andrade, 3504-510 Viseu, Portugal

[3] Department of Wood Engineering, Polytechnic Institute of Viseu, Av. Cor. José Maria Vale de Andrade, 3504-510 Viseu, Portugal

[4] CI&DEI Research Centre, Polytechnic Institute of Viseu, 3504-510 Viseu, Portugal

* Correspondence: lvalente@estgv.ipv.pt; Tel.: +351-232-480-500

Abstract: Invasive plants spread in such a way that they are threats to native species and to biodiversity. In this context, this work aims to determine possible valorizations of Scotch Broom *Cytisus scoparius* (L.) Link. This species harvested in the Viseu region was used in the present study. The eco-valorization of these renewable resources was made by conversion into liquid mixtures that can later be used in the manufacture of valuable products. For a better understanding of the results obtained, a chemical characterization of the *Cytisus scoparius* branches (CsB) was made. The ash content, extractives in dichloromethane, ethanol and water, lignin, cellulose and hemicellulose of the initial material were determined. Liquefaction was made in a reactor with different granulometry, temperatures and time. Results show that Broom is mainly composed of cellulose (36.1%), hemicelluloses (18.6%) and lignin (14.6%) with extractives mainly soluble in ethanol, followed by water and a small amount in dichloromethane. Ashes were around 0.69%, mainly composed of potassium and calcium. Generally, smaller size, higher solvent ratio, higher temperature and higher time of liquefaction lead to higher liquefaction. The highest percentage of liquefaction was 95% which is better than most of the lignocellulosic materials tested before.

Keywords: *Cytisus scoparius*; chemical composition; liquefied; optimization; eco-valorization; agro-industrial residues

Citation: Cruz-Lopes, L.; Almeida, D.; Dulyanska, Y.; Domingos, I.; Ferreira, J.; Fragata, A.; Esteves, B. Chemical Composition and Optimization of Liquefaction Parameters of *Cytisus scoparius* (Broom). *Forests* **2022**, *13*, 1772. https://doi.org/10.3390/f13111772

Academic Editor: Antonios Papadopoulos

Received: 22 September 2022
Accepted: 20 October 2022
Published: 27 October 2022

Publisher's Note: MDPI stays neutral with regard to jurisdictional claims in published maps and institutional affiliations.

1. Introduction

Many plants that surround us have not always existed in our territory; they came from other places brought by man from their native habitat. Certain species remain only in places where they have been embedded with native species in a stable way, but others prosper rapidly without the support of man, expanding their populations. These are called invasive plants. The proliferation of these species promotes environmental changes and economic damage so that they can be a threat to natural ecosystems, food production, human health and the economy itself [1]. Therefore, Broom, also called Scotch broom or English broom (*Cytisus scoparius* (L.) Link), is an example of an invasive plant.

In Portugal, communities are dominated by shrubs of the genera *Cytisus*, *Genista*, *Adenocarpus* and *Retama*. In accordance with Costa et al. [2], the area of occupation of the class *Cytisetea scopario-striati* has increased in recent decades, mainly due to the abandonment of agricultural land. Its development in the Iberian Peninsula resulted in fragmented landscapes, determining spatial patterns of richness, composition and specific abundance or the periods of regression of climatic forests [2]. The class *Cytisetea scopario-striati* develops well on siliceous, poor on bases, non-hydomorphic soils and presents its greatest evolution and diversity in the west of the Iberian Peninsula in the Atlantic and Ibero-Atlantic Mediterranean provinces [2]. According to the national forest inventory, forests represent 40% of Portugal's mainland, which is equivalent to almost 3.5 million hectares, 1 million more hectares than agriculture. The forests with the highest territorial expression are eucalypts (*Eucalyptus globulus* Labill) (26%), cork oak (*Quercus suber* L.) (23%) and pine (*Pinus pinaster* Ait.) (22%), and the invasive species represent 0.4% [3,4].

Cytisus scoparius can be found in almost all of Europe and Macaronesia. It was introduced in North America, Australia and South Africa. It is common in bushes, scrubs and riparian and the edge of woods and paths. It is used as a medicinal plant, in agroforestry management, as fuel and in the manufacture of outdoor brooms [5]. *Cytisus scoparius* is a shrub 1 to 2 m high. The stem is green, angled, striated longitudinally and hard, with consistent and flexible branches, usually with five well-defined streaks in an inverted V-shape. The space between the stretch marks is greater than their width. This plant is known to have several antioxidant compounds such as flavones, isoflavones, flavonols and carotenoids [6–8].

The emergence of liquefaction techniques at low pressure and temperatures has increased the interest in invasive plants that can be a cheap lignocellulosic alternative to other more valuable materials and constitute an alternative for petroleum-derived fuels. This alternative also has the advantage of contributing to the reduction of greenhouse gas emissions, thus preventing global warming [9]. Liquefaction of moderate to low temperatures has been done using two different types of solvents, phenol [10–13] or polyalcohols [14–18], with basic (NaOH, KOH) or acid catalysis (sulphuric acid, paratuluenosulphonic acid, hydrochloric acid, oxalic acid or phosphoric acid). Phenol is mostly used when the objective is to prepare resins such as epoxy resins since there is a higher content of phenols in the liquefied material [19]. On the other hand, phenol is more toxic than the polyalcohols used and less green since most of the polyalcohols can be obtained from crude glycerol or other non-petroleum materials. Several lignocellulosic materials have been liquefied with polyalcohols such as agricultural wastes, for example, wheat straw [20], cornstarch [21] or rice straw, oilseed rape straw and corn stover [22], forest management residues, such as several woods [15,23,24], barks [16,25,26] or shells [14,27,28] and also some industrial residues such as orange peel [29]. The lignocellulosic material is composed of carbohydrate polymers (cellulose and hemicelluloses), lignin and a small part of other compounds (extractives, salts and minerals). Cellulose and hemicelluloses, which usually account for two-thirds of the dry mass, are polysaccharides that can be hydrolyzed into sugars and, eventually, by fermentation, converted into ethanol. Lignin cannot be used for ethanol production, so that it can be used for thermal energy production by combustion [9]. To improve the knowledge and evaluation of possible applications, the chemical composition of Broom, *Cytisus scoparius*, was determined.

This study intends to determine if *Cytisus scoparius* is a suitable material to produce high quality polyols to be later used for the production of polyurethane foams or adhesives. The knowledge of the chemical composition of the material and the FTIR analysis of initial, liquefied material and solid residue will allow us to understand the chemical changes involved and give some clues on how to improve the liquefaction procedure.

2. Materials and Methods

2.1. Materials

The branches of *Cytisus scoparius* (CsB) come from the district of Viseu (São Pedro do Sul). The samples were dried at room temperature, crushed in a Retsh SKI mill (Retsch GmbH, Haan, Germany) and sieved using a Retsch AS200 for 30 min at 50 rpm. Five fractions, >35 mesh (>0.425 mm); 35–40 mesh (0.425–0.450 mm); 40–60 mesh (0.425–0.250 mm); 60–80 mesh (0.250–0.180 mm) and <80 mesh (<0.180 mm) fraction was used for the tests. The fractions obtained were dried in a greenhouse at 100 °C for 24 h before each test. The chemical reagents used were analytical-grade reagents.

2.2. Chemical Composition

The CsB were characterized for their ash content, extractives (in dichloromethane, ethanol and hot water), α-cellulose, lignin and hemicelluloses. The 40–60 mesh fraction was dried at 105 °C for at least 24 h and afterward used for the chemical analyses according to Tappi T 264 om-97 [30]. The average chemical composition of each sample was determined in triplicate. The extractives were determined by extraction with different solvents in sequential order of ascending polarity.

The ash content was determined by calcination at 525 °C in accordance to Tappi T 211 om-93 [31]. The inorganic composition was determined by ICP after ash wet digestion in a Leco CHNS-932 Elemental Analyzer (St. Joseph, MI, USA).

Extractives were determined by sequential Soxhlet extraction with 150 mL of dichloromethane, ethanol and hot water extractives in accordance with Tappi T 204 om-88 [32]. A total of 10 g of dried material was placed in a filter paper cartridge inside the Soxhlet and refluxed until extractives were removed. Extractions lasted for 6 h for dichloromethane and 16 h for both ethanol and water. Extractive content was determined in relation to the dry material.

Alkaline extraction was done by reflux with 0.3% (m/v) NaOH under a nitrogen atmosphere.

Lignin was determined by the Klason method in extractive free material. Two hydrolyses were performed, the first in a water bath at 30 °C with 72% H_2SO_4 for 1 h followed by a second with 3% H_2SO_4 in an autoclave at 120 °C for 1 h (according to modified Tappi T 204 om-88) [33]. The insoluble residue was filtered in a G2 glass crucible, washed with warm water and acetone, and dried at 100 °C until constant weight. The lignin percentage was determined according to Equation (1).

$$\text{Lignin (\%)} = \frac{\text{Insoluble residue}}{\text{Dried material}} \times 100 \tag{1}$$

Soluble lignin was determined by measuring the absorbance at 205 nm.

Cellulose *Kürschner–Hoffer* was determined by using 1 g of material that was refluxed with 100 mL of nitric acid: ethanol 20:80 solution for 1 h. This procedure was repeated again, and the final insoluble material was filtered using a G2 crucible and by washing with ethanol and hot water.

Hemicelluloses were determined by difference.

2.3. Liquefaction

In order to study the optimal conditions of liquefaction for the branches of *Cytisus scoparius* shrubs (CsB), different liquefaction reaction times, temperature and size of the samples were tested.

The liquefaction process was conducted in a double shirt reactor (600 mL) heated with oil (Reactor Parr LKT PED). The samples were introduced in the reactor with a mixture of glycerol and ethylene glycol 1:1, catalyzed with sulfuric acid (3%). Liquefied samples were dissolved in methanol and filtered in a G3 crucible.

The liquefaction yield was determined by the insoluble material retained in the crucible according to Equation (2).

$$\text{Liquefaction yield (\%)} = \frac{\text{Insoluble residue}}{\text{Dried initial material}} \times 100 \tag{2}$$

The effect of particle size was studied for five fractions, >35 mesh (>0.425 mm); 35–40 mesh (0.425–0.450 mm); 40–60 mesh (0.425–0.250 mm); 60–80 mesh (0.250–0.180 mm) and 80 mesh (<0.180 mm). The temperature ranged between 140 °C and 180 °C and liquefaction time from 15 to 60 min. Studies were made using different ratios of CsB:solvent, 1:5, 1:10 and 1:12.

3. Results and Discussion

3.1. Chemical Composition

Table 1 presents the chemical composition of CsB, where it can be observed that it is composed of 0.69% ashes, similar to the 0.78% obtained by González et al. with Broom from Ourense (north-west Spain) [34]. Ashes are essentially composed by potassium (0.694%) and calcium (0.046%) followed by sodium (0.042%), magnesium (0.030%) and phosphor (0.028%). This ash, given its composition rich in potassium and calcium, can be used in the glass industry to reduce the melting temperature of silica.

Table 1. Chemical composition of CsB.

Parameters		Content (% Dry Matter, *w/w*)
Ashes		0.694 ± 0.012
	K	0.194 ± 0.001
	Ca	0.046 ± 0.001
	Na	0.042 ± 0.001
	Mg	0.030 ± 0.001
	P	0.028 ± 0.001
	Fe	<0.01 ± 0.000
	Zn	<0.01 ± 0.000
Extractives	Dichloromethane	0.689 ± 0.274
	Ethanol	6.638 ± 1.466
	Hot water	1.977 ± 0.794
	Total	9.304 ± 1.466
1% NaOH extract [a]		20.77 ± 0.15
Klason Lignin [b]		14.57 ± 0.06
Cellulose *Kürscher and Höffer*		36.05 ± 1.15
Hemicelluloses		18.61 ± 1.47

[a] Corrected for the extractive content and ash. [b] Corrected for the extractive content, ash and alkaline extract.

Total extractives represented around 9.3%, most of which were soluble in ethanol (6.64%) and hot water (1.98%) and only 0.69% in dichloromethane. This means that extractives from Broom are most likely phenolic compounds and some small sugars that are obtained in ethanol and water extracts, respectively. These extractives are important since they proved to be good for the preparation of formulations for topical application to protect skin against oxidative damage [6,35]. Dichloromethane removes mostly non-polar extractives such as fatty acids, alkanes, waxes, terpenes and terpenoids, as stated before [36]. On the other hand, ethanol extractives are generally composed of lignans, flavonoid, stilbenes and mostly tannins (hydrolyzable and condensed). Some of these tannins have larger molecules and can only be removed by more aggressive extraction procedures such as 1% NaOH solutions. Results show that Broom has a high amount of such compounds. Klason lignin represents 14.57% of Broom, while cellulose *Kürscher and Höffer* represent around 36.1% and hemicelluloses 18.6%. The determination of the chemical composition will allow us to understand the possible uses of CsB better.

Cytisus scoparius chemical composition is somewhat different from the one presented by González et al. [34]. These authors reported similar ash (0.78%) and extractives content (9.13%) but a higher lignin percentage (26.6%); however, no alkaline extraction was made before lignin determination. Therefore, the existent polyphenols were most likely counted as lignin. In this study, the 20.8% alkaline extract has removed most of the polyphenols reducing the Klason lignin obtained. This extraction might also remove some other non-tannin material, such as lignin [37] and possibly some hemicelluloses, as stated before [38].

On the other hand, the lignin content is not that far from the lignin content (16%) from Broom obtained from agricultural wastelands located in Zachodniopomorskie Voivodeship in Poland [39]. In the study by González et al. [34], cellulose was estimated by glucan to be around 40% and hemicelluloses by their sugar constituents to be around 19% which is similar to the results obtained here. Moreover, some chemical differences are expected since the origin of the plants is different. Results obtained for *Cytisus striatus*, generally called Portuguese Broom, are not that different, with 0.8% ash content, 4.7% ethanol toluene extractives, 22.4% lignin and 70% holocellulose [40].

3.2. Liquefaction Optimization

Figure 1 presents the variation of the liquefaction yield with the size of particles. Results show that for the larger particles >35 mesh, the percentage of liquefaction was only around 62.5% which can mean that the size of some of these particles was too big for them to be solubilized. On the other hand, for lower-size particles, the differences between the different sizes are small with 35–40 mesh, 40–60 mesh and <80 mesh (powder) with a very similar value ranging from 92.4 to 95.1%. In relation to 60–80 mesh particles, the percentage of liquefaction was slightly lower (80.9%), which can be due to the different chemical composition in this particle size has observed before [14,17,41]. The different chemical composition of each fraction is particularly important in heterogeneous materials such as in this case where there is a mixture of wood, bark and even flowers with very different chemical composition between each other. This result is extremely important as it allows industries to use larger particles (below 35 mesh), excluding the time-consuming process and complex transformation of *Cytisus scoparius* bushes to dust.

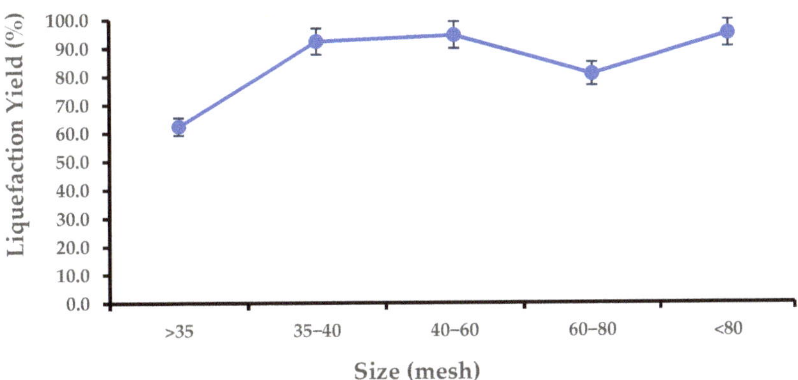

Figure 1. Liquefaction yield for different particle size for the *Cytisus scoparius* branches (CsB) (constant parameters: 180 °C, time 60 min and ratio CsB:solvent of 1:10).

To perform an optimization of the CsB/solvent ratios, tests with different ones were performed using 1:5, 1:10, 1:12 for 60 min (min) at 180 °C. The results obtained are given in Figure 2.

The results show an increase in liquefaction yield, between 54.8%, 55.8% and 87.2% for a CsB/solvent ratio of 1:5, 1:10 and 1:12, respectively. In fact, this shows that using a higher amount of solvent leads to better liquefaction yields, as proven before [17]; however, it should be noted that more solvent corresponds to a higher cost. Thus, in the choice of solvent ratio, economic and environmental issues should be considered. It was decided to use the 1:10 ratio rather than the 1:5 ratio for the subsequent tests since, although here the difference is small, some earlier tests have shown that 1:5 was not enough in some conditions, and choosing this ratio could affect the study of the other parameters [17].

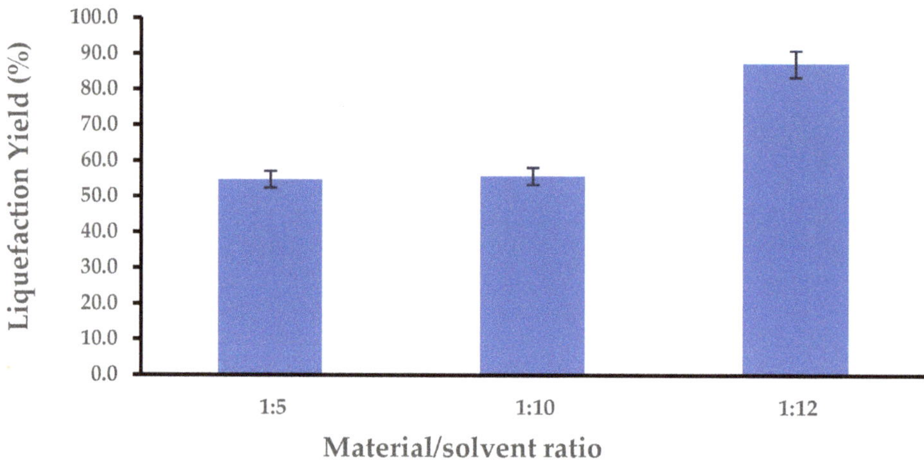

Figure 2. Study of the ratio CsB:solvent for the samples of the branches (CsB) of *Cytisus scoparius* (constant parameters: 180 °C, time 60 min and size for the samples <80 mesh).

Temperature optimization was performed by varying the temperature between 140 °C and 180 °C, keeping all other parameters constant. The results obtained are shown in Figure 3.

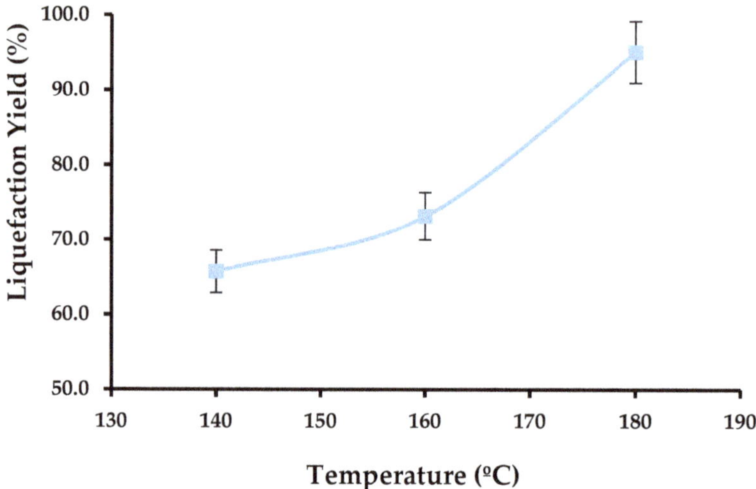

Figure 3. Study of the temperature for the samples of the branches (CsB) of *Cytisus scoparius* (constant parameters: time 60 min, size for the samples <80 mesh and ratio CsB:solvent of 1:10).

The liquefaction yield increased with the increase of temperature liquefaction, achieving a promising value of 95.1% for 180 °C. At lower temperatures, the liquefaction yield was only 65% and 73.5% for 140 °C and 160 °C, respectively. The high percentage of liquefaction for 180 °C is higher than that obtained for other materials in similar conditions, such as, for example, *Eucalyptus globulus* bark (61.6%) and branches (62.2%) [26]. Until 180 °C, no recondensation reactions were observed, or at least it was not significant, probably due to the use of a co-solvent system by the addition of a low-molecular-weight glycol (glycerol) and ethylene glycol, as mentioned before [15].

The optimization of liquefaction time was performed by varying the time between 15 and 60 min, keeping all other parameters constant. The results obtained are presented in Figure 4.

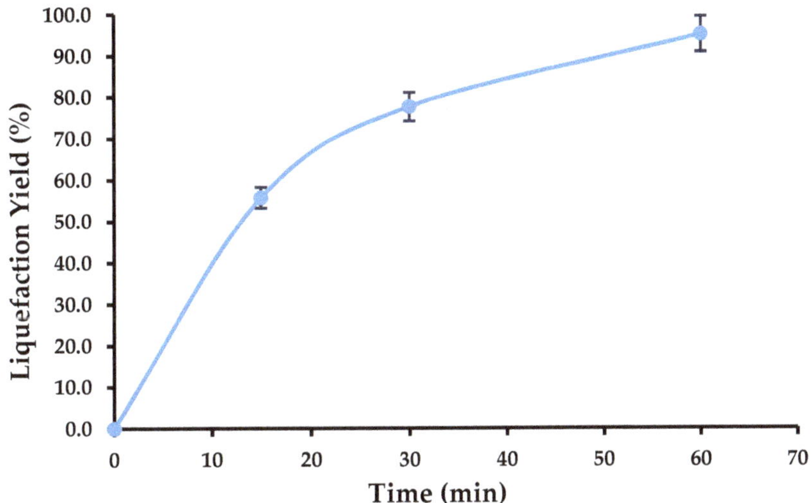

Figure 4. Study of liquefaction yield along liquefaction time for Broom branches for the fraction > 80 mesh (powder) at 180 °C with acid catalysis.

Results show that it takes only 30 min to obtain a good liquefaction yield of 77.7% but also, that higher liquefaction can be obtained for 60 min (95.1%). The liquefaction yield was higher than the obtained for walnut shells (85%) before [42]. No decrease was observed with the increase in temperature, as reported, for instance, by Soares et al. [23] with cork liquefaction. These authors observed a decrease in higher temperatures which was attributed to polycondensation reactions between the liquefaction intermediates. Depending on the final product to be produced, a lower or higher liquefaction can be obtained.

The high percentage of liquefaction (95%) shows that broom can be successfully liquefied bi polyalcohols with acid catalysis. The percentage of liquefaction is higher than the one obtained before for several other lignocellulosic materials such as rice straw, oilseed rape straw, wheat straw or corn stover [22] or orange peel wastes [43]. The low amount of residue allows using the obtained polyol without any separation between the solid residue and the liquefied material, which can be costly. Therefore, Broom has a nice potential to be used for the production of polyurethane foams, as was done before for several lignocellulosic materials, for instance, wood, barks or agricultural wastes [22,43–46].

3.3. FTIR of Liquefaction

Figure 5 presents the variation of the FTIR spectrum between the initial material and the liquefied polyol at 180 °C for 15 min, 30 min and 60 min. There are no major differences between the spectra of polyols produced with 15 min, 30 min or 60 min reactions. Big differences are observed between the initial and the liquefied material. There is a clear increase in the hydroxyl peak and a shifting of the peak maximum from around 3350 cm^{-1} to around 3250 cm^{-1}. This is to be expected since the solvents used, glycerol and ethyleneglycol, are alcohols and, therefore, have several hydroxyl groups in their constitution. The shifting for lower wavenumbers can be due to the different functional groups present in liquefied material. Generally, alcohols absorb at higher wavenumbers (3600–3300 cm^{-1}) than carboxylic acids (3300–2500 cm^{-1}), therefore, this shifting to lower wavenumbers could be due to the increase in carboxylic acids. Nevertheless, generally, the

C=O of carboxylic acids has a strong absorption at about 1725–1700 cm^{-1}, which is not seen in liquefied spectra.

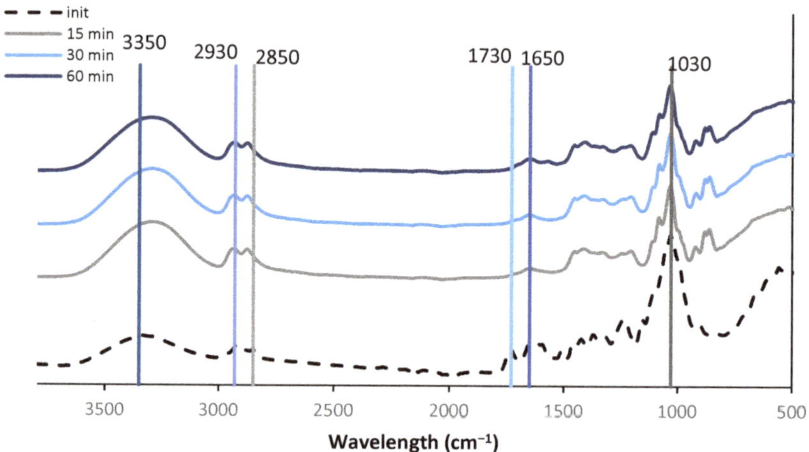

Figure 5. FTIR spectra of the initial material and the liquefied polyol at 180 °C for 15 min, 30 min and 60 min at 180 °C with acid catalysis.

The CH peaks are also very different in the polyol and in the initial material. These bands result from the overlapping of -CH$_2$- (2935–2915 cm^{-1}) and -CH$_3$ (2970–2950 cm^{-1}) stretching asymmetric vibrations and -CH$_2$- (2865–2845 cm^{-1}) and -CH$_3$ (2880–2860 cm^{-1}) stretching symmetric vibrations [47]. The original peaks are sharper, and there is an increase in the 2850 cm^{-1} peak in relation to 2930 cm^{-1} shifting the maximum to around 2860 cm^{-1}. Similar behavior has been observed before for liquefied cherry seeds [48].

The peak at 1730 cm^{-1}, corresponding to non-conjugated C=O linkages, completely disappeared in the spectra of the liquefied material, and the 1650 cm^{-1} band was significantly reduced and narrowed. This might be due to the low amount of lignin in the initial material and to the lower resistance of Broom polysaccharides to hydrolysis. This is in line with the increase observed in this peak in the FTIR spectrum of the solid residue, especially after 60 min liquefaction. Similarly, the peak at 1510 cm^{-1} also disappears in the polyol spectra, and this peak has been associated with benzene ring stretching vibrations [47,49]. The 1230 cm^{-1} peak also decreases in the liquefied material. Although there is an apparent decrease in the 1030 cm^{-1} peak, this might be due to the high increase in the 3350 cm^{-1} band that proportionally decreases the remaining peaks. There is, however, a narrowing of the peak and the appearance of a more pronounced shoulder at 1100 cm^{-1}. There is a new visible peak appearing at around 850 cm^{-1} which can be due to stretching vibrations in the pyranose ring [49]. The changes in FTIR spectra for the liquefied material are very different from those observed for sweet cherry seeds, which can be due to the significantly different chemical composition between them.

Figure 6 presents the variation of the FTIR spectrum between the initial material and the liquefied polyol for a 60 min reaction at 140 °C, 160 °C and 180 °C. Similar to the results presented in Figure 6, the main differences are observed between the initial and the liquefied material. Nevertheless, at 140 °C, there is a smaller increase in the 3350 cm^{-1} band and a much higher C-O-C peak at 1030 cm^{-1} in relation to the other temperatures.

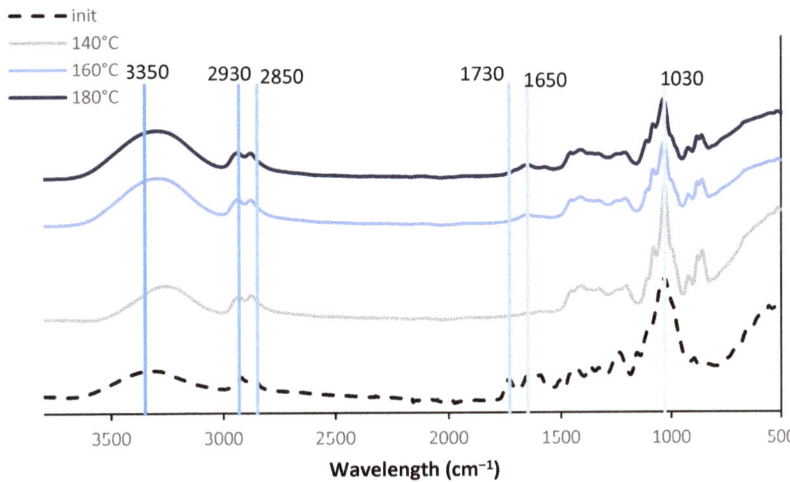

Figure 6. FTIR spectra of the initial material and the liquefied polyol for a 60 min reaction at 140 °C, 160 °C and 180 °C with acid catalysis.

Figure 7 presents the FTIR spectra of the initial material and of the solid residue obtained by liquefaction at 180 °C for 15 min, 30 min and 60 min reaction time. The solid residue spectrum changes rapidly from the initial material and along the reaction time. There is a slight narrowing of the 3350 cm^{-1} band in the initial of the liquefaction and a decrease for 60 min reaction time. In relation to the CH peaks, there is a decrease of the 2930 cm^{-1} peak followed by an increase for longer reaction times. The same seems to happen to the 1730 cm^{-1} peak and somewhat to the 1650 cm^{-1} band. These changes might be due to an initial cleavage of acetyl or acetoxy groups in hemicelluloses. The fingertip region is much less defined in the solid residue, and there seems to be a decrease at the 1226 cm^{-1} peak. The C-O-C peak also seems to increase and decrease later.

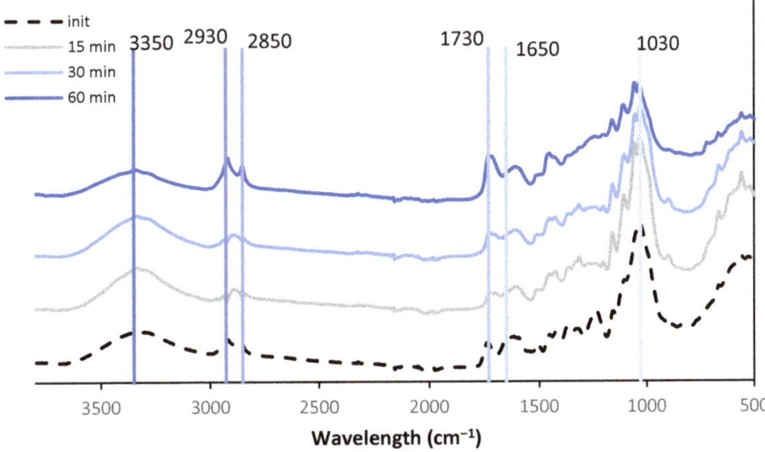

Figure 7. FTIR spectra of the initial material and of the solid residue obtained by liquefaction at 180 °C for 15 min, 30 min and 60 min reaction time with acid catalysis.

The changes observed in the solid residue along the liquefaction time are similar to those presented in Figure 8 with the increase in liquefaction temperature.

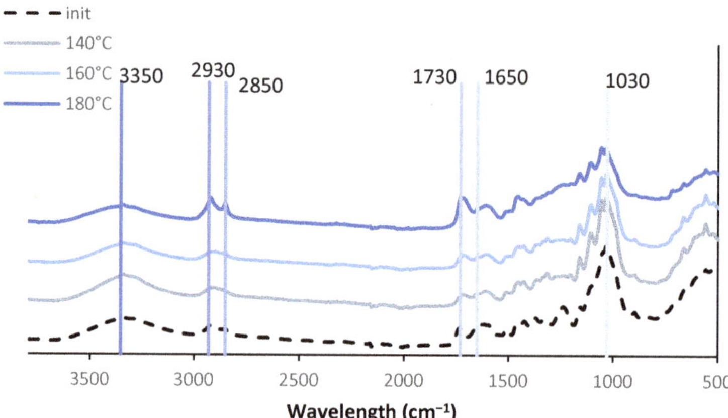

Figure 8. FTIR spectra of the initial material and of the solid residue obtained by liquefaction for 60 min at temperatures 140 °C, 160 °C and 180 °C with acid catalysis.

FTIR analysis shows that it is not only one of the macromolecular compounds that is being liquefied but rather that all the chemical compounds are affected by polyalcohol liquefaction with acid catalysis.

4. Conclusions

Results have shown that in relation to macromolecular compounds, Broom is mainly composed of cellulose, followed by hemicelluloses and lignin. Alkaline extract represented more than 20% and is probably composed of tannins. Extractives were more than 9%, with ethanol soluble representing more than half, followed by water soluble and dichloromethane extractives. These phenolic compounds can be retrieved before liquefaction in a biorefinery approach. Ashes were around 0.69%. No significative differences were found for the different fractions, although the 40–60 mesh fraction presented the best results. A higher amount of solvent improved the liquefaction percentage. The same was observed for higher temperatures and the time of liquefaction. At the ideal conditions, a high percentage of liquefaction (95%) could be achieved, better than most lignocellulosic materials. These conditions were 180 °C, 60 min for the 40–60 mesh fraction. The results have proven that Broom can be a valuable material that, once liquefied, can be later used for the manufacture of goods without any further separation to substitute petroleum-based products such as, for instance, polyurethane foams.

Author Contributions: Conceptualization, L.C.-L. and B.E.; methodology, L.C.-L., Y.D. and B.E.; formal analysis, L.C.-L. and Y.D.; investigation, L.C.-L., Y.D., D.A. and B.E.; resources, L.C.-L., Y.D., I.D., J.F., A.F. and B.E.; writing—original draft preparation, Y.D.; writing—review and editing, L.C.-L. and B.E.; supervision, L.C.-L. All authors have read and agreed to the published version of the manuscript.

Funding: This work is funded by National Funds through the FCT—Foundation for Science and Technology. I.P. within the scope of the project Ref UIDB/00681/2020 (CERNAS) and Ref UIDB/05507/ 2020 (CI&DEI).

Acknowledgments: We would like to thank the CERNAS and CI&DEI Researchs Centres and the Polytechnic Institutes of Viseu for their support and the PROJ/IPV/ID&I/021 "Valorização de resíduos: potencial de aproveitamento do caroço de cereja (VALCER)" by Yulia's Research Grant.

Conflicts of Interest: The authors declare no conflict of interest. The funders had no role in the design of the study; in the collection, analyses, or interpretation of data; in the writing of the manuscript, or in the decision to publish the results.

References

1. Marchante, H.; Morais, M.; Freitas, H.; Marchante, E. *Guia Prático para a Identificação de Plantas Invasoras em Portugal*; Imprensa da Universidade de Coimbra/Coimbra University Press: Lisboa, Portugal, 2014; ISBN 978-989-26-0785-6.
2. Costa, J.C.; Aguiar, C.; Capelo, J.; Lousã, M.; Antunes, J.H.S.; Honrado, J.J.; Izco, J.; Ladero, M. A Classe Cytisetea Scopario-Striati Em Portugal Continental. *Quercetea* 2003, *4*, 45–70.
3. o Inventário Florestal Nacional—PDF Free Download. Available online: https://docplayer.com.br/146300038-6-o-inventario-florestal-nacional.html (accessed on 17 July 2021).
4. Caetano, M.; Igreja, C.; Marcelino, F.; Costa, H. *Estatísticas e Dinâmicas Territoriais Multiescala de Portugal Continental 1995–2007–2010 Com Base Na Carta de Uso e Ocupação Do Solo (COS)*; Direção-Geral do Território: Lisbon, Portugal, 2017.
5. Pinela, J.; Barros, L.; Carvalho, A.M.; Ferreira, I.C. Influence of the Drying Method in the Antioxidant Potential and Chemical Composition of Four Shrubby Flowering Plants from the Tribe Genisteae (Fabaceae). *Food Chem. Toxicol.* 2011, *49*, 2983–2989. [CrossRef] [PubMed]
6. González, N.; Otero, A.; Conde, E.; Falqué, E.; Moure, A.; Domínguez, H. Extraction of Phenolics from Broom Branches Using Green Technologies. *J. Chem. Technol. Biotechnol.* 2017, *92*, 1345–1352. [CrossRef]
7. Nirmal, J.; Babu, C.S.; Harisudhan, T.; Ramanathan, M. Evaluation of Behavioural and Antioxidant Activity of Cytisus Scoparius Link in Rats Exposed to Chronic Unpredictable Mild Stress. *BMC Complement. Altern. Med.* 2008, *8*, 15. [CrossRef]
8. Raja, S.; Ahamed, K.N.; Kumar, V.; Mukherjee, K.; Bandyopadhyay, A.; Mukherjee, P.K. Antioxidant Effect of Cytisus Scoparius against Carbon Tetrachloride Treated Liver Injury in Rats. *J. Ethnopharmacol.* 2007, *109*, 41–47. [CrossRef]
9. Gil, N.R.S. Pré-Tratamento de Materiais Lenhocelulósicos para Produção de Bioetanol. Ph.D. Thesis, Universidade da Beira Interior, Covilha, Portugal, 2008.
10. Lin, R.; Sun, J.; Yue, C.; Wang, X.; Tu, D.; Gao, Z. Study on Preparation and Properties of Phenol-Formaldehyde-Chinese Fir Liquefaction Copolymer Resin. *Maderas. Cienc. Tecnol.* 2014, *16*, 159–174. [CrossRef]
11. Mun, S.P.; Gilmour, I.A.; Jordan, P.J. Effect of Organic Sulfonic Acids as Catalysts during Phenol Liquefaction of Pinus Radiata Bark. *J. Ind. Eng. Chem.* 2006, *12*, 720–726.
12. Zhang, Q.; Zhao, G.; Jie, S. Liquefaction and Product Identification of Main Chemical Compositions of Wood in Phenol. *For. Stud. China* 2005, *7*, 31–37. [CrossRef]
13. Li, G.; Hse, C.; Qin, T. Wood Liquefaction with Phenol by Microwave Heating and FTIR Evaluation. *J. For. Res.* 2015, *26*, 1043–1048. [CrossRef]
14. Cruz-Lopes, L.P.; Domingos, I.; Ferreira, J.; Esteves, B. Chemical Composition and Study on Liquefaction Optimization of Chestnut Shells. *Open Agric.* 2020, *5*, 905–911. [CrossRef]
15. Kurimoto, Y.; Tamura, Y. Species Effects on Wood-Liquefaction in Polyhydric Alcohols. *Holzforschung* 1999, *53*, 617–622. [CrossRef]
16. D'Souza, J.; Wong, S.Z.; Camargo, R.; Yan, N. Solvolytic Liquefaction of Bark: Understanding the Role of Polyhydric Alcohols and Organic Solvents on Polyol Characteristics. *ACS Sustain. Chem. Eng.* 2015, *4*, 851–861. [CrossRef]
17. Esteves, B.; Dulyanska, Y.; Costa, C.; Ferreira, J.V.; Domingos, I.; Pereira, H.; de Lemos, L.T.; Cruz-Lopes, L.V. Cork Liquefaction for Polyurethane Foam Production. *BioResources* 2017, *12*, 2339–2353. [CrossRef]
18. Jin, Y.; Ruan, X.; Cheng, X.; Lü, Q. Liquefaction of Lignin by Polyethyleneglycol and Glycerol. *Bioresour. Technol.* 2011, *102*, 3581–3583. [CrossRef] [PubMed]
19. Kobayashi, M.; Tukamoto, K.; Tomita, B. Application of Liquefied Wood to a New Resin System-Synthesis and Properties of Liquefied Wood/Epoxy Resins. *Holzforschung* 2000, *54*, 93–100. [CrossRef]
20. Chen, F.; Lu, Z. Liquefaction of Wheat Straw and Preparation of Rigid Polyurethane Foam from the Liquefaction Products. *J. Appl. Polym. Sci.* 2009, *111*, 508–516. [CrossRef]
21. Huang, Y.; Zheng, Y.; Zheng, Z.; Zhou, Q.; Feng, H. Study on Liquefaction Technology of Cornstarch in Polyhydric Alcohols. *J. Southwest For. Univ.* 2011, *31*, 72–74.
22. Zhang, J.; Hori, N.; Takemura, A. Optimization of Agricultural Wastes Liquefaction Process and Preparing Bio-Based Polyurethane Foams by the Obtained Polyols. *Ind. Crops Prod.* 2019, *138*, 111455. [CrossRef]
23. Pan, H.; Zheng, Z.; Hse, C.Y. Microwave-Assisted Liquefaction of Wood with Polyhydric Alcohols and Its Application in Preparation of Polyurethane (PU) Foams. *Eur. J. Wood Wood Prod.* 2012, *70*, 461–470. [CrossRef]
24. Zheng, Z.; Pan, H.; Huang, Y.; Chung, Y.; Zhang, X.; Feng, H. Rapid Liquefaction of Wood in Polyhydric Alcohols under Microwave Heating and Its Liquefied Products for Preparation of Rigid Polyurethane Foam. *Open Mater. Sci. J.* 2011, *5*, 1–8. [CrossRef]
25. Esteves, B.; Cruz-Lopes, L.; Ferreira, J.; Domingos, I.; Nunes, L.; Pereira, H. Optimizing Douglas-Fir Bark Liquefaction in Mixtures of Glycerol and Polyethylene Glycol and KOH. *Holzforschung* 2018, *72*, 25–30. [CrossRef]
26. Fernandes, A.; Cruz-Lopes, L.; Dulyanska, Y.; Domingos, I.; Ferreira, J.; Evtuguin, D.; Esteves, B. Eco Valorization of *Eucalyptus globulus* Bark and Branches through Liquefaction. *Appl. Sci.* 2022, *12*, 3775. [CrossRef]

27. Ye, L.; Zhang, J.; Zhao, J.; Tu, S. Liquefaction of Bamboo Shoot Shell for the Production of Polyols. *Bioresour. Technol.* **2014**, *153*, 147–153. [CrossRef] [PubMed]

28. Zhang, J.; Du, M.; Hu, L. Factors Influencing Polyol Liquefaction of Nut Shells of Different Camellia Species. *BioResources* **2016**, *11*, 9956–9969. [CrossRef]

29. Duarte, J.; Cruz-Lopes, L.; Dulyanska, Y.; Domingos, I.; Ferreira, J.; de Lemos, L.T.; Esteves, B. Orange Peel Liquefaction Monitored by FTIR. *J. Int. Sci. Publ.* **2017**, *5*, 309–313.

30. *264 Om-97*; Preparation of Wood in Chemical Analysis. Tappi Press: Atlanta, GA, USA, 1997.

31. *211 Om–93*; Ash in Wood, Pulp, Paper and Paperboard: Combustion at 525. Tappi Press: Atlanta, GA, USA, 1993.

32. Prozil, S.O.; Evtuguin, D.V.; Lopes, L.P.C. Chemical Composition of Grape Stalks of *Vitis vinifera* L. from Red Grape Pomaces. *Ind. Crops Prod.* **2012**, *35*, 178–184. [CrossRef]

33. Neto, C.P.; Seca, A.; Fradinho, D.; Coimbra, M.A.; Domingues, F.; Evtuguin, D.; Silvestre, A.; Cavaleiro, J.A.S. Chemical Composition and Structural Features of the Macromolecular Components of Hibiscus Cannabinus Grown in Portugal. *Ind. Crops Prod.* **1996**, *5*, 189–196. [CrossRef]

34. González, D.; Campos, A.R.; Cunha, A.M.; Santos, V.; Parajó, J.C. Manufacture of Fibrous Reinforcements for Biodegradable Biocomposites from Citysus Scoparius. *J. Chem. Technol. Biotechnol.* **2011**, *86*, 575–583. [CrossRef]

35. González, N.; Ribeiro, D.; Fernandes, E.; Nogueira, D.R.; Conde, E.; Moure, A.; Vinardell, M.P.; Mitjans, M.; Domínguez, H. Potential Use of Cytisus Scoparius Extracts in Topical Applications for Skin Protection against Oxidative Damage. *J. Photochem. Photobiol. B Biol.* **2013**, *125*, 83–89. [CrossRef]

36. Esteves, B.; Ayata, U.; Cruz-Lopes, L.; Brás, I.; Ferreira, J.; Domingos, I. Changes in the Content and Composition of the Extractives in Thermally Modified Tropical Hardwoods. *Maderas. Cienc. Tecnol.* **2022**, *24*, 1–14. [CrossRef]

37. Chupin, L.; Motillon, C.; Charrier-El Bouhtoury, F.; Pizzi, A.; Charrier, B. Characterisation of Maritime Pine (*Pinus pinaster*) Bark Tannins Extracted under Different Conditions by Spectroscopic Methods, FTIR and HPLC. *Ind. Crops Prod.* **2013**, *49*, 897–903. [CrossRef]

38. Kofujita, H.; Ettyu, K.; Ota, M. Characterization of the Major Components in Bark from Five Japanese Tree Species for Chemical Utilization. *Wood Sci. Technol.* **1999**, *33*, 223–228. [CrossRef]

39. Smuga-Kogut, M.; Szymanowska, D.; Markiewicz, R.; Piskier, T.; Kobus-Cisowska, J.; Cielecka-Piontek, J.; Schöne, H. Evaluation of the Potential of Fireweed (*Epilobium angustifolium* L.), European Goldenrod (*Solidago virgaurea* L.), and Common Broom (*Cytisus scoparius* L.) Stems in Bioethanol Production. *Energy Sci. Eng.* **2020**, *8*, 3244–3254. [CrossRef]

40. Gil, N.; Domingues, F.C.; Amaral, M.E.; Duarte, A.P. Optimization of diluted acid pretreatment of Cytisus striatus and Cistus ladanifer for bioethanol production. *J. Biobased Mater. Bioenergy* **2012**, *6*, 292–298. [CrossRef]

41. Sen, A.; Miranda, I.; Esteves, B.; Pereira, H. Chemical Characterization, Bioactive and Fuel Properties of Waste Cork and Phloem Fractions from *Quercus cerris* L. Bark. *Ind. Crops Prod.* **2020**, *157*, 112909. [CrossRef]

42. Domingos, I.; Ferreira, J.; Esteves, B.; Cruz-Lopes, L. Liquefaction and Chemical Composition of Walnut Shells. *Open Agric.* **2022**, *7*, 249–256. [CrossRef]

43. Domingos, I.; Ferreira, J.; Cruz-Lopes, L.; Esteves, B. Polyurethane Foams from Liquefied Orange Peel Wastes. *Food Bioprod. Process.* **2019**, *115*, 223–229. [CrossRef]

44. Gama, N.V.; Soares, B.; Freire, C.S.; Silva, R.; Brandão, I.; Neto, C.P.; Barros-Timmons, A.; Ferreira, A. Rigid Polyurethane Foams Derived from Cork Liquefied at Atmospheric Pressure: Rigid Polyurethane Foams Derived from Cork Liquefied. *Polym. Int.* **2015**, *64*, 250–257. [CrossRef]

45. Cinelli, P.; Anguillesi, I.; Lazzeri, A. Green Synthesis of Flexible Polyurethane Foams from Liquefied Lignin. *Eur. Polym. J.* **2013**, *49*, 1174–1184. [CrossRef]

46. Kahlerras, Z.; Irinislimane, R.; Bruzaud, S.; Belhaneche-Bensemra, N. Elaboration and Characterization of Polyurethane Foams Based on Renewably Sourced Polyols. *J. Polym. Environ.* **2020**, *28*, 3003–3018. [CrossRef]

47. Esteves, B.; Velez Marques, A.; Domingos, I.; Pereira, H. Chemical Changes of Heat Treated Pine and Eucalypt Wood Monitored by FTIR. *Maderas. Cienc. Tecnol.* **2013**, *15*, 245–258. [CrossRef]

48. Cruz-Lopes, L.; Dulyanska, Y.; Domingos, I.; Ferreira, J.; Fragata, A.; Guiné, R.; Esteves, B. Influence of Pre-Hydrolysis on the Chemical Composition of Prunus Avium Cherry Seeds. *Agronomy* **2022**, *12*, 280. [CrossRef]

49. Michell, A.; Higgins, H. *Infrared Spectroscopy in Australian Forest Products Research*; CSIRO Forestry: Victoria, Australia, 2002.

Article

Chemical-Anatomical Characterization of Stems of Asparagaceae Species with Potential Use for Lignocellulosic Fibers and Biofuels

Agustín Maceda [1], Marcos Soto-Hernández [2] and Teresa Terrazas [1,*]

[1] Instituto de Biología, Universidad Nacional Autónoma de México, Mexico City 04510, Mexico
[2] Programa de Botánica, Colegio de Postgraduados, Montecillo 56230, Mexico
* Correspondence: tterrazas@ib.unam.mx; Tel.: +52-55-5622-9116

Abstract: During the last decades, the possibility of using species resistant to droughts and extreme temperatures has been analyzed for use in the production of lignocellulosic materials and biofuels. Succulent species are considered to identify their potential use; however, little is known about Asparagaceae species. Therefore, this work aimed to characterize chemically-anatomically the stems of Asparagaceae species. Stems of 10 representative species of Asparagaceae were collected, and samples were divided into two. One part was processed to analyze the chemical composition, and the second to perform anatomical observations. The percentage of extractives and lignocellulose were quantified, and crystalline cellulose and syringyl/guaiacyl lignin were quantified by Fourier transform infrared spectroscopy. Anatomy was observed with epifluorescence microscopy. The results show that there were significant differences between the various species ($p < 0.05$) in the percentages of extractives and lignocellulosic compounds. In addition, there were anatomical differences in fluorescence emission that correlated with the composition of the vascular tissue. Finally, through the characterization of cellulose fibers together with the proportion of syringyl and guaiacyl, it was obtained that various species of the Asparagaceae family have the potential for use in the production of lignocellulosic materials and the production of biofuels.

Keywords: Asparagaceae; lignocellulose; crystalline cellulose; syringyl/guaiacyl; anatomy

Citation: Maceda, A.; Soto-Hernández, M.; Terrazas, T. Chemical-Anatomical Characterization of Stems of Asparagaceae Species with Potential Use for Lignocellulosic Fibers and Biofuels. *Forests* **2022**, *13*, 1853. https://doi.org/10.3390/f13111853

Academic Editors: Vicelina Sousa, Helena Pereira, Teresa Quilhó and Isabel Miranda

Received: 3 October 2022
Accepted: 3 November 2022
Published: 6 November 2022

Publisher's Note: MDPI stays neutral with regard to jurisdictional claims in published maps and institutional affiliations.

1. Introduction

Asparagaceae is one of the most important families in Mexico due to its biological, economic, and cultural importance [1]. Several species are used in the production of fibers, intoxicating drinks, food preparation, and the consumption of plant parts (flower, stem, and leaf) [2]. Asparagaceae has several subfamilies, including Agavoideae with *Agave*, *Furcraea*, *Manfreda*, and *Polianthes*; Yuccoideae including *Yucca*, *Hesperaloe*, and *Hesperoyucca* [3], and the subfamily Nolinoideae that includes the genera *Beaucarnea* and *Nolina* [4]. Most of the species of Agavoideae and Nolinoideae are distributed in arid and semiarid, and warm temperate regions of North and Central America [3] and in other parts of the world naturally or introduced. Several species of *Agave* have been used and studied the most because ethanol is produced in the form of intoxicating beverages such as *mezcal* and *tequila* [5].

In recent years, mainly the fibers and bagasse waste of several agave species, mainly *A. tequilana* [6,7], *A. angustifolia* [8], and *A. salmiana* [9] have been studied because a large amount of waste is produced annually from the production of *tequila* and *mezcal*. In addition, the subfamilies Agavoideae and Nolinoideae present acid metabolism of the crassulacean (CAM), which is considered raw material for the production of biofuel [10,11]. Furthermore, these species are part of the second generation of plants focused on biofuels. They are not part of the plants essential for human consumption [12], and they tolerate drought conditions and high temperatures [13].

Other species within the group of the second generation are cacti, such as *Opuntia* spp. [14], which also withstand extreme drought conditions and have CAM metabolism [15]. However, except for some genera such as *Opuntia* spp. and *Selenicereus* spp. [16,17], the other cacti species have slow vegetative development or very small sizes [18], so they could not be profitable in their use as biofuels or production of paper, while in species of the genus *Agave* growth and yield are higher [19].

However, even though many species of Asparagaceae exist in Mexico, there is not much information on the composition of the main lignocellulosic structural components, the anatomical distribution, or the potential use for farmers to cultivate and protect the plants in their natural environment [1]. Therefore, the objectives were to characterize the different Asparagaceae species with the extractives and lignocellulosic percentages, obtain crystallinity indexes, syringyl/guaiacyl (S/G) lignin ratio, and the anatomical distribution, with which it will be possible to identify the potential use of the different species as biofuels or in the paper industry, in addition to the possible biological implications.

2. Materials and Methods

2.1. Plant Materials and Extractives

Healthy adult plants were donated from the *Universidad Nacional Autónoma de México* (UNAM) Botanical Garden collection (Table 1), located at 19°18′44″ N, 99°11′46″ O and 2320 m a. s. l. The climate of the area is temperate, with rain in summer, with the rainy season from June to October, and the dry season from November to May. It has an average annual temperature of 15.6 °C and a rainfall of 833 mm. The plants were collected in their natural populations and grew in the garden. The leaves and stems of the ten species were cut, and only the stem was selected for the study. Stem samples were cut into pieces and dried in an oven for two weeks at 70 °C. Subsequently, the samples were ground (40–60 mesh size, Cyclone Sample Mill, (UDY Corporation, Fort Collins, CO, USA) until a particle size of 0.4 mm was obtained.

Table 1. Asparagaceae Species.

Species	Type of Stem	Life Forms	Size Category	Natural Distribution
Agave attenuata Salm-Dyck	Fibrous	Herbaceous	Medium	Mexico
Agave celsii Hook.	Fibrous	Herbaceous	Medium	Mexico
Agave convallis Trel.	Fibrous	Herbaceous	Medium	Mexico
Agave striata Zucc.	Fibrous	Herbaceous	Medium	Mexico
Beaucarnea gracilis Lem.	Fibrous	Arborescent	Tall	Mexico
Furcraea longaeva Karw. & Zucc.	Fibrous	Arborescent	Tall	Mexico
Nolina excelsa García-Mend. & E. Solano	Fibrous	Arborescent	Tall	Mexico
Yucca filifera Chabaud	Fibrous	Arborescent	Tall	Mexico
Yucca gigantea Lem.	Fibrous	Arborescent	Tall	Mexico, Guatemala
Yucca periculosa Baker	Fibrous	Arborescent	Tall	Mexico

The samples were analyzed in triplicate based on the TAPPI T-222 om-02 standard and based on the method proposed by Maceda et al. [20,21]. From each ground sample, 2 g were taken, which were placed in filter paper cartridges to carry out successive extractions for six hours in a Soxhlet with ethanol-benzene (1:2 *v/v*) and subsequently in ethanol (96%). After each extraction, the cartridges were allowed to dry for 24 h at 70 °C to record their constant weight.

Subsequently, the cartridges were discarded, and the samples were kept in a reflux system for 1 h in water at 90 °C. The samples were filtered through a medium pore Büchner filter and dried at 70 °C for 24 h to record constant dry weight. The formula used was the following:

$$\text{Total extractives (\%)} = [(A + B + C)/W_0] \times 100$$

where A is the weight lost (g) after extraction with ethanol:benzene, B is the weight lost (g) after extraction with ethanol (96%), C is the weight lost (g) after extraction with water at

90 °C, and W_0 is the initial weight of each sample. The percentage of extractive-free lignocellulose was obtained by subtracting from 100% the initial weight of the total percentage of extractives.

2.2. Lignocellulosic Purification

Klason lignin. From the extractive-free lignocellulose of each species, 0.2 g were taken, and 15 mL of concentrated sulfuric acid (72%) was added at a temperature of 2 °C. The mixture was kept under constant stirring and at room temperature (18 °C) for 2 h. Then, 560 mL of distilled water was added, and the mixture was refluxed and boiled for 4 h. The samples were filtered through a fine-pore Büchner filter and dried at 105 °C for 24 h to record constant dry weight. Lignin was quantified as follows:

$$\text{Klason lignin (\%)} = (W_L/W_W) \times 100$$

where W_L is the obtained weight of lignin (g), and W_W is the extractives-free lignocellulose (g).

Cellulose. From the extractive-free lignocellulose, 0.2 g were taken to purify the cellulose using the Kûshner-Höffer method [21]. Twenty-five mL of HNO_3/ethanol (1:4 *v/v*) were added to each sample and kept in a reflux system, and boiled for one hour. The sample was allowed to decant to discard the HNO_3/ethanol solution, and another 25 mL was added again. This cycle was repeated three more times, and in the last cycle, 25 mL of an aqueous solution of KOH at 1% was added and kept for 30 min at reflux and boiling to finally filter the sample through a fine-pore Büchner filter. The sample was left to dry at 70 °C for 12 h to record the constant dry weight and obtain the percentage of cellulose based on the following formula:

$$\text{Cellulose (\%)} = (W_C/W_W) \times 100$$

where W_C is the obtained weight of cellulose(g), and $/W_W$ is the extractives-free lignocellulose (g).

Hemicellulose. The purification was carried out based on the methodology proposed by Li. et al. [22]. From the extractive-free lignocellulose, 0.5 g were taken and placed in a reflux system with 10 mL of water for 3 h (solid-to-liquid ratio 1:20 g/mL). The system was cooled to room temperature and filtered. The filtrate was concentrated at 1.25 mL and purified into 3.75 mL of ethanol (95%) with stirring. The mixture was held for 1 h without stirring, and the hemicellulose precipitated. In order to obtain the dry weight (H_0), the sample was centrifuged at $4500\times g$ for 4 min, and then lyophilized. The residue insoluble in water was dried at 60 °C for 16 h, then successive extractions were performed with different concentrations of KOH (0.6, 1.0, 1.5, 2.0, and 2.5%) in a ratio of 1:20 (g/mL) at 75 °C for 3 h in each extraction. In the last concentration of 2.5% KOH, ethanol (99.7%) was added in a ratio of 2:3. The five mixtures were filtered and acidified to pH 5.5 with glacial acetic acid and concentrated to 1.25 mL. The mixtures were poured into 3.75 mL of ethanol (95%) with constant stirring. The mixtures were kept for 1 h and finally were centrifuged ($4500\times g$ for 4 min) and lyophilized. The constant dry weight was recorded in each extraction stage ($H_{0.6}$, $H_{1.0}$, $H_{1.5}$, $H_{2.0}$, $H_{2.5}$), and the percentage of cellulose was obtained with the following formula:

$$\text{Hemicellulose (\%)} = (W_H/W_W) \times 100$$

where W_H is the sum of $H_{0.6}$ to $H_{2.5}$ and W_W is the extractive-free lignocellulose (g).

2.3. Fourier Transform Infrared Spectroscopy Analysis

Lignin analysis. The ratio of syringyl/guaiacyl monomers (S/G) was obtained by Fourier transform infrared spectroscopy (FTIR) analysis. Klason lignin samples were kept dry until analyzed by FTIR (30 scans with a resolution of 4 cm^{-1}, 15 s per repeat). Three FTIR readings (Agilent Cary 630 FTIR) were made from each sample, and then the baseline correction was performed to separate the peaks of the fingerprints (wavelength of 800–1800 cm^{-1}) [23] in the MicroLab PC program (Agilent Technologies). The peaks of 1269 to 1272 cm^{-1} and 1328 to 1330 cm^{-1} were used to quantify the proportion of guaiacyl (G) and syringyl (S), respectively [24]. The value of each peak was obtained by drawing a line connecting the lowest values and a similar line for the highest values of each peak. A vertical line was drawn from the base of the $X-$axis to the highest part of the peak. The portion of the line between the top and the base is the value of each peak, so the S/G ratio was calculated by dividing the values of each peak [24].

Cellulose analysis. The proportion of crystalline cellulose was obtained by analyzing the dried samples with FTIR [21]. From each sample and doing the analyzes in triplicate, a small portion of the sample was placed in the FTIR Spectrometer (Agilent Cary 630 FTIR), and the spectrum was obtained in a range of 400–650 cm^{-1} (30 scans with a resolution of 4 cm^{-1}, 15 s per repeat). Samples were converted from transmittance to absorbance, and spectra were averaged using the Resolution Pro FTIR Software program (Agilent Technologies, Santa Clara, CA, United States).

The crystallinity indices used were: Total crystallinity index (TCI) proposed by Nelson and O'Connor [25] or also called the crystallinity ratio [26,27]. The lateral order index (LOI) [25,27] or second proportion of crystallinity [26]; and hydrogen bonding intensity (HBI) [28]. TCI was calculated with the ratio between the absorption intensity of the peaks 1370 cm^{-1} and 2900 cm^{-1} [27], LOI was calculated from the ratio between the absorption intensity of the peaks 1430 cm^{-1} and 893 cm^{-1} [26], while HBI was calculated with the ratio between 3350 cm^{-1} and 1315 cm^{-1} [28].

2.4. Statistical Analysis

The data obtained from the percentages of extractives and lignocellulosic components were analyzed with the non-parametric Kruskal-Wallis test and Dunn's post hoc analysis since the values did not present normality based on the results of Kolmogorov-Smirnov and Shapiro$-$Wilk, even when they were transformed with the square root of the arc sine. In addition, a multivariate principal component analysis was performed to separate the groups based on the values of the structural components.

2.5. Lignocellulosic Anatomical Distribution

Stem fragments were saved from each sample and were fixed, embedded, and cut based on the procedures of Arias and Terrazas [29] for succulent hardwood species. The transverse sections were stained with acridine orange and calcofluor [30] to observe the distribution of cellulose and lignin in the stem, in addition to comparing the anatomical results with the chemical ones.

3. Results

3.1. Extractives and Lignocellulosic Structural Compounds

The Asparagaceae species had significant differences between species (Table 2) in the variables of extractives and lignocellulosic components. The percentages of extractives were heterogeneous between the species of the same genus; however, the differences occurred mainly between *Agave* and *Yucca-Nolina*. In Tables 3 and 4, the means and standard deviation of the extractives are presented, and the different superscript capital letters show the species that are significantly different. In the ethanol extractives, *A. striata* had the lowest percentage and was significantly different from *Y. gigantea*, which had the highest percentage (Table 3). In hot water extractives, a similar situation was shown; *A. convallis* had the highest percentage, while *N. excelsa* and *Y. periculosa* had the lowest percentage. In the ethanol:benzene extractives, the statistical differences were presented between *N. excelsa*, with the lowest percentage, and *F. longaeva*, with the highest percentage. The species with the highest content of extractives and the lowest content of lignified tissue were *A. convallis* and *Y. gigantea*. On the contrary, the species with the lowest content of extractives and the highest amount of lignified tissue were *N. excelsa* and *Y. periculosa*.

Table 2. Kruskal-Wallis analysis for the lignocellulosic and extractives variables.

Variables	$\chi-$Square	Df	Significance
Ethanol-benzene	21.16344	9	0.01194
Ethanol 96%	22.17698	9	0.00833
Water 90 °C	24.08278	9	0.00417
Total extractives	26.28172	9	0.00184
Extractive-free lignocellulose	26.28172	9	0.00184
Lignin	20.94839	9	0.01288
Cellulose	26.64249	9	0.00160
Hemicelluloses	24.54409	9	0.00352

Table 3. Extractives percentage from the 10 Asparagaceae species.

Species	Extractive Compounds (%)				Extractive−Free Lignocellulose (%)
	Ethanol−Benzene	Ethanol 96%	Water 90 °C	Total Extractives	
Agave attenuata	7.5 + 1.8 AB	9.9 + 1.1 AB	4.7 + 1.1 AB	22.0 ± 1.8 ABC	78.0 ± 1.8 ABC
Agave celsii	6.9 ± 1.8 AB	6.5 ± 1.1 AB	3.9 ± 0.8 AB	17.4 ± 0.8 ABC	82.6 ± 0.8 ABC
Agave convallis	9.4 ± 1.9 AB	6.4 ± 1.2 AB	10.2 ± 1.5 B	26.0 ± 0.5 C	74.0 ± 0.5 A
Agave striata	7.6 ± 1.6 AB	4.6 ± 0.7 A	3.0 ± 0.7 AB	15.3 ± 2.5 ABC	84.7 ± 2.5 ABC
Beaucarnea gracilis	5.9 ± 0.5 AB	8.3 ± 1.9 AB	5.3 ± 0.5 AB	19.5 ± 1.8 ABC	80.5 ± 1.8 ABC
Furcraea longaeva	9.7 ± 0.6 B	6.9 ± 0.7 AB	3.7 ± 0.7 AB	20.2 ± 0.8 ABC	79.8 ± 0.8 ABC
Nolina excelsa	4.5 ± 0.9 A	5.1 ± 0.6 AB	1.7 ± 0.6 A	11.3 ± 1.1 A	88.7 ± 1.1 C
Yucca filifera	9.6 ± 0.2 AB	6.2 ± 1.1 AB	3.2 ± 0.2 AB	18.9 ± 0.9 ABC	81.1 ± 0.9 ABC
Yucca gigantea	8.4 ± 1.1 AB	11.9 ± 1.2 B	3.8 ± 0.9 AB	24.1 ± 2.2 BC	75.9 ± 2.2 AB
Yucca periculosa	6.1 ± 1.8 AB	4.9 ± 1.0 AB	1.8 ± 0.6 A	12.9 ± 2.5 AB	87.1 ± 2.5 BC

Different letters in each column indicate significant differences ($p < 0.05$). Mean ± standard deviation (SD).

Table 4. Lignin, cellulose, and hemicellulose percentage of dry biomass of Asparagaceae species.

Species	Lignin (%)	Cellulose (%)	Hemicellulose (%)
Agave attenuata	20.4 ± 2.7 AB	35.4 ± 0.8 ABC	22.1 ± 1.6 ABC
Agave celsii	23.8 ± 0.9 AB	31.6 ± 0.8 A	27.1 ± 2.3 BC
Agave convallis	18.2 ± 1.8 AB	34.9 ± 1.4 AB	20.9 ± 1.6 ABC
Agave striata	24.2 ± 6.1 AB	37.1 ± 1.9 ABC	23.4 ± 5.9 ABC
Beaucarnea gracilis	20.2 ± 1.1 AB	38.1 ± 1.7 ABC	22.3 ± 2.8 ABC
Furcraea longaeva	26.3 ± 0.8 B	37.3 ± 0.9 ABC	16.2 ± 2.4 ABC
Nolina excelsa	24.5 ± 3.2 AB	52.2 ± 2.1 C	12.0 ± 4.1 AB
Yucca filifera	11.4 ± 1.1 A	38.9 ± 1.2 ABC	30.7 ± 0.9 C
Yucca gigantea	24.9 ± 1.7 AB	45.3 ± 1.8 BC	5.7 ± 1.5 A
Yucca periculosa	24.5 ± 0.5 AB	41.6 ± 1.8 ABC	21.0 ± 2.3 ABC

Different letters in each column indicate significant differences ($p < 0.05$). Mean ± standard deviation (SD).

The percentages of lignocellulosic components had significant differences in the percentages of lignin between *Y. filifera* and *F. longaeva* (Table 4). In cellulose, the species that had the lowest percentage were *A. celsii* and *A. convallis*, and *N. excelsa* was the species with the highest percentage. Finally, in the hemicelluloses, *Y. gigantea* presented the least quantity and *Y. filifera* the largest.

In the Principal Component (PC) analysis, the first two PCs had eigenvalues above 1, while PC3 was less than 1; however, PC3 was considered in the analysis due that it explained 11% of the variance, so the three PCs explained 84.5% of the total variation (Table 5). In each PC, the highest negative or positive values were those that influenced the separation of each species, as shown in Figure 1. For PC1, the variables that determined the separation of the different groups were hot water extractives and extractive-free lignocellulose. In PC2, the determinant variables of the variation were hemicelluloses and ethanol extractives, while in PC3 was the cellulose percentage (Table 5).

Table 5. Vectors, eigenvalues, and cumulative proportion of the variation are explained by each variable.

Variables	PC1	PC2	PC3
Ethanol-Benzene	0.439	0.058	0.013
Ethanol	0.194	**0.553**	−0.169
Water	**0.453** *	0.096	0.125
Extractive-free lignocellulose	−0.492	−0.346	0.015
Cellulose	−0.327	0.302	**0.788**
Lignin	−0.410	0.272	−0.578
Hemicellulose	0.216	**−0.630**	−0.008
Eigenvalue	3.127	2.011	0.771
Variance (%)	44.7	28.7	11.0
Accumulative variance (%)	44.7	73.4	**84.4**

* The highest values are in bold on each PC.

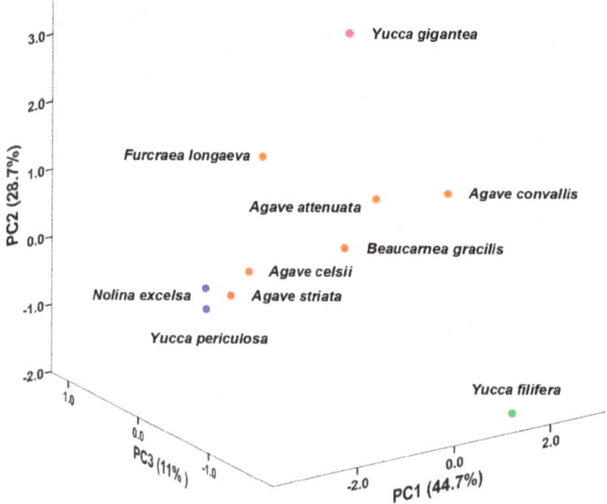

Figure 1. The three-dimension plot of the principal components from Asparagaceae stems. Blue points: group 1, green point: group 2, pink point: group 3, brown point: group 4.

When plotting the species based on the first three PCs and the variables that influenced each PC (Figure 1), the species were separated into four groups. The first group (blue points) included the species of *N. excelsa* and *Y. periculosa*, which were the species with the lowest content of hot water extractives and higher extractive–free lignocellulose content. The second group (green point), represented by *Y. filifera* had the highest percentage of hemicelluloses. The third group (pink point), conformed by *Y. gigantea* was separated from the other Yuccas and the other species because they had lower percentages of hemicelluloses, but it was one of the species with the highest percentage of cellulose and percentage of total extractives. In the fourth group (orange points), the remaining species that belong to *Agave*, *F. longaeva*, and *B. gracilis* were clustered (Figure 1).

3.2. Cellulose Crystallinity

In the cellulose spectra (Figure 2) and Table 6, the main cellulose peaks are observed. In order to determine the purity of the cellulose, the absence of xylans and hemicellulose was obtained by not detecting the peak at 1735 cm^{-1}. Lignin was not detected with the peaks 1595 cm^{-1}, 1512 cm^{-1}, and 1463 cm^{-1}. Only weak lignin signals were observed in *A.*

striata and *Y. periculosa*. In addition, the absence of hemicellulose was observed without the presence of the 1269 cm^{-1} peak, except for *A. celsii*, which had a weak peak.

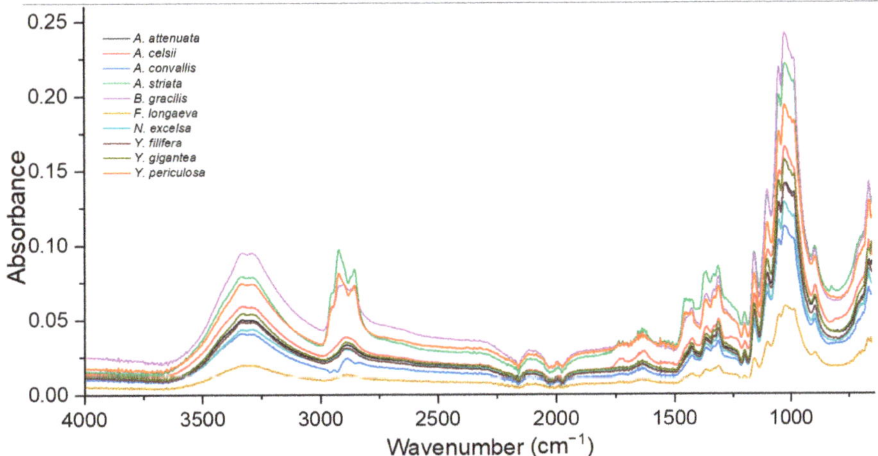

Figure 2. Cellulose FTIR spectra of Asparagaceae species.

Table 6. Cellulose peaks of Asparagaceae species.

Wavenumber (cm^{-1})	Assignments
3000–3600	OH stretching
2900	CH stretching
1430	CH$_2$ symmetric bending (crystalline and amorphous cellulose)
1370	C-H and C-O bending vibration bonds
1336	C-O-H in-plane bending (amorphous cellulose)
1315	CH$_2$ wagging vibration (crystalline cellulose)
1163	C-O-C asymmetrical stretching
893	Out-of-plane asymmetrical stretching of cellulose ring
670	C-O-H out-of-plane stretching

In the crystallinity indexes (Table 7), the TCI values showed that most of the species had values above one because they had a higher percentage of crystalline cellulose, except *A. striata*, *B. gracilis*, and *Y. periculosa*, which presented a higher percentage of amorphous cellulose. In LOI, the species that had the highest value was *A. striata*, while the species that had the lowest value was *F. longaeva*, which had the highest value in TCI. In HBI, similarly, *A. striata* had the lowest value, while *B. gracilis* had the highest value.

Table 7. Crystallinity indexes of Asparagaceae species.

Species	TCI (A1370/A2900)	LOI (A1430/A893)	HBI (A3400/A1315)
Agave attenuata	1.12 ± 0.01	0.49 ± 0.01	1.16 ± 0.02
Agave celsii	1.15 ± 0.05	0.49 ± 0.03	1.18 ± 0.09
Agave convallis	1.27 ± 0.03	0.56 ± 0.28	1.17 ± 0.08
Agave striata	0.85 ± 0.17	0.64 ± 0.03	0.92 ± 0.12
Beaucarnea gracilis	0.9 ± 0.01	0.59 ± 0.01	1.21 ± 0.02
Furcraea longaeva	1.18 ± 0.05	0.47 ± 0.02	1.09 ± 0.04
Nolina excelsa	1.09 ± 0.1	0.54 ± 0.02	1.09 ± 0.01
Yucca filifera	1.16 ± 0.02	0.52 ± 0.01	1.11 ± 0.02
Yucca gigantea	1.15 ± 0.03	0.51 ± 0.01	1.11 ± 0.02
Yucca periculosa	0.78 ± 0.05	0.59 ± 0.02	1.03 ± 0.05

Mean ± standard deviation (SD).

3.3. Lignin S/G Ratio

Figure 3 shows the FTIR spectra for the 10 species of Asparagaceae. Representative peaks of lignin were 1501 cm^{-1} which showed the C=C aromatic ring vibration of syringyl and guaiacyl monomers. The 1325 cm^{-1} peak reflected the breathing of the ring of syringyl in addition to C-O stretching. The 1271 cm^{-1} and 1225 cm^{-1} peaks reflected the symmetric vibration of C-O and the glucopyranose cycle of guaiacyl and syringyl, respectively. The 1030 cm^{-1} peak reflected the C-H in-plane deformation of guaiacyl and C-O deformation in primary alcohol; finally, the 913 cm^{-1} peaks showed the =CH out-of-plane deformation in the aromatic ring of syringyl and guaiacyl monomers (Figure 3).

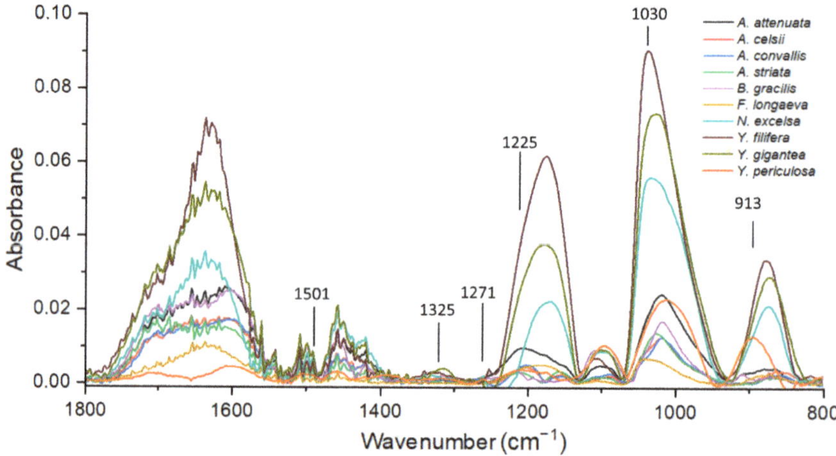

Figure 3. FTIR spectra of the lignin fingerprint of the Asparagaceae species.

Whit the peaks 1325 and 1271 cm^{-1} the S/G ratio was calculated (Table 8). The species with the lowest proportion of S/G was *A. attenuata* because it had a higher percentage of guaiacyl in its structure, while in the genus *Yucca*, the three species presented high values of syringyl, for which the proportion of S/G was in the range of 2.8 to 3.9. The other species had similar proportions, so the syringyl monomer prevailed except in *A. convallis*.

Table 8. Percentages of syringyl and guaiacyl, and S/G ratio of Asparagaceae species.

Species	Guaiacyl (%)	Syringyl (%)	S/G Ratio
Agave attenuata	63.1	36.9	0.6
Agave celsii	47.0	53.0	1.1
Agave convallis	52.0	48.0	0.9
Agave striata	44.0	56.0	1.3
Beaucarnea gracilis	42.3	57.7	1.4
Furcraea longaeva	39.1	60.9	1.6
Nolina excelsa	49.0	51.0	1.0
Yucca filifera	26.1	73.9	2.8
Yucca gigantea	23.0	77.0	3.3
Yucca periculosa	20.5	79.5	3.9

3.4. Anatomical Distribution

The species of Asparagaceae studied showed the same type of vascular tissue, with closed vascular bundles forming isolated patches or two or more patches joined through lignified parenchyma. In the vascular bundle, the presence of tracheary elements with mainly reticular type of secondary wall thickenings was observed, with completely lignified fibers and non-lignified phloem with contents inside with fluorescence emission in bluish tones (Figure 4). In most species, patches of non-lignified parenchyma were observed on the edges of the stem, while in the center, the parenchyma was completely lignified. In the species *A. attenuata* and *A. convallis*, the fluorescence tones of the tracheary elements and fibers predominated in yellow-green tones, while in the *Yucca* species, the fluorescence emission was observed in green to bluish-green tones. In all species, the presence of crystals was observed, mainly raphides and prisms. Cellulose fluoresced in bluish tones but differed from lignin due to the intensity of fluorescence emission since the cellulose had lower intensity compared to lignin.

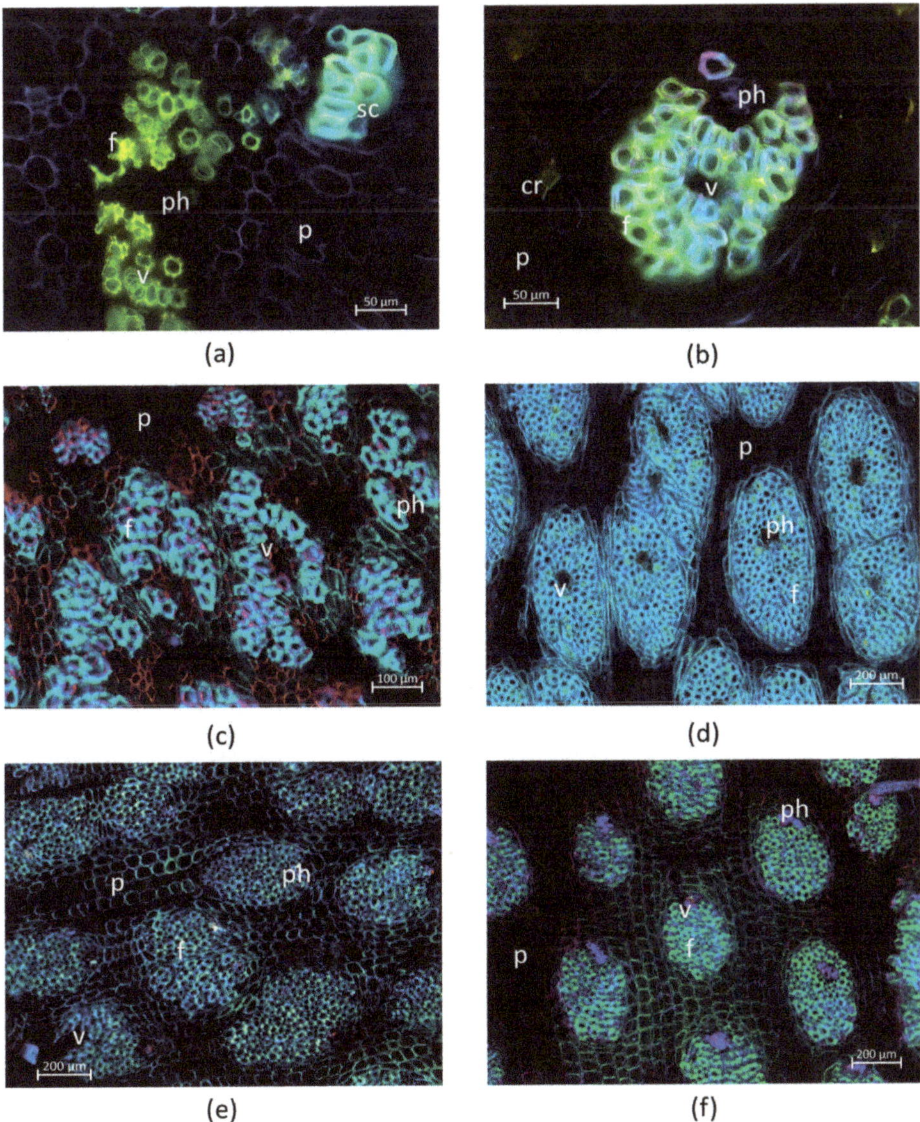

Figure 4. Fluorescence images from representative Asparagaceae species. (**a**) *A. attenuata*. (**b**) *A. convallis*. (**c**) *B. gracilis*. (**d**) *N. excelsa*. (**e**) *Y. gigantea*. (**f**) *Y. periculosa*. f: fiber, v: vessel, p: parenchyma, ph: phloem, cr: crystal.

4. Discussion

The Asparagaceae family presented heterogeneity in the percentages of the structural components that allow its grouping based on the values of extractive compounds, hemicellulose, and cellulose. In addition, the presence of high percentages of cellulose, a majority proportion of crystalline cellulose, and the presence of S/G ratios greater than one make it possible to use in biofuels and cellulosic compounds.

4.1. Extractives and Lignocellulose

Asparagaceae species had similar percentages of extractives when extracted with ethanol: benzene and ethanol. However, hot water extractives were low except for *A. convallis* (Table 3). Compared with the literature on other *Agave* species (Table 9), the extractive percentages in water were similar to the values reported here. *A. tequilana, A. angustifolia*, and *A. salmiana* have values between 4.4%–6.0% (Table 9). On the contrary, the values reported in *Agave* for the percentages of extractives in ethanol: benzene and ethanol (Table 9) were low (1.5%–4.0% in ethanol: benzene and 1.3%–5.0% in ethanol) compared to the values reported here (Table 3). For the genera *Beaucarnea, Furcraea*, and *Nolina*, there are no reports of extractive percentages in their stems or leaves. In the *Yucca* genus, the literature report that *Y. gloriosa* presents a percentage of total extractives of 1.1 [31], which is low compared to the percentages obtained here (Table 3).

Although extractives in ethanol: benzene and in ethanol are hardly reported in the literature for the species of the genera analyzed here, they have been reported for other succulent species such as cacti [20,21], which had lower percentages of ethanol: benzene (2.5%–4.2%), similar in ethanol (1.0%–9.1%), higher in hot water (8.2%–44.5%) and total extractives (16.7–49.2 %).

In the percentages of lignocellulosic compounds, Table 4 shows that there was heterogeneity in the results within the same genus. In the species that are considered arborescent, it was obtained that the percentages of lignin were low. However, the percentage of cellulose was higher (37.3%–52.2%) and in herbaceous species was lower (31.6%–37.1%). In the hemicelluloses, there is homogeneity between the agaves (20.9%–27.1%), while in the *Yucca* genus, notable differences are observed between *Y. gigantea* (5.7%) concerning the other two species of *Yucca* (21.0% and 30.7%). The differences possibly were due to the type of environment in which they live; *Y. gigantea* is distributed in regions with higher humidity [32], while *Y. filifera* [33,34] and *Y. periculosa* [35] grow in arid zones and semi-desert, respectively. In celluloses, the percentages reported in the literature (Table 9) were high compared to those obtained here (Table 4), possibly due to the differences in the species or the conditions in which they have been developed [36,37].

When comparing the results obtained with succulent species such as cacti, it was observed that the percentages of lignocellulose, in general, are higher in agave species than in non-fibrous wood cacti species [20], while in fibrous species, the percentages were similar [38].

Table 9. Percentages of extractives and lignocellulosic compounds of *Agave, Furcraea*, and *Yucca*.

Species	S *	Ce	He	Li	TE	W	E−B	Et	Ref
A. angustifolia									
A. lechuguilla									
A. salmiana	L	33.2–44.3	2.6–3.5	2.1–2.9	–	–	–	–	[9]
A. tequilana									
A. americana	L	68.4	15.7	4.9	–	–	–	–	[39]
A. americana	St	65.0	32.0	3.0	–	–	–	–	[40]
A. angustifolia	F	55.0	34.1	20.7	5.3	4.4	1.5	–	[8]
A. angustifolia	F	67.0	25.2	6.3	–	–	–	–	[41]
A. angustifolia	F	64.0	25.0	6.5	2.5	–	–	–	[42]
A. lechuguilla	St	17.7	17.5	7.3	45.3				[43]
A. lechuguilla	St	–	–	9.1	25.7	–	–	–	[44]

Forests **2022**, 13, 1853

Table 9. *Cont.*

Species	S *	Ce	He	Li	TE	W	E−B	Et	Ref
A. lechuguilla	L	79.8	3–6	15.3	—	—	—	—	[45]
A. fourcroudes	L	77.6	5.0–7.0	13.1					[45]
A. salmiana	F	48.9	—	8.5	—	—	—	—	[46]
A. salmiana	B	47.3	12.8	4.9	—	—	—	—	[47]
A. salmiana *A. americana* *A. tequilana*	L St	39.7–45.0		7.3–11.9	—	6.0–15.1	—	1.3–3.2	[48]
A. salmiana	L St	—	—	9.8	—	—	—	—	[49]
A. tequilana	L St	—	—	11.9	—	—	—	—	[49]
A. americana	L St	—	—	8.2	—	—	—	—	[49]
A. sisalana	—	43.0	32.0	15.0	—	—	—	—	[50]
A. tequilana	F	49.4	—	21.1	—	10.6	2.9	—	[51]
A. tequilana	F	41.9	—	7.2	—	5.8	3.1	—	[52]
A. tequilana	L St	40.0 and 51.0	—	—	—	—	—	—	[53]
A. tequilana	B	56.0–69.0	—	—	—	—	—	—	[54]
A. tequilana	B	40.9	—	—	—	—	—	—	[55]
A. tequilana	B	42.0	18.5	14.0	—	—	—	—	[56]
A. tequilana	B	—	—	—	—	—	—	—	[57]
A. tequilana	B	44.5	20.1	25.3	—	3.7	3.6	—	[58]
A. tequilana	L	24.7–33.5	10.7–15.2	15.6–19.5	—	—	—	—	[59]
A. tequilana	L	47.0	16.0	9.0	—	—	—	—	[6]
A. americana	L	50.0	22.0	13.0	—	—	—	—	[6]
A. lechuguilla	L	46.0–48.0	30.0	11.0	—	4.0	4.0	5.0	[60]
Agave spp.	B	70.0–80.0	5.0–10.0	15.0–20.0	—	—	—	—	[61]
Furcraea foetida	F	68.4	11.5	12.3	—	—	—	—	[62]
Yucca aloifolia	L	52.5	20.5	20.0	—	—	—	—	[63]
Yucca gloriosa	F	66.4	17.5	6.7	1.1	—	—	—	[31]

S: sample; Ce: cellulose, He: hemicellulose, Li: lignin, TE: total extractives, W: hot water extractives, E−B: ethanol−benzene extractives, Et: ethanol, Ref: references. * B: Bagasse, F: Fiber, L: Leaf, St: Stem.

The differences in the extractive and lignocellulosic compounds between species allowed us to identify four groups through the principal component analysis. The first group with *N. excelsa* and *Y. periculosa* presented a lower amount of hot water and total extractives. This could have implications for their use because various authors consider the hot water and total extractives to represent the nonstructural sugars that are used in fermentation processes for the production of ethanol [8,48,51]. However, in the case of *N. excelsa*, the percentage of cellulose presented in the stem would allow it to be used to obtain cellulosic products (Table 4). The characteristic that distinguishes *Y. filifera* from the other species studied was the presence of a large amount of hemicellulose. This hemicellulose can be transformed into usable sugars either through enzymatic hydrolysis processes or with temperature [9], in addition to the fact that *Y. filifera* presented lower percentages of lignin so that the purification of cellulose would be more efficient [64]. The main characteristic of *Y. gigantea* was the lower amount of hemicelluloses; however, it had a high percentage of cellulose that could be used to obtain cellulosic fibers. The three species of *Yucca* studied here revealed the higher differences in the extractives and lignocellulosic compounds between them, and other species of *Yucca* should be studied in the future to support these findings. All agave species plus *Beaucarnea* and *Furcraea* were grouped together by similar values in the extractive compounds (Table 4). In the lignocellulosic components, the agaves had low percentages of cellulose and lignin; however, they had a higher percentage of hemicelluloses. Finally, chromatographic analyzes can be carried out on all the extractives compounds to identify components with potential use, such as flavonoids and triterpenes, which have the antioxidant capacity and can be used mainly in the food and cosmetic industries [65]. In addition, lignocellulose can be treated in different ways to obtain various derivatives, as mentioned by Palomo-Briones et al. [66] for *Agave tequilana*: through the elimination of lignin by alkaline, organosolv or enzymatic methods, lignin can be solubilized and fermented to obtain ethanolic derivatives. Hemicelluloses and celluloses can be degraded by acids or enzymes to obtain insoluble fractions and use microcrystalline and nanocrystalline cellulose, while the solubilized carbohydrates can be used in the fermentation to obtain ethanol. Solubilized carbohydrates can also be fermented by anaerobic digestion or dark fermentation to obtain biogas and H_2 which can be used to obtain energy [66]. Therefore, the species analyzed in this work could be used in different ways, both energetically and in the production of paper or cellulose components [21].

4.2. Cellulose Crystallinity

The crystallinity indexes allowed identifying the species that had the highest proportion of crystalline cellulose. The purity of the cellulose could affect these indexes by presenting hemicellulose or lignin residues that alter the peaks [67]. The FTIR spectra can determine the purity of the samples and calculate the proportion of crystalline cellulose [68,69] by the absence of peaks belonging to hemicelluloses [70] and lignin [71].

The most commonly used indexes in conjunction with FTIR are TCI, LOI, and HBI [68,72]. The TCI index provides information about the amount of crystalline or amorphous cellulose in a sample. The peak indicating the presence of crystalline cellulose is 1370 cm^{-1}, while the peak of 2900 cm^{-1} [27] indicates the presence of amorphous cellulose. Therefore, if the value of the ratio is greater than one, there will be a greater amount of crystalline cellulose [28]. The LOI index is related to the order of crystalline cellulose, in addition to the fact that the 1430 cm^{-1} peak shows the presence of crystalline cellulose of type I and the peak 893 cm^{-1} the presence of type II crystalline cellulose and cellulose amorphous [73]. The order of the crystalline cellulose and LOI peaks can be altered by the type of chemical extraction, and the type of purification used [74]. The HBI index reflects the crystallinity of the sample and its water absorption. Low values reflect a greater amount of crystalline cellulose, while high values indicate the presence of cellulose II or amorphous [27], however, TCI and HBI values are related in terms of cellulose structure and stability, so if they are similar, the cellulose structure has greater stability [28].

Generally, the samples had a higher proportion of crystalline cellulose; however, the type of extraction modifies the order of the cellulose, so the low LOI values reflect the presence of type II crystalline cellulose [75], in addition to possibly the presence of NaOH during purification and the temperature used would alter the order of the crystalline [73]. Furthermore, LOI is correlated with the overall degree of cellulose order, while TCI is directly proportional to the percentage of crystalline cellulose [28]. The similar values of HBI with TCI confirm that the highest proportion of crystalline cellulose predominates among the Asparagaceae species, except the species *A. striata*, *B. gracilis*, and *Y. periculosa*, which had lower crystalline cellulose, and this is reflected in the LOI values that were also the highest [76].

The presence of a greater amount of crystalline cellulose (between 50 and 56% so that the proportion of TCI is 1 to 1.27) in most of the Asparagaceae species agrees with that reported for stems (Table 10). Furthermore, the percentages were also similar for other structures such as fibers in *Agave* (50.07 [46]), *Furcraea* (52.6% [62]), and *Yucca* spp. (55–56% [77]). The other structures have a higher percentage, possibly due to the species analyzed, the type of sample, such as bagasse that no longer has extractives, and some structural sugars, such as hemicelluloses. However, in future studies, X−ray diffraction analysis [78] could be used to confirm the proportion of crystalline cellulose in samples of Asparagaceae species.

Table 10. Crystalline index *Agave, Furcraea,* and *Yucca.*

Species	Sample	Crystalline Index	Reference
A. salmiana	Fiber	50.07	[46]
A. tequilana	Bagasse	63–68	[54]
A. salmiana			
A. americana	Leaf and stem	45–55	[48]
A. tequilana			
A. tequilana	Bagasse	71	[57]
A. tequilana	Bagasse	60.5	[58]
A. americana	Stem	50.1–64.1	[40]
Yucca spp.	Leaf	76	[77]
Yucca aloifolia	Leaf	69.43	[63]
Yucca spp.	Fiber	55–56	[79]
Furcraea foetida	Fiber	52.6	[62]

4.3. Lignin S/G Ratio

The proportion of S/G obtained for the 10 species reflected that most of the species presented high percentages of syringyl monomers, except for *A. attenuata* (36.9%) and *A. convallis* (48%). The species with the highest percentage was of the *Yucca* genus since it had percentages of 73.9 to 79.5%. The presence of higher percentages of syringyl in species with stem potential allows the purification of cellulose and the degradation of lignin by hydrolytic processes to be more efficient [80] by presenting more bonds of the β-O-4 type and being less condensed [81]. In species with higher percentages of guaiacyl, the stem tissue is hard, and the lignin is difficult to degrade, which is why it is called recalcitrant lignin [82].

Therefore, in the *Yucca* species and generally in the other species of *Agave, Beaucarnea, Furcraea,* and *Nolina,* by presenting a higher percentage of syringyl, the hydrolyzation process with Kraft would be more efficient [83]. In the literature, there are few reports on the proportion of S/G for *A. fourcroydes,* proportions are reported in fibers (1.05) and spines (1.2) [84], in leaves of *A. sisalana* (2.0) [85], and leaf fibers (3.0–3.5) [86]. In stems, S/G values have been reported for *A. americana* (1.27), *A. angustifolia* (1.29), *A. fourcroydes* (1.40), *A. salmiana* (1.33) y *A. tequilana* (1.57) in untreated samples [11], while for *A. tequilana* S/G values of 4.3 have also been reported in untreated samples [87] while in *A. sisalana* the proportion of S/G in stem fibers is 3.6 [88]. In the other genera analyzed here, there are no reports on the composition of lignin. In general, it is observed that the values obtained are similar to those reported by the aforementioned authors.

The presence of high percentages of syringyl in lignin has also been reported in various fibrous species where it provides resistance to and cellular support [89–91]. However, in succulent species such as cacti, the presence of syringyl monomers is more associated with tissues with non−lignified parenchyma and not fibers [38,92,93], so the presence of syringyl could be associated with a defense mechanism against pathogens [93] as it has been reported for bryophytes, conifers and angiosperms [94–98].

4.4. Cellulose and Lignin Anatomical Distribution

Analyzing the anatomical distribution of lignin and cellulose in the vascular bundles of Asparagaceae species was useful in identifying the location of the main structural components, in addition to explaining the results obtained both in the percentages of lignocellulosic components and in the proportion of S/G [49,99].

In Figure 4, the fluorescence emitted by the tracheary elements, the fibers, and the non−lignified and lignified parenchyma showed that lignin had different emissions in its fluorescence since it ranged from bluish tones to yellow tones. In the species of *A. attenuata* and *A. convallis,* the presence of yellow tones was observed, which would be related to the type of lignin present in the lignified walls. As proposed by Maceda et al. [93], based on a tone scale, yellow to green tones would reflect the presence of guaiacyl-type lignin, while syringyl−type lignin would have shades of lime-green to blue. The difference in the tones in the emission of fluorescence has already been reported for various species with similar tones [100]. The presence of blue tones in fibers in species with S/G ratios above two [93], as for *Ferocactus hamatacanthus* and *F. pilosus*, whose S/G ratio is 11.7 and 3.5, respectively [38]. Therefore, in the species of Yuccas, whose proportion ranges from 2.8 to 3.9, the presence of fibers and tissue in blue tones responds to the presence of lignin of the syringyl type.

The presence of lignified parenchyma in the stems of the *Yucca* species (Figure 4c–f) agrees with the lignin percentages shown in Table 4 since these species, except for *B. gracilis*, had high lignin values, in contrast to the species of *A. convallis* and *A. attenuata* (Figure 4a,b). Therefore, the presence of lignin is a structural support factor for species of tall size and arborescent shape. However, these species also had high percentages of cellulose (Table 4), with a predominance of crystalline cellulose (Table 7). Cellulose and lignin not only provide structural rigidity but also the proportion of S/G, and the presence of crystalline cellulose could improve water conduction in tracheary elements [101] in addition to protecting against pathogens [91,102]. The species analyzed in this study are naturally distributed in Mexico and *Y. guatemalensis* in Mexico and Guatemala, so it would be interesting to expand the number of genera to determine if there is heterogeneity in a greater number of species of the same genus or if they present homogeneity in the composition of cellulose and lignin.

5. Conclusions

The presence of high percentages of cellulose, the predominance of syringyl−type lignin, percentages above 10% of total extractive components, and the high percentages of holocellulose (structural sugars) show that Asparagaceae species have potential in the use of both productions as biofuels as in the production of paper. In addition, the chemical composition that Asparagaceae species present would be related to biological implications such as conduction, support, and protection against pathogens.

Author Contributions: Conceptualization, A.M. and T.T.; methodology, A.M., M.S.-H. and T.T.; validation, A.M. and T.T.; investigation, A.M. and T.T.; resources, T.T.; writing—review and editing, A.M., T.T. and M.S.-H. All authors have read and agreed to the published version of the manuscript.

Funding: Funding was provided by DGAPA−UNAM postdoctoral fellowship (document number: CJ IC/CTIC I5OO7I2O2I) to AM.

Data Availability Statement: Raw data and FTIR spectra are available in the Figshare repository: https://doi.org/10.6084/m9.figshare.21259230.v1 https://figshare.com/articles/dataset/Chemical-anatomical_characterization_of_stems_of_Asparagaceae_species_with_potential_use_for_lignocellulosic_fibers_and_biofu-els/21259230 (accessed on 3 November 2022)

Acknowledgments: The authors thank Abisaí Josué García for providing the plants from the Botanic Garden, UNAM; thanks to Rubén San Miguel−Chávez for allowing us to use the FTIR in COLPOS. Thanks to Elizabeth Navarro Cerón for allowing us to use the laboratory LANISAF, thanks to Pedro Mercado Ruaro from Laboratorio de Morfo−Anatomía y Citogenética (LANABIO, UNAM), and thanks to Steffany Aguilar Moreno for the support and the laboratory glassware provided.

Conflicts of Interest: The authors declare no conflict of interest.

References

1. Delgado-Lemus, A.; Casas, A.; Téllez, O. Distribution, abundance and traditional management of *Agave potatorum* in the Tehuacán Valley, Mexico: Bases for sustainable use of non−timber forest products. *J. Ethnobiol. Ethnomed.* **2014**, *10*, 63. [CrossRef] [PubMed]
2. Martínez Jiménez, R.; Ruiz-Vega, J.; Caballero Caballero, M.; Silva Rivera, M.E.; Montes Bernabé, J.L. Agaves silvestres y cultivados empleados en la elaboración de mezcal en Sola de vega, Oaxaca, México. *Trop. Subtrop. Agroecosyst.* **2019**, *22*, 477–485.
3. García-Mendoza, A.J.; Franco Martínez, I.S.; Sandoval Gutiérrez, D. Cuatro especies nuevas de *Agave* (Asparagaceae, Agavoideae) del sur de México. *Acta Bot. Mex.* **2019**, *126*, e1461. [CrossRef]
4. Seberg, O.; Petersen, G.; Davis, J.I.; Chris Pires, J.; Stevenson, D.W.; Chase, M.W.; Fay, M.F.; Devey, D.S.; Jørgensen, T.; Sytsma, K.J.; et al. Phylogeny of the Asparagales based on three plastid and two mitochondrial genes. *Am. J. Bot.* **2012**, *99*, 875–889. [CrossRef]
5. Pérez Hernández, E.; Del Carmen Chávez Parga, M.; González Hernández, J.C. Revisión del agave y el mezcal. *Rev. Colomb. Biotecnol.* **2016**, *18*, 148–164. [CrossRef]

6. Corbin, K.R.; Byrt, C.S.; Bauer, S.; Debolt, S.; Chambers, D.; Holtum, J.A.M.; Karem, G.; Henderson, M.; Lahnstein, J.; Beahan, C.T.; et al. Prospecting for energy−rich renewable raw materials: *Agave* leaf case study. *PLoS ONE* **2015**, *10*, e0135382. [CrossRef]

7. Flores-Gómez, C.A.; Escamilla Silva, E.M.; Zhong, C.; Dale, B.E.; Da Costa Sousa, L.; Balan, V. Conversion of lignocellulosic agave residues into liquid biofuels using an AFEX^{TM}–based biorefinery. *Biotechnol. Biofuels* **2018**, *11*, 1–18. [CrossRef]

8. Hidalgo-Reyes, M.; Caballero-Caballero, M.; Hernández-Gómez, L.H.; Urriolagoitia-Calderón, G. Chemical and morphological characterization of *Agave angustifolia* bagasse fibers. *Bot. Sci.* **2015**, *93*, 807–817. [CrossRef]

9. Jiménez-Muñóz, E.; Prieto-García, F.; Prieto-Méndez, J.; Acevedo-Sandoval, O.A.; Rodríguez-Laguna, R. Physicochemical characterization of four species of agaves with potential in obtaining pulp for paper making. *DYNA* **2016**, *83*, 232–242. [CrossRef]

10. Davis, S.C.; Dohleman, F.G.; Long, S.P. The global potential for *Agave* as a biofuel feedstock. *GCB Bioenergy* **2011**, *3*, 68–78. [CrossRef]

11. Pérez-Pimienta, J.A.; Mojica-Álvarez, R.M.; Sánchez-Herrera, L.M.; Mittal, A.; Sykes, R.W. Recalcitrance assessment of the agro−industrial residues from five *Agave* species: Ionic liquid pretreatment, saccharification and structural characterization. *BioEnergy Res.* **2018**, *11*, 551–561. [CrossRef]

12. Hill, J.; Nelson, E.; Tilman, D.; Polasky, S.; Tiffany, D. Environmental, economic, and energetic costs and benefits of biodiesel and ethanol biofuels. *Proc. Natl. Acad. Sci. USA* **2006**, *103*, 11206–11210. [CrossRef] [PubMed]

13. Sims, R.E.H.; Mabee, W.; Saddler, J.N.; Taylor, M. An overview of second generation biofuel technologies. *Bioresour. Technol.* **2010**, 1570–1580. [CrossRef] [PubMed]

14. Comparetti, A.; Febo, P.; Greco, C.; Mammano, M.M.; Orlando, S. Potential production of biogas from prinkly pear (*Opuntia ficus−indica* L.) in Sicilian uncultivated areas. *Chem. Eng. Trans.* **2017**, *58*, 559–564. [CrossRef]

15. Gilman, I.S.; Edwards, E.J. Crassulacean acid metabolism. *Curr. Biol.* **2019**, *30*, 51–63. [CrossRef]

16. López Collado, C.J.; Vázquez, A.M.; López-Collado, J.; García-Pérez, E.; Sánchez, Á.S. Crecimiento de *Opuntia ficus−indica* (L.) Mill. en la zona central de Veracruz. *Rev. Mex. Cienc. Agríc.* **2013**, *4*, 1005–1014. [CrossRef]

17. Garbanzo-León, G.; Chavarría-Pérez, G.; Vega-Villalobos, E.V. Correlaciones alométricas en *Hylocereus costaricensis* y H. monocanthus (pitahaya): Una herramienta para cuantificar el crecimiento. *Agron. Mesoam.* **2019**, *30*, 425–436. [CrossRef]

18. Loza-Cornejo, S.; Terrazas, T.; López-Mata, L.; Trejo, C. Características morfo−anatómicas y metabolismo fotosintético en plántulas de Stenocereus queretaroensis (Cactaceae): Su significado adaptativo. *Interciencia* **2003**, *28*, 83–89.

19. Zúñiga-Estrada, L.; Rosales Robles, E.; Yáñez-Morales, M.d.J.; Jacques-Hernández, C. Características y productividad de una planta MAC, *Agave tequilana* desarrollada con fertigación en Tamaulipas, México. *Rev. Mex. Cienc. Agríc.* **2018**, *9*, 553–564.

20. Maceda, A.; Soto-Hernández, M.; Peña-Valdivia, C.B.; Terrazas, T. Chemical composition of cacti wood and comparison with the wood of other taxonomic groups. *Chem. Biodivers.* **2018**, *15*, e1700574. [CrossRef]

21. Maceda, A.; Soto-Hernández, M.; Peña-Valdivia, C.B.; Trejo, C.; Terrazas, T. Characterization of lignocellulose of *Opuntia* (Cactaceae) species using FTIR spectroscopy: Possible candidates for renewable raw material. *Biomass Convers. Biorefin.* **2020**. [CrossRef]

22. Li, R.; Yang, G.; Chen, J.; He, M. The characterization of hemicellulose extract from corn stalk with stepwise alkali extraction. *Palpu Chongi Gisul/J. Korea Tech. Assoc. Pulp Pap. Ind.* **2017**, *49*, 29–40. [CrossRef]

23. Popescu, C.-M.; Popa, V.I. Analytical methods for lignin characterization. II. Spectroscopic studies. *Cellul. Chem. Technol.* **2006**, *40*, 597–621.

24. Pandey, K.K. Study of the effect of photo−irradiation on the surface chemistry of wood. *Polym. Degrad. Stab.* **2005**, *90*, 9–20. [CrossRef]

25. Nelson, M.L.; O'Connor, R.T. Relation of certain infrared bands to cellulose crystallinity and crystal latticed type. Part I. Spectra of lattice types I, II, III and of amorphous cellulose. *J. Appl. Polym. Sci.* **1964**, *8*, 1311–1324. [CrossRef]

26. Ciolacu, D.; Ciolacu, F.; Popa, V.I. Amorphous cellulose−structure and characterization. *Cellul. Chem. Technol.* **2011**, *45*, 13–21.

27. Poletto, M.; Ornaghi, H.L.; Zattera, A.J. Native cellulose: Structure, characterization and thermal properties. *Materials* **2014**, *7*, 6105–6119. [CrossRef]

28. Cichosz, S.; Masek, A. IR study on cellulose with the varied moisture contents: Insight into the supramolecular structure. *Materials* **2020**, *13*, 4573. [CrossRef]

29. Arias, S.; Terrazas, T. Variación en la anatomía de la madera de *Pachycereus pecten−aboriginum* (Cactaceae). *An. Inst. Biol. Univ. Nac. Autón. Méx.* **2001**, *72*, 157–169.

30. Nakaba, S.; Kitin, P.; Yamagishi, Y.; Begum, S.; Kudo, K.; Nugroho, W.D.; Funada, R. Three−dimensional imaging of cambium and secondary xylem cells by confocal laser scanning microscopy. *Plant Microtech. Protoc.* **2015**, 431–465. [CrossRef]

31. Taban, E.; Mirzaei, R.; Faridan, M.; Samaei, E.; Salimi, F.; Tajpoor, A.; Ghalenoei, M. Morphological, acoustical, mechanical and thermal properties of sustainable green *Yucca* (*Y. gloriosa*) fibers: An exploratory investigation. *J. Environ. Health Sci. Eng.* **2020**, *18*, 883–896. [CrossRef]

32. Nataren-Velazquez, J.; Del Angel-Pérez, A.L.; Megchún-García, J.V.; Ramírez-Herrera, E.; Ibarra-Pérez, F. Colecta y caracterización morfologica de izote (*Yucca elephantipes*) y cruceta (*Acanthocereus tetragonus*), del estado de Veracruz. In *Prospectiva, de la Investigación Agrícola en el Siglo XXI en México*; Avendaño Ruiz, B.D., Bautista Ortega, J., Del Angel-Pérez, A.L., Ireta Paredes, A.d.R., Martinez-Trejo, G., Pérez Hernández, P., Schwentesius Ridermann, R., Eds.; Universidad Autónoma de Chapingo and Plaza y Valdés, S.L.: Texcoco, México, 2020; pp. 137–149.

33. Mora-Olivo, A.; Hurtado-González, M.; Gaona-García, G.; Treviño-Carreón, J. Las flores comestibles del desierto. *CienciaUAT* **2009**, *4*, 10–13.

34. Granados-Sánchez, D.; Sánchez-González, A.; Granados-Victorino, R.L.; de la Rosa, A.B. Ecología de la vegetación del desierto chihuahuense. *Rev. Chapingo Ser. Cienc. For. Y Ambient.* **2011**, *18*, 111–130. [CrossRef]

35. Hernández-Moreno, M.M.; Téllez-Valdés, O.; Martínez-Meyer, E.; Islas-Saldaña, L.A.; Salazar-Rojas, V.M.; Macías-Cuéllar, H. Distribución de la cobertura vegetal y del uso del terreno del municipio de Zapotitlán, Puebla, México. *Rev. Mex. Biodivers.* **2021**, *92*, e923649. [CrossRef]

36. Moura, J.C.M.S.; Bonine, C.A.V.; de Oliveira Fernandes Viana, J.; Dornelas, M.C.; Mazzafera, P. Abiotic and biotic stresses and changes in the lignin content and composition in plants. *J. Integr. Plant Biol.* **2010**, *52*, 360–376. [CrossRef] [PubMed]

37. Lipiec, J.; Doussan, C.; Nosalewicz, A.; Kondracka, K. Effect of drought and heat stresses on plant growth and yield: A review. *Int. Agrophys.* **2013**, *27*, 463–477. [CrossRef]

38. Reyes-Rivera, J.; Canché-Escamilla, G.; Soto-Hernández, M.; Terrazas, T. Wood chemical composition in species of Cactaceae the relationship between lignification and stem morphology. *PLoS ONE* **2015**, *10*, e0123919. [CrossRef]

39. Mylsamy, K.; Rajendran, I. Investigation on physio-chemical and mechanical properties of raw and alkali-treated *Agave americana* fiber. *J. Reinf. Plast. Compos.* **2010**, *29*, 2925–2935. [CrossRef]

40. Krishnadev, P.; Subramanian, K.S.; Janavi, G.J.; Ganapathy, S.; Lakshmanan, A. Synthesis and characterization of nano-fibrillated cellulose derived from green *Agave americana* L. Fiber. *BioResources* **2020**, *15*, 2442–2458. [CrossRef]

41. Rosli, N.A.; Ahmad, I.; Abdullah, I. Isolation and characterization of cellulose nanocrystals from *Agave angustifolia* fibre. *BioResources* **2013**, *8*, 1893–1908. [CrossRef]

42. Teli, M.D.; Jadhav, A.C. Effect of alkali treatment on the properties of *Agave augustifolia* v. *marginata* fibre. *Int. Res. J. Eng. Technol.* **2016**, *3*, 2754–2761.

43. Carmona, J.E.; Morales-Martínez, T.K.; Mussatto, S.I.; Castillo-Quiroz, D.; Ríos-González, L.J. Chemical, structural and functional properties of lechuguilla (*Agave lechuguilla* Torr.). *Rev. Mex. Cienc. For.* **2017**, *8*, 100–122.

44. Ortíz-Méndez, O.H.; Morales-Martínez, T.K.; Rios-González, L.J.; Rodríguez-De La Garza, J.A.; Quintero, J.; Aroca, G. Bioethanol production from *Agave lechuguilla* biomass pretreated by autohydrolysis. *Rev. Mex. Ing. Quím.* **2017**, *16*, 467–476.

45. Vieira, M.C.; Heinze, T.; Antonio-Cruz, R.; Mendoza-Martinez, A.M. Cellulose derivatives from cellulosic material isolated from *Agave lechuguilla* and *A. fourcroydes*. *Cellulose* **2002**, *9*, 203–212. [CrossRef]

46. De Dios Naranjo, C.; Alamilla-Beltrán, L.; Gutiérrez-Lopez, G.F.; Terres-Rojas, E.; Solorza-Feria, J.; Romero-Vargas, S.; Yee-Madeira, H.T.; Areli, F.-M.; Mora-Escobedo, R. Aislamiento y caracterización de celulosas obtenidas de fibras de *Agave salmiana* aplicando dos métodos de extracción ácido-alcali. *Rev. Mex. Cienc. Agríc.* **2016**, *7*, 31–43.

47. Bernardo, G.R.R.; Rene, R.M.J. Contribution of agro-waste material main components (hemicelluloses, cellulose, and lignin) to the removal of chromium (III) from aqueous solution. *J. Chem. Technol. Biotechnol.* **2009**, *84*, 1533–1538. [CrossRef]

48. Li, H.; Foston, M.B.; Kumar, R.; Samuel, R.; Gao, X.; Hu, F.; Ragauskas Cd, A.J.; Wyman, C.E. Chemical composition and characterization of cellulose for Agave as a fast-growing, drought-tolerant biofuels feedstock. *RSC Adv.* **2012**, *2*, 4951–4958. [CrossRef]

49. Li, H.; Pattathil, S.; Foston, M.B.; Ding, S.Y.; Kumar, R.; Gao, X.; Mittal, A.; Yarbrough, J.M.; Himmel, M.E.; Ragauskas, A.J.; et al. *Agave* proves to be a low recalcitrant lignocellulosic feedstock for biofuels production on semi-arid lands. *Biotechnol. Biofuels* **2014**, *7*, 50. [CrossRef]

50. McDougall, G.J.; Morrison, I.M.; Stewart, D.; Weyers, J.D.B.; Hillman, J.R. Plant fibres: Botany, chemistry and processing for industrial use. *J. Sci. Food Agric.* **1993**, *62*, 1–20. [CrossRef]

51. Kestur G., S.; Flores-Sahagun, T.H.S.; Dos Santos, L.P.; Dos Santos, J.; Mazzaro, I.; Mikowski, A. Characterization of blue agave bagasse fibers of Mexico. *Compos. Part A Appl. Sci. Manuf.* **2013**, *45*, 153–161. [CrossRef]

52. Iñiguez, G.; Acosta, N.; Martínez, L.; Parra, J.; González, O. Utilización de supbroductos de la industria tequilera. Parte 7. Compostaje de bagazo de agave y vinazas tequileras. *Rev. Int. Contam. Ambient.* **2005**, *21*, 37–50.

53. Robles, E.; Fernández-Rodríguez, J.; Barbosa, A.M.; Gordobil, O.; Carreño, N.L.V.; Labidi, J. Production of cellulose nanoparticles from blue agave waste treated with environmentally friendly processes. *Carbohydr. Polym.* **2018**, *183*, 294–302. [CrossRef] [PubMed]

54. Hernández, J.A.; Romero, V.H.; Escalante, A.; Toriz, G.; Rojas, O.J.; Sulbarán, B.C. *Agave tequilana* bagasse as source of cellulose nanocrystals via organosolv treatment. *BioResources* **2018**, *13*, 3603–3614. [CrossRef]

55. Robles-García, M.Á.; Del-Toro-Sánchez, C.L.; Márquez-Ríos, E.; Barrera-Rodríguez, A.; Aguilar, J.; Aguilar, J.A.; Reynoso-Marín, F.J.; Ceja, I.; Dórame-Miranda, R.; Rodríguez-Félix, F. Nanofibers of cellulose bagasse from *Agave tequilana* Weber var. *azul* by electrospinning: Preparation and characterization. *Carbohydr. Polym.* **2018**, *192*, 69–74. [CrossRef] [PubMed]

56. Ramírez-Cortina, C.; Alonso-Gutiérrez, M.S.; Rigal, L. Tratamiento alcalino de los residuos agroindustriales de la producción del tequila, para su uso como complemento de alimento de rumiantes. *Rev. AIDIS Ing. Y Cienc. Ambient.* **2012**, *5*, 69–77.

57. Espino, E.; Cakir, M.; Domenek, S.; Román-Gutiérrez, A.D.; Belgacem, N.; Bras, J. Isolation and characterization of cellulose nanocrystals from industrial by-products of *Agave tequilana* and barley. *Ind. Crop. Prod.* **2014**, *62*, 552–559. [CrossRef]

58. Palacios Hinestroza, H.; Hernández Diaz, J.A.; Esquivel Alfaro, M.; Toriz, G.; Rojas, O.J.; Sulbarán-Rangel, B.C. Isolation and characterization of nanofibrillar cellulose from *Agave tequilana* Weber bagasse. *Adv. Mater. Sci. Eng.* **2019**, *2019*, 1342547. [CrossRef]

59. Rijal, D.; Walsh, K.B.; Subedi, P.P.; Ashwath, N. Quality estimation of *Agave tequilana* leaf for bioethanol production. *J. Near Infrared Spectrosc.* **2017**, *24*, 453–465. [CrossRef]

60. Márquez, A.; Cazaurang, N.; González, I.; Colunga-GarciaMarin, P. Extraction of chemical cellulose from the fibers of *Agave lechuguilla* Torr. *Econ. Bot.* **1996**, *50*, 465–468. [CrossRef]

61. Ponce-Reyes, C.E.; Chanona-Pérez, J.J.; Garibay-Febles, V.; Palacios-González, E.; Karamath, J.; Terrés-Rojas, E.; Calderón-Domínguez, G. Preparation of cellulose nanoparticles from agave waste and its morphological and structural characterization. *Rev. Mex. Ing. Quím.* **2014**, *13*, 897–906.

62. Manimaran, P.; Senthamaraikannan, P.; Sanjay, M.R.; Marichelvam, M.K.; Jawaid, M. Study on characterization of *Furcraea foetida* new natural fiber as composite reinforcement for lightweight applications. *Carbohydr. Polym.* **2018**, *181*, 650–658. [CrossRef] [PubMed]

63. do Nascimento, H.M.; dos Santos, A.; Duarte, V.A.; Bittencourt, P.R.S.; Radovanovic, E.; Fávaro, S.L. Characterization of natural cellulosic fibers from *Yucca aloifolia* L. leaf as potential reinforcement of polymer composites. *Cellulose* **2021**, *28*, 5477–5492. [CrossRef]

64. Razali, N.A.M.; Sohaimi, R.M.; Othman, R.N.I.R.; Abdullah, N.; Demon, S.Z.N.; Jasmani, L.; Yunus, W.M.Z.W.; Ya'acob, W.M.H.W.; Salleh, E.M.; Norizan, M.N.; et al. Comparative study on extraction of cellulose fiber from rice straw waste from chemo-mechanical and pulping method. *Polymers* **2022**, *14*, 387. [CrossRef] [PubMed]

65. Álvarez-Chávez, J.; Villamiel, M.; Santos-Zea, L.; Ramírez-Jiménez, A.K. *Agave* by-products: An overview of their nutraceutical value, current applications, and processing methods. *Polysaccharides* **2021**, *2*, 720–743. [CrossRef]

66. Palomo-Briones, R.; López-Gutiérrez, I.; Islas-Lugo, F.; Galindo-Hernández, K.L.; Munguia-Aguilar, D.; Rincón Pérez, J.A.; Cortés-Carmona, M.A.; Alatriste-Mondragón, F.; Razo-Flores, E. *Agave* bagasse biorefinery: Processing and perspectives. *Clean Technol. Environ. Policy* **2018**, *20*, 1423–1441. [CrossRef]

67. Fuller, M.E.; Andaya, C.; McClay, K. Evaluation of ATR-FTIR for analysis of bacterial cellulose impurities. *J. Microbiol. Methods* **2018**, *144*, 145–151. [CrossRef]

68. Zaltariov, M.-F. FTIR investigation on crystallinity of hydroxypropyl methyl cellulose-based polymeric blends. *Cellul. Chem. Technol. Cellul. Chem. Technol.* **2021**, *55*, 981–988. [CrossRef]

69. Kruer-Zerhusen, N.; Cantero-Tubilla, B.; Wilson, D.B. Characterization of cellulose crystallinity after enzymatic treatment using Fourier transform infrared spectroscopy (FTIR). *Cellulose* **2018**, *25*, 37–48. [CrossRef]

70. Chen, C.J.; Luo, J.J.; Huang, X.P.; Zhao, S.K. Analysis on cellulose crystalline and FTIR spectra of artocarpus heterophyllus Lam wood and its main chemical compositions. *Adv. Mater. Res.* **2011**, *236–238*, 369–375. [CrossRef]

71. Lionetto, F.; Del Sole, R.; Cannoletta, D.; Vasapollo, G.; Maffezzoli, A. Monitoring wood degradation during weathering by cellulose crystallinity. *Materials* **2012**, *5*, 1910–1922. [CrossRef]

72. Poletto, M.; Pistor, V.; Santana, R.M.C.; Zattera, A.J. Materials produced from plant biomass: Part II: Evaluation of crystallinity and degradation kinetics of cellulose. *Mater. Res.* **2012**, *15*, 421–427. [CrossRef]

73. Široký, J.; Blackburn, R.S.; Bechtold, T.; Taylor, J.; White, P. Attenuated total reflectance Fourier-transform Infrared spectroscopy analysis of crystallinity changes in lyocell following continuous treatment with sodium hydroxide. *Cellulose* **2010**, *17*, 103–115. [CrossRef]

74. Hofmann, D.; Fink, H.P.; Philipp, B. Lateral crystallite size and lattice distortions in cellulose II samples of different origin. *Polymer* **1989**, *30*, 237–241. [CrossRef]

75. Kljun, A.; Benians, T.A.S.; Goubet, F.; Meulewaeter, F.; Knox, J.P.; Blackburn, R.S. Comparative analysis of crystallinity changes in cellulose i polymers using ATR-FTIR, X-ray diffraction, and carbohydrate-binding module probes. *Biomacromolecules* **2011**, *12*, 4121–4126. [CrossRef]

76. Colom, X.; Carrillo, F. Crystallinity changes in lyocell and viscose-type fibres by caustic treatment. *Eur. Polym. J.* **2002**, *38*, 2225–2230. [CrossRef]

77. Kamali Moghaddam, M.; Torabi, T. Cellulose microfibers isolated from *Yucca* leaves: Structural, chemical, and thermal properties. *J. Nat. Fibers* **2022**. [CrossRef]

78. Beluns, S.; Gaidukovs, S.; Platnieks, O.; Gaidukova, G.; Mierina, I.; Grase, L.; Starkova, O.; Brazdausks, P.; Thakur, V.K. From wood and hemp biomass wastes to sustainable nanocellulose foams. *Ind. Crop. Prod.* **2021**, *170*, 113780. [CrossRef]

79. Moghaddam, M.K.; Karimi, E. Structural and physical characteristics of the *yucca* fiber. *J. Ind. Text.* **2020**, *51*, 8018S–8034S. [CrossRef]

80. Alves, A.; Simoes, R.; Stackpole, D.J.; Vaillancourt, R.E.; Potts, B.M.; Schwanninger, M.; Rodrigues, J. Determination of the syringyl/guaiacyl ratio of *Eucalyptus globulus* wood lignin by near infrared-based partial least squares regression models using analytical pyrolysis as the reference method. *J. Near Infrared Spectrosc.* **2011**, *19*, 343–348. [CrossRef]

81. Wang, C.; Li, H.; Li, M.; Bian, J.; Sun, R. Revealing the structure and distribution changes of *Eucalyptus* lignin during the hydrothermal and alkaline pretreatments. *Sci. Rep.* **2017**, *7*, 593. [CrossRef]

82. Li, M.; Pu, Y.; Ragauskas, A.J. Current understanding of the correlation of lignin structure with biomass recalcitrance. *Front. Chem.* **2016**, *4*, 45. [CrossRef] [PubMed]

83. Ralph, J.; Landucci, L.L. NMR of Lignins. In *Lignin and Lignans; Advances in Chemistry*; Heitner, C., Dimmel, D.R., Schmidt, J.A., Eds.; Taylor & Francis Group: Boca Raton, FL, USA, 2010; pp. 137–234.

84. Morán-Velázquez, D.C.; Monribot-Villanueva, J.L.; Bourdon, M.; Tang, J.Z.; López-Rosas, I.; Maceda-López, L.F.; Villalpando-Aguilar, J.L.; Rodríguez-López, L.; Gauthier, A.; Trejo, L.; et al. Unravelling chemical composition of agave spines: News from *Agave fourcroydes* Lem. *Plants* **2020**, *9*, 1642. [CrossRef] [PubMed]

85. Del Río, J.C.; Rencoret, J.; Marques, G.; Gutiérrez, A.; Ibarra, D.; Santos, J.I.; Jiménez-Barbero, J.; Zhang, L.; Martínez, Á.T. Highly Acylated (acetylated and/or p−coumaroylated) native lignins from diverse herbaceous plants. *J. Agric. Food Chem.* **2008**, *56*, 9525–9534. [CrossRef] [PubMed]

86. Del Río, J.C.; Prinsen, P.; Cadena, E.M.; Ngel, A.; Martínez, T.; Gutiérrez, A.; Rencoret, J. Lignin−carbohydrate complexes from sisal (*Agave sisalana*) and abaca (*Musa textilis*): Chemical composition and structural modifications during the isolation process. *Planta* **2016**, *243*, 1143–1158. [CrossRef]

87. Perez-Pimienta, J.A.; Flores-Gómez, C.A.; Ruiz, H.A.; Sathitsuksanoh, N.; Balan, V.; da Costa Sousa, L.; Dale, B.E.; Singh, S.; Simmons, B.A. Evaluation of *Agave* bagasse recalcitrance using AFEX^{TM}, autohydrolysis, and ionic liquid pretreatments. *Bioresour. Technol.* **2016**, *211*, 216–223. [CrossRef]

88. Rencoret, J.; Marques, G.; Gutiérrez, A.; Ibarra, D.; Li, J.; Gellerstedt, G.; Santos, J.I.; Jiménez-Barbero, J.; Martínez, Á.T.; Del Río, J.C. Structural characterization of milled wood lignins from different eucalypt species. *Holzforschung* **2008**, *62*, 514–526. [CrossRef]

89. Lee, Y.; Voit, E.O. Mathematical modeling of monolignol biosynthesis in Populus xylem. *Math. Biosci.* **2010**, *228*, 78–89. [CrossRef]

90. Barros, J.; Serk, H.; Granlund, I.; Pesquet, E. The cell biology of lignification in higher plants. *Ann. Bot.* **2015**, *115*, 1053–1074. [CrossRef]

91. Li, L.; Cheng, X.F.; Leshkevich, J.; Umezawa, T.; Harding, S.A.; Chiang, V.L. The last step of syringyl monolignol biosynthesis in angiosperms is regulated by a novel gene encoding sinapyl alcohol dehydrogenase. *Plant Cell* **2001**, *13*, 1567–1585. [CrossRef]

92. Reyes-Rivera, J.; Soto-Hernández, M.; Canché-Escamilla, G.; Terrazas, T. Structural characterization of lignin in four cacti wood: Implications of lignification in the growth form and succulence. *Front. Plant Sci.* **2018**, *871*, 1518. [CrossRef]

93. Maceda, A.; Reyes-Rivera, J.; Soto-Hernández, M.; Terrazas, T. Distribution and chemical composition of lignin in secondary xylem of Cactaceae. *Chem. Biodivers.* **2021**, *18*, e2100431. [CrossRef] [PubMed]

94. Weng, J.-K.; Banks, J.A.; Chapple, C. Parallels in lignin biosynthesis: A study in *Selaginella moellendorffii* reveals convergence across 400 million years of evolution. *Commun. Integr. Biol.* **2008**, *1*, 20–22. [CrossRef] [PubMed]

95. Weng, J.-K.; Li, X.; Stout, J.; Chapple, C. Independent origins of syringyl lignin in vascular plants. *Proc. Natl. Acad. Sci. USA* **2008**, *105*, 7887–7892. [CrossRef] [PubMed]

96. Lange, B.M.; Lapierre, C.; Sandermann, H. Elicitor-induced spruce stress lignin: Structural similarity to early developmental lignins. *Plant Physiol.* **1995**, *108*, 1277–1287. [CrossRef] [PubMed]

97. Menden, B.; Kohlhoff, M.; Moerschbacher, B.M. Wheat cells accumulate a syringyl−rich lignin during the hypersensitive resistance response. *Phytochemistry* **2007**, *68*, 513–520. [CrossRef] [PubMed]

98. Skyba, O.; Douglas, C.J.; Mansfield, S.D. Syringyl−Rich lignin renders poplars more resistant to degradation by wood decay fungi. *Appl. Environ. Microbiol.* **2013**, *79*, 2560–2571. [CrossRef]

99. Li, Y.; Mai, Y.W.; Ye, L. Sisal fibre and its composites: A review of recent developments. *Compos. Sci. Technol.* **2000**, *60*, 2037–2055. [CrossRef]

100. De Micco, V.; Aronne, G. Combined histochemistry and autofluorescence for identifying lignin distribution in cell walls. *Biotech. Histochem.* **2007**, *82*, 209–216. [CrossRef]

101. Pereira, L.; Domingues-Junior, A.P.; Jansen, S.; Choat, B.; Mazzafera, P. Is embolism resistance in plant xylem associated with quantity and characteristics of lignin? *Trees Struct. Funct.* **2018**, *32*, 349–358. [CrossRef]

102. Martone, P.T.; Estevez, J.M.; Lu, F.; Ruel, K.; Denny, M.W.; Somerville, C.; Ralph, J. Discovery of lignin in seaweed reveals convergent evolution of cell−wall architecture. *Curr. Biol.* **2009**, *19*, 169–175. [CrossRef]

forests

MDPI

Article

Branchwood Properties of Two *Tilia* Species Collected from Natural Secondary Forests in Northeastern China

Pingping Guo [1], Xiping Zhao [1,*], Qi Feng [1] and Yongqiang Yang [2]

[1] College of Horticulture and Plant Protection, Henan University of Science and Technology, Luoyang 471023, China
[2] Pingjiang County Forestry Bureau, Yueyang 414599, China
* Correspondence: zhaoxiping1977@126.com; Tel.: +86-1378-314-8985

Abstract: *Tilia amurensis* Rupr. and *Tilia mandshurica* Rupr. and Maxim. are two essential commercial species, though there is surprisingly little concern about whether their branches can be used in the current situation of a wood shortage in China. In this study, tissue proportions and fiber morphology, physical and mechanical properties, and chemical composition of the branchwood were studied and compared with stemwood to evaluate the potential for papermaking. The branchwood and stemwood showed similar cell arrangement but different tissue proportions and fiber morphology. The branchwood had more than 40% fiber proportion, 90%–97% below 0.9 mm in length, 75%–90% less than 33 in slenderness ratio, and 80% less than 1 in Runkel ratio. The branchwood was as light and soft as stemwood with a density of 0.32–0.36 g/cm^3 and a compressive strength of about 30 MPa. The branchwood had 6% water extractives, 66% holocellulose, and 22% lignin for *T. amurensis*, 58% holocellulose and 30% lignin for *T. mandshurica*. The results suggest the branchwood is favorable for mechanical chipping, has the potential to obtain high pulp yield and its fibers can be mixed with wide, long and thick fibers from other tree species to produce specific paper products. In contrast, *T. mandshurica* branchwood may not be suitable for chemical pulping.

Keywords: *Tilia amurensis* Rupr.; *Tilia mandshurica* Rupr. & Maxim.; branch; wood properties; papermaking material

Citation: Guo, P.; Zhao, X.; Feng, Q.; Yang, Y. Branchwood Properties of Two *Tilia* Species Collected from Natural Secondary Forests in Northeastern China. *Forests* **2023**, *14*, 760 . https://doi.org/10.3390/f14040760

Academic Editors: Vicelina Sousa, Helena Pereira, Teresa Quilhó and Isabel Miranda

Received: 17 February 2023
Revised: 31 March 2023
Accepted: 1 April 2023
Published: 7 April 2023

1. Introduction

Branches are an essential part of a tree, the proportion of branchwood to the volume of the whole tree is about 20%, which varies with species, tree height, and stand condition [1]. Unfortunately, many branches are discarded when harvesting operations concentrate on stems. Using branchwood for pulping and papermaking can be an additional measure to ease the conflict between timber supply and demand [2,3]. Branchwood has been used in pulp and paper making in China since the mid-1960s [4]. With improvements in harvesting methods [5] and technological advances in wood pulping [6], the prospects for increased use of branchwood for papermaking materials are promising.

The qualities of pulp and paper products highly depend on the properties of raw materials, e.g., wood density, fiber dimensions, and chemical composition. Higher pulp yield is consistent with higher cellulose and hemicellulose content and lower lignin and extractive content [7,8]. Species with low wood density are indicated to produce printing and writing sheets, while high-density wood is favorable to make tissue paper [9]. Long fibers are beneficial for improving the fracture toughness of paper, and short fibers give superior tear resistance [10,11]. However, wood properties closely related to pulping and papermaking are lacking for branch, especially for broadleaves grown in China. Branches and stems of the same tree may differ in wood properties. For example, branchwood has a smaller cell size than stemwood, because some types of cells are more or less abundant in branchwood than stemwood [2,12–14]. These differences are important factors affecting the utilization of wood as a papermaking material [15–17].

Tilia amurensis Rupr. and *Tilia mandshurica* Rupr Maxim., two native linden trees in the temperate zone of the northern hemisphere, are not easily visually distinguishable from each other [18,19]. They are of commercial importance for honey production and as essential timber trees in China [20]. The wood of the two linden trees is diffuse-porous in its gross structure and exhibits well-differentiated growth rings, but there are no color differences between heartwood and sapwood [21,22]. Linden wood has a minor use in construction and is commercialized mainly for the manufacture of furniture, craft items, and plywood [23–25]. A survey shows that most pencil boards are made of linden wood in China [26].

The work described in this paper will examine tissue proportions and fiber morphology, physical and mechanical properties, and chemical composition of the branchwood of the two *Tilia* species. We expect that the branches will have similar wood properties to the stem and could meet the basic requirements of pulping and papermaking.

2. Materials and Methods

2.1. Materials

For each species, three healthy straight trees were randomly selected from the Mao-ershan Forest Ecosystem Research Station in Heilongjiang Province, northeastern China (127°30′–34′ E, 45°20′–25′ N, 400 m elevation). The site has a continental monsoon climate with a warm summer, cold, dry winter, and temperate species-rich deciduous broadleaf forests [27]. A live branch was sampled from each tree's upper, middle, and lower canopy. The characteristics of the trees and branches sampled were measured (Table 1). A segment (about 1 m long) was cut from each sample branch. It should be stressed that efforts to ensure sustainable management in the station are crucial; only one 5 mm increment core (from pith to bark) was collected at 1.3 m stem height (b.h., breast height) as non-destructively as possible to the stem.

Table 1. Characteristics of the trees and branches sampled [1].

Species	Tree Age (yr)	Tree Height (m)	Tree Diameter at Breast Height (cm)	Branch Length (m)	Branch Diameter (cm)
T. amurensis	60 ± 11.1	21.7 ± 0.6	30.8 ± 5.5	4.4 ± 0.4	8.3 ± 0.7
T. mandshurica	63 ± 5.0	20.3 ± 1.4	31.8 ± 2.5	4.5 ± 0.2	9.1 ± 1.5

[1] The data is represented as Means ± SD.

2.2. Xylem Anatomy

A strip (located in the middle of the branches) with 20 (radial) × 10 (tangential) × 10 (longitudinal) mm and a chip with 20 (radial) × 2 (tangential) × 10 (longitudinal) mm were cut from the lateral wood (mid-way between the tension wood and the opposite wood) of each branch segment. Two 20 mm long segments (including 3–5 growth rings) were cut from the middle of each tree core, one for slicing and the other for macerating. A 15-μm-thick transverse section was cut from each strip and core segment (embedded in paraffin blocks) using a rotary microtome (RM 2235, Leica Microsystems, Wetzlar, Germany) equipped with a knife holder, stained with safranin, and then fixed on microscopic slides [28]. The wood chip and core segment were macerated using the chromic acid-nitric acid method described by Jeffrey [29]. The macerated material was rinsed and placed on microscopic slides. All microscopic slides were photographed using a digital light microscopy (Mshot-MD50; Microshot Technology Limited, Guangzhou, China) to measure tissue proportions with an image analysis system (TDY5.2; Beijing Tian Di Yu Technology Co., Ltd., Beijing, China) [30]. Fiber dimensions were measured manually. At least 60 measurements were done per parameter. Two derived indices were calculated using fiber dimension: slenderness ratio as fiber length/fiber diameter and Runkel ratio as two times fiber cell wall thickness/lumen diameter [31].

2.3. Physical and Mechanical Properties

Five defect-free cubic test pieces with 20 (radial) × 20 (tangential) × 20 (longitudinal) mm and prismatic test pieces with 20 (radial) × 20 (tangential) × 30 (longitudinal) mm were cut from each branch segment to test basic wood density (hereafter referred to as "wood density") and compressive strength parallel to grain (hereafter referred to as "compressive strength") by the Chinese Standard GB/T1933-2009 [32] and GB/T 1935–2009 [33], respectively. The green volume and absolute weight of each cubic test piece were measured using the immersion and electronic balance weighing methods, respectively. Wood density was calculated based on weight and volume values. The prismatic test pieces were adjusted to a moisture level of 12%, and then the compressive strength was tested using a universal testing machine (WDW-50B, Jinan Shengfeng Testing Machine Co., Ltd., Jinan, China) with a loading capacity of 50 kN and a testing speed of 25,000 ± 5000 N/min.

2.4. Chemical Composition

After mechanical tests, all damaged columns for each branch were ground to 40 to 60 mesh wood powder.

The extractives were determined according to the Chinese Standard GB/T 35816–2018 [34]. Cold water-soluble extraction was carried out at room temperature (23 ± 2 °C) for 48 h. Hot-water-soluble and 1% sodium hydroxide (NaOH) extraction were carried out by boiling water bath for 3 and 1 h, respectively. The extractive content of the wood powder was calculated by weight lost when extracted.

The holocellulose, cellulose, and lignin were determined by glacial acetic acid-sodium chlorite, nitric acid-ethanol, and sulphuric acid hydrolysis methods, respectively, according to the Chinese Standard GB/T 35818–2018 [35]. The contents were calculated based on the final residue. The hemicellulose content was calculated as the subtraction of cellulose content from holocellulose content.

2.5. Data Analysis

Data processing and analysis were performed using IBM SPSS Statistics software (Version 24.0, International Business Machines Corporation, Armonk, NY, USA). Interspecific and differences between branch and stem, and the two species in wood properties were evaluated by variance analyses with alpha significance less than 0.05 and 0.01. The fitting curves of the fiber length and derived parameter distributions were developed using a normal distribution function, and skewness and kurtosis were used to check the normality of the data sets.

3. Results and Discussion

3.1. Wood Tissue Proportions

The stemwood of the *T. amurensis* and *T. mandshurica* was diffuse-porous, with vessels arranged as clusters, multiples, or solitary (Figure 1a,b). Growth rings were distinct. There were many, but relatively narrow, rays. Axial parenchyma was sparce, axial parenchyma apotracheal short tangential lines, (terminal) and in marginal or in seemingly marginal lines, as reported in previous studies [22], *T. mandshurica* showed a more frequent diagonal pattern in vessels arrangement than the *T. amurensis*. Compared to stemwood, branchwood showed diffuse to echelon arrangement in vessel groups; latewood vessels were about half the diameter of earlywood vessels (Figure 1c,d).

The mean fiber proportion of stemwood was over 50% and above the value reported by Fang et al. [36]. Fang et al. also measured the proportions of vessels (26.6%), rays (13.6%) and axial parenchyma (13%) that sightly differed from the present results (Table 2). In the study by Fang et al., the linden trees were sampled from southwest China, which is about 3300 km from the sampling site (northeast China). Northeast China is a sub-humid region in the cold temperate zone, the mean annual precipitation is 629 mm, and January and July air temperature are −18 and 22 °C, respectively [27]. Southwest is a humid region in the subtropical zone, the mean annual precipitation is 800–1600 mm, and January and July air temperature are 5–12 and 20–30 °C, respectively [37]. Thus, variation in provenance may be the main reason for the difference [38]. Interspecific differences in tissue proportions were significant for stemwood ($p < 0.05$, except ray proportion) but not for branchwood. These differences in tissue proportions can lead to differences in porosity, shrinkage, and treatment capacity.

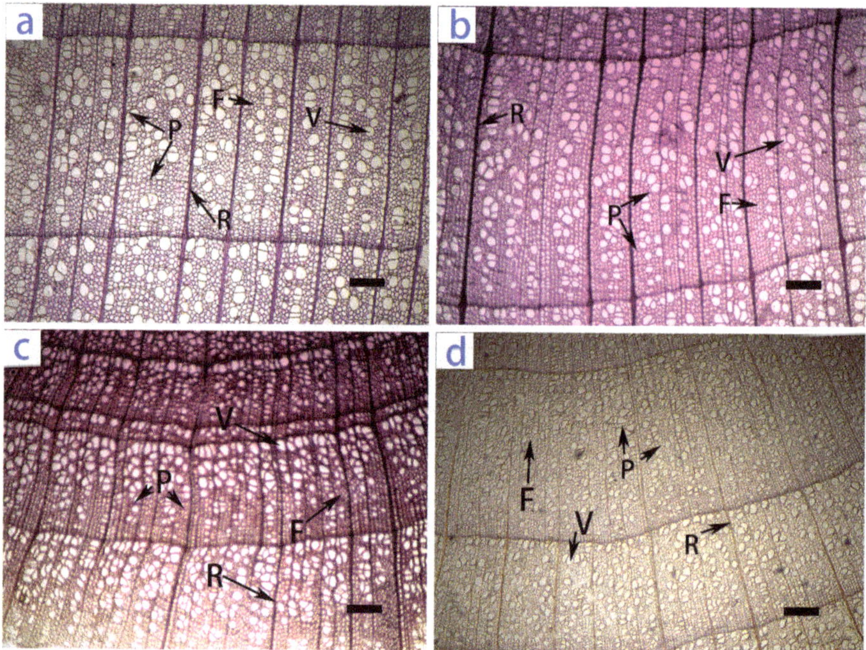

Figure 1. Cross-sections of the stemwood of *T. amurensis* (**a**) and *T. mandshurica* (**b**), and of the branchwood of *T. amurensis* (**c**) and *T. mandshurica* (**d**): axial parenchyma (P), ray parenchyma (R), vessels (V) and fibers (F). The scale bar represents 200 μm.

Table 2. Comparison of tissue proportions in the branchwood and stemwood from two *Tilia* species [1].

Tissue	Species	Stemwood	Branchwood	
Fiber (%)	*T. amurensis*	60.35 ± 0.73	**42.50 ± 4.75** [2]	*
	T. mandshurica	53.77 ± 0.80	**50.65 ± 1.71**	*
Vessel (%)	*T. amurensis*	17.60 ± 0.78	**30.01 ± 2.37**	**
	T. mandshurica	25.38 ± 0.96	**30.64 ± 1.73**	*
Ray (%)	*T. amurensis*	13.12 ± 0.53	19.75 ± 1.90	**
	T. mandshurica	12.04 ± 0.47	17.00 ± 1.54	**
Axial parenchyma (%)	*T. amurensis*	9.20 ± 0.14	**7.93 ± 0.61**	*
	T. mandshurica	11.95 ± 0.29	**9.24 ± 0.99**	*

[1] The data is represented as Means ± SE. [2] Data in bold indicate significant differences between species at $p = 0.01$. * and ** indicate significant differences between branchwood and stemwood at $p = 0.05$ and 0.01, respectively.

The proportion of fiber in branchwood was lower than the stemwood, but it was more than 40%, indicating the potential of branchwood to obtain a high pulp yield [39]. Specified proportions of parenchyma are good for press-dried paper, increasing the bond strength in the paper [40]. However, redundant vessel elements in the sheets reduce mechanical properties and cause linting problems [16].

3.2. Fiber Morphology

Interspecific differences in fiber dimensions were significant ($p < 0.01$) for the stemwood of the two species (Table 3). The average fiber length for stemwood was similar (0.7–1.6 mm) to most of the hardwood species [41], and the mean length for *T. amurensis* stemwood was similar to the medium-length fibers (0.9–1.6 mm) according to the IAWA [42]. The stemwood fibers were wide (average 31–41 µm) and similar to *Paulownia fortune* and *Alniphyllum fortunei* [43], and were up to a particular wide grade (>30 µm) proposed by Cheng et al. [44]. The average lumen diameter and double wall thickness of the fibers were about 20 µm for *T. amurensis* stemwood, and 15 µm for *T. mandshurica*. The average fiber lumen diameter of the linden stemwood was similar to *Betulaplaty phylla* [45] and *Alnus sibirica* [31], but the fiber wall was too thick. For example, a thicker cell wall would make the fiber more flexible but lead to a massive void in the paper produced in *Prunus domestica* [46].

Table 3. Comparison of fiber dimensions and their derived indices of branchwood and stemwood from two *Tilia* species [1].

	Species	**Stemwood**	**Branchwood**	
Fiber length (mm)	*T. amurensis*	**0.96 ± 0.20** [2]	**0.59 ± 0.16**	**
	T. mandshurica	**0.83 ± 0.22**	**0.73 ± 0.17**	**
Fiber width (µm)	*T. amurensis*	**40.58 ± 0.43**	**23.95 ± 0.33**	**
	T. mandshurica	**31.02 ± 0.38**	**25.65 ± 0.45**	**
Lumen diameter (µm)	*T. amurensis*	**20.51 ± 0.35**	**13.80 ± 0.26**	**
	T. mandshurica	**15.91 ± 0.29**	**15.38 ± 0.31**	n.s.
Double wall thickness (µm)	*T. amurensis*	**20.08 ± 0.27**	**10.15 ± 0.13**	**
	T. mandshurica	**15.11 ± 0.24**	**10.28 ± 0.28**	**
Slenderness ratio	*T. amurensis*	**24.73 ± 0.43**	**25.86 ± 0.52**	n.s.
	T. mandshurica	**29.64 ± 0.30**	**30.21 ± 0.51**	**
Runkel ratio	*T. amurensis*	1.07 ± 0.01	0.79 ± 0.02	**
	T. mandshurica	1.03 ± 0.02	0.74 ± 0.02	**

[1] The data is represented as Means ± SE. [2] Data in bold indicate significant differences between species at $p = 0.01$. ** indicate significant differences between branchwood and stemwood at $p = 0.01$, respectively. n.s. indicate no significant differences.

Average fiber dimensions (except lumen diameter) for branchwood were significantly lower (i.e., shorter and narrower fibres) compared to stemwood ($p < 0.01$). Similar results were found in many hardwoods [14,31]. These may be ascribed to cambial age and distance from apical meristem [28,47]. Short and narrow fibers could have resulted from a faster growth rate during wood formation in the branches and the extent of invasive growth of the tip of the fibers during their differentiation [48]. The average fiber dimensions were significantly lower in branchwood of *T. amurensis* than those of *T. mandshurica* ($p < 0.01$), except for cell wall thickness. Compared to long and wide fibers, short and narrow fibers make it challenging in specific usage of lignocellulosic materials [38,49]. For example, longer fibers are preferred to shorter fibers due to their capacity to produce paper with greater tensile strength and toughness [10]. However, shorter fibers could give superior tear resistance at higher levels of sheet density [11].

The average fiber length for *T. amurensis* branchwood was 0.59 mm, which does not fall in the range values (0.7–1.6 mm) for most of the hardwood species [41]. Even the average fibers in *T. mandshurica* branchwood that were 0.72 mm long only met the IAWA category for the shorter fibers [42]. Figure 2a–d shows the distribution of fiber length. Table 4 shows that the skewness and kurtosis of distributions appeared to meet the normality assumption (|skewness| < 2.1 and |kurtosis| < 7.1) set by West et al. [50]. However, the fact was that all distributions were not standard but slightly skewed. About 50% and 30% of the fibers had a length greater than 0.9 mm in *T. amurensis* and *T. mandshurica* stemwood, respectively.

Only 10% of the fibers in *T. mandshurica* branchwood with a length of more than 0.9 mm, and less *T. amurensis*, only 3%. About 80% of the fibers showed 0.4–0.8 mm and 0.5–0.9 mm in length in *T. amurensis* and *T. mandshurica* branchwood, respectively. These results showed that the short fibers of the branchwood would be a factor limiting its application in pulp and papermaking.

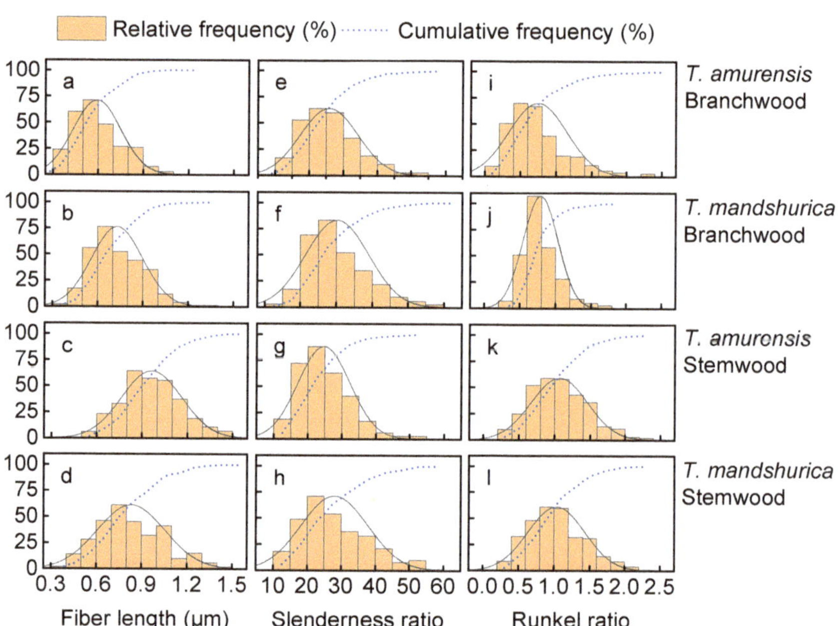

Figure 2. Relative frequency (histogram), cumulative frequency (dotted line), and fitted normal distribution (solid line) of fiber length, slenderness ratio and Runkel ratio. (**a**) Fiber length of *T. amurensis* branchwood. (**b**) Fiber length of *T. amurensis* stemwood. (**c**) Fiber length of *T. mandshurica* branchwood. (**d**) Fiber length of *T. mandshurica* stemwood. (**e**) Slenderness ratio of fibers in *T. amurensis* branchwood. (**f**) Slenderness ratio of fibers in *T. amurensis* stemwood. (**g**) Slenderness ratio of fibers in *T. mandshurica* branchwood. (**h**) Slenderness ratio of fibers in *T. mandshurica* stemwood. (**i**) Runkel ratio of fibers in *T. amurensis* branchwood. (**j**) Runkel ratio of fibers in *T. amurensis* stemwood. (**k**) Runkel ratio of fibers in *T. mandshurica* branchwood. (**l**) Runkel ratio of fibers in *T. mandshurica* stemwood.

Table 4. Skewness and kurtosis (data in brackets) of normal distribution curves for fiber length, slenderness ratio and Runkel ratio.

	T. amurensis		*T. mandshurica*	
	Branchwood	**Stemwood**	**Branchwood**	**Stemwood**
Fiber length (mm)	0.61 (−0.20)	0.32 (−0.25)	0.51 (0.01)	0.42 (−0.23)
Slenderness ratio	0.76 (0.55)	0.96 (1.18)	0.77 (0.46)	0.70 (−0.07)
Runkel ratio	1.13 (2.02)	0.48 (−0.11)	1.12 (1.38)	0.50 (−0.15)

The average slenderness ratio of the fibers for both branchwood and stemwood was smaller than the acceptable value (33) in papermaking [51]. About 10% of the fibers with the slenderness ratio were above 33 for *T. mandshurica*, and 25% for *T. amurensis* (Figure 2e–h). Therefore, it is difficult for both species to produce high-quality pulp. In fact, many paper products on the market have slenderness ratio of less than 33 due to a severe shortage of wood resources, as witnessed, for example, by study of paper and paper egg trays used in Southwestern Nigeria by Amoo et al. [52].

The average Runkel ratio of fibers in the branchwood was less than 1, which implied that the fibers would collapse and provide a large surface area for bonding during paper-making [46]. The distribution of the Runkel ratio were skewed to the right (Figure 2i–l), indicating that most of the data were below its average value. The distribution of the Runkel ratio in *T. mandshurica* branchwood was steep, with kurtosis up to 2.02 (Table 4), indicating that the data were relatively centralized. More than 55% of the fibers in stemwood had a Runkel ratio less than 1, while up to 80 % in branchwood. Therefore, the short fibers of branchwood can be mixed with some long fibers (for example, softwood fibers) in different proportions to produce particular products, such as newsprint, packaging, and hygienic tissue products [48,53].

3.3. Physical and Mechanical Properties

Wood density is generally believed to reflect fiber wall thickness [54]. However, these results showed that linden trees with thick-walled fibers (Table 3) did not appear to produce high wood density (Table 5). Despite the low density of wood, the values still met the density requirement (0.3 to 0.5 g/cm^3) of papermaking raw material [55]. The correlation between wood density and pulp yield is still controversial, but the correlation between wood density and pulp and paper quality is recognized [7,56]. Colodette et al. [57] suggest that species with low density wood should be directed towards manufacturing refined paper (printing and writing grades). Referring to the research of Kennedy et al. [58], pulp produced from linden wood is suitable for manufacturing fine paper. Wood density was similar between stemwood and branchwood. Similar wood density creates a favorable condition for mixing stemwood and branchwood in cooking to optimize the production process and pulp quality [56].

Table 5. Comparison of physical and mechanical properties of the branchwood and stemwood from two *Tilia* species [1].

	Species	Branchwood	Stemwood [59]
Wood density (g/cm^3)	*T. amurensis*	0.36 ± 0.09	0.36
	T. mandshurica	0.32 ± 0.01	0.33
Compressive strength parallel to grain (MPa)	*T. amurensis*	30.57 ± 1.05	34.9
	T. mandshurica	30.93 ± 1.03	32.7

[1] The data of branchwood is represented as Means ± SE.

Generally, the mechanical strength is weak in species with low wood density [60]. The compressive strength of linden wood is below 45 MPa, and even below the values of fast-growing *Populus deltoides* (36.46 MPa) reported by Feng [61], belonging to the low strength range according to the classification of wood mechanical properties [62]. Although having similar densities, the branchwood has a slightly lower compressive strength than the stemwood's value [59]. The low density and strength of linden wood confirm that it is rarely used in construction (see Section 1). Still, it also implies less energy consumption in mechanical chipping [63] and an enormous potential to produce printing and writing sheets.

3.4. Chemical Properties

Table 6 shows that the branchwood of *T. amurensis* had more cold and hot water extractives than the stemwood reported by Lu [64], suggesting that branchwood contained more water-soluble material, such as starch and soluble sugar. This suggestion has already been demonstrated in previous studies [27,65]. The cellulose content in *T. amurensis* branchwood was lower (36.73%) and hemicellulose content was higher (30.93%) than in stemwood, and the contents exceeded the reference 45%–50% for cellulose content and 20%–25% for hemicellulose content [66]. Similar results were found in other tree species [45,67]. A high proportion of sapwood in the branches may be the main reason [45]. Still, branchwood contained a large amount of holocellulose (total cellulose and hemicellulose) and a small amount of lignin (22.34%), which was beneficial for pulping [8,68]. The branchwood of *T. mandshurica* showing cold and hot water extractives were not much different from those of *T. amurensis*. However, NaOH extractive in the branchwood of *T. mandshurica* was about 6% less than that of *T. amurensis*. This may be because the branchwood of *T. mandshurica* had less hemicellulose (20.98%) than that of *T. amurensis*. Some studies [69,70] have shown that although hemicellulose is a structural carbohydrate, it is not as stable as cellulose. Wood treated with 1% NaOH can dissolve a portion of hemicellulose in addition to water-soluble substances. Thus, low hemicellulose content in *T. mandshurica* branchwood leads to low 1% NaOH extractives. The cellulose content in *T. mandshurica* branchwood was close to that of *T. amurensis* branchwood. However, *T. mandshurica* branchwood had a high lignin content (30.48%), and the content was almost close to the critical reference 20%–30% [66], which was not beneficial for pulping [68].

Table 6. Comparison of chemical composition of branchwood and stemwood from two *Tilia* species [1].

	Species	Branchwood	Stemwood [64]
Cold water extractives content (%)	*T. amurensis*	6.45 ± 1.05 [2]	2.81
	T. mandshurica	6.01 ± 0.65	—
hot water extractives content (%)	*T. amurensis*	6.88 ± 0.16	2.77
	T. mandshurica	6.45 ± 1.05	—
1%NaOH content (%)	*T. amurensis*	**18.0 ± 2.05**	24.61
	T. mandshurica	**11.92 ± 0.90**	—
Cellulose content (%)	*T. amurensis*	36.73 ± 3.10	49.64
	T. mandshurica	38.64 ± 3.43	—
Hemi- cellulose content (%)	*T. amurensis*	30.93 ± 1.59	22.59
	T. mandshurica	20.88 ± 2.28	—
Lignin content (%)	*T. amurensis*	**22.34 ± 4.69**	24.68
	T. mandshurica	**30.48 ± 4.50**	—

[1] The data is represented as Means ± SE. [2] Data in bold indicate significant differences between species at $p = 0.01$. — indicate no data for reference.

4. Conclusions

The branchwood and stemwood showed significant differences in their tissue proportions and fiber dimensions despite showing similar cell arrangements ($p < 0.05$). The branchwood have more than 40% fiber proportion, indicating there is the potential to obtain high pulp yield. The branchwood have a short, narrow, and thin fiber dimensions and small slenderness ratio, but a suitable Runkel ratio to papermaking. Branchwood fibers can be mixed with large fibers from other tree species to produce specific paper products. The density and compressive strength of branchwood and the reported stemwood values are similar and relatively weak, which would be beneficial for mechanical chipping. Compared to stemwood, *T. amurensis* branchwood has less cellulose but more holocellulose (about 67%) and less lignin (22%), which was beneficial for pulping. *T. mandshurica* branchwood may not be suitable for chemical pulping due to its high lignin content (up to 30%).

Author Contributions: Conceptualization, P.G., X.Z. and Q.F.; methodology, P.G. and X.Z.; software, P.G., Q.F. and Y.Y.; validation, P.G., X.Z. and Q.F.; formal analysis, P.G. and X.Z.; investigation, Q.F. and Y.Y.; resources, X.Z.; data curation, P.G. and Y.Y.; writing—original draft preparation, P.G.; writing— review and editing, X.Z., Q.F. and Y.Y.; visualization, P.G.; supervision, X.Z.; project administration, P.G. and X.Z.; funding acquisition, X.Z. All authors have read and agreed to the published version of the manuscript.

Funding: This research was funded by the National Science Foundation of China, grant number 32171701.

Data Availability Statement: The data used in the study are published in this paper.

Acknowledgments: The authors would like to thank Xingchang Wang of the Maoershan Forest Ecosystem Research Station for collecting samples. We thank a thoughtful reviewer and editor for helping to improve this work.

Conflicts of Interest: The authors declare no conflict of interest.

References

1. Nilsson, P.O.; Wernius, S. Whole-tree utilization: A method of increasing the wood supply. *Ecol. Bull.* **1976**, *21*, 131–136. [CrossRef]
2. Dadzie, P.K.; Amoah, M.; Ebanyenle, E.; Frimpong-Mensah, K. Characterization of density and selected anatomical features of stemwood and branchwood of *E. cylindricum*, *E. angolense* and *K. ivorensis* from natural forests in Ghana. *Eur. J. Wood Wood Prod.* **2018**, *76*, 655–667. [CrossRef]
3. Guo, P.; Zhao, X.; Liu, Z.; Peng, H. Fiber morphology in walnut branchwood: Relation to the branch diameter, branching Level, and tension wood. *BioResources* **2022**, *17*, 2214–2227. [CrossRef]
4. Gansu Light Industry Science Research Institute. Research and production report on pulping and papermaking of branchwood from broad-leaved trees. *China Pulp Pap.* **1971**, *4*, 23–27. (In Chinese)
5. Strandgard, M.; Turner, P.; Shillabeer, A. Optimizing operational-level forest biomass logistic costs for storage, chipping and transportation through roadside drying. *Forests* **2022**, *13*, 138. [CrossRef]
6. Rullifank, K.; Roefinal, M.; Kostanti, M.; Sartika, L.; Evelyn, E. Pulp and paper industry: An overview on pulping technologies, factors, and challenges. *IOP Conf. Ser. Mater. Sci. Eng.* **2020**, *845*, 012005. [CrossRef]
7. Santos, A.; Anjos, O.; Amaral, M.E.; Gil, N.; Pereira, H.; Simões, R. Influence on pulping yield and pulp properties of wood density of *Acacia melanoxylon*. *J. Wood Sci.* **2012**, *58*, 479–486. [CrossRef]
8. Ramírez, M.; Rodríguez, J.; Balocchi, C.; Peredo, M.; Elissetche, J.; Mendonça, R.; Valenzuela, S. Chemical composition and wood anatomy of *Eucalyptus globulus* clones: Variations and relationships with pulpability and handsheet properties. *J. Wood Chem. Technol.* **2009**, *29*, 43–58. [CrossRef]
9. Dos Santos, S.R.; Sansígolo, C.A. Wood basic density effect of *Eucalyptus grandis* x *Eucalyptus urophylla* clones on bleached pulp quality. *Ciência Florest.* **2007**, *17*, 53–63. [CrossRef]
10. Larsson, P.T.; Lindström, T.; Carlsson, L.A.; Fellers, C. Fiber length and bonding effects on tensile strength and toughness of kraft paper. *J. Mater. Sci.* **2018**, *53*, 3006–3015. [CrossRef]
11. Wangaard, F.G.; Woodson, G.E. Fiber length strength interrelationship for slash pine and its effect on pulp-sheet properties. *Wood Sci.* **1973**, *5*, 235–240.
12. Zhao, X.; Guo, P.; Zhang, Z.; Peng, H. Anatomical features of branchwood and stemwood of *Betula costata* Trautv. from natural secondary forests in China. *BioResources* **2019**, *14*, 1980–1991. [CrossRef]
13. Zhao, X.; Guo, P.; Zhang, Z.; Wang, X.; Peng, H.; Wang, M. Wood density and fiber dimensions of root, stem, and branch wood of *Populus ussuriensis* Kom. trees. *BioResources* **2018**, *13*, 7026–7036. [CrossRef]
14. Dong, H.; Bahmani, M.; Humar, M.; Rahimi, S. Fiber morphology and physical properties of branch and stem wood of hawthorn (*Crataegus Azarolus* L) grown in Zagros forests. *Wood Res-Slovak.* **2021**, *66*, 391–402. [CrossRef]
15. Hunt, K.; Keays, J.L. Kraft pulping of trembling Aapen tops and branches. *Can. J. Forest Res.* **1973**, *3*, 535–542. [CrossRef]
16. Law, K.N.; Lapointe, M. Chemimechanical pulping of boles and branches of white spruce, white birch, and trembling aspen. *Can. J. Forest Res.* **1983**, *13*, 412–418. [CrossRef]
17. Hassan, K.; Kandeel, E.-S.; Kherallah, I.; Abou-Gazia, H.; Hassan, F. *Pinus halepensis* and *Eucalyptus camaldulensis* grown in Egypt: A comparison between stem and branch properties for pulp and paper making. *BioResources* **2020**, *15*, 7598–7614. [CrossRef]
18. Wang, C. *Tilia amurensis* Rupr. and *Tilia mandshurica* Rupr. & Maxim. from northeast in China. *J. Syst. Evol.* **1976**, *14*, 36–37. (In Chinese)
19. Ma, Y.; Wang, H.; Mu, L.; Zou, Q. Anatomy structure and analysis of branches and leaves of *Tilia amurensis* and *T. mandshurica*. *J. Qiqihar Univ.* **2006**, *22*, 93–96. [CrossRef]
20. Long, Z.; Hu, Y. Current situation and countermeasure of *Tilia* Linn. *For. By-Prod. Spec. China* **2013**, *28*, 88–90. (In Chinese)
21. Hather, J.G. *The Identification of the North European Woods. A Guide for Archaeologist and Conservators*; Archetype Publication: London, UK, 2000.
22. Ni, F.; Ma, B.; Li, C.; Wang, Z.; Liu, Q. Study on xylem comparison of *Tulia mandshuarica* and *Tilia amurensis*. *Jilin Norm. Univ. J.* **2013**, *34*, 99–100. (In Chinese)
23. Dai, C. Furniture materials: Linden, birch, Maple. *Furniture* **1980**, *1*, 39–40. (In Chinese)
24. Yu, Z.; Chang, S.; Hu, J.; Tan, Y.; Shang, J.; Xu, D.; Liu, Y. Study on the main physical and mechanical properties of *Tilia* sp. used for shutters. *J. For. Eng.* **2019**, *4*, 159–164. [CrossRef]
25. Wang, H.; Shen, F.; Gong, Z.; Li, Q.; Yuan, S.; Zhang, J.; Yang, S. Study on the dyeing properties of poplar and basswood veneer. *Shandong For. Sci. Technol.* **2020**, *50*, 4. (In Chinese)
26. Yong, C.; Li, Y. Wood structure and pencil board processing. *Chin. Pen Mak.* **1995**, *16*, 4–5. (In Chinese)
27. Zhang, H.; Wang, C.; Wang, X. Spatial variations in non-structural carbohydrates in stems of twelve temperate tree species. *Trees-Struct. Funct.* **2014**, *28*, 77–89. [CrossRef]

28. Zhao, X. Effects of cambial age and flow path-length on vessel characteristics in birch. *J. Forest Res-Jpn.* **2015**, *20*, 175–185. [CrossRef]
29. Jeffrey, E.C. *The Anatomy of Woody Plants*; The University of Chicago Press: Chicago, IL, USA, 1917; p. 478. [CrossRef]
30. Yu, H.; Liu, Y.; Han, G.; Cui, Y. Comparison of image analysis and conventional methods for cellular tissue proportion measurement of wood. In Proceedings of the International Conference on Information and Automation, 2009. ICIA'09, Zhuhai/Macau, China, 22–24 June 2009; pp. 133–137. [CrossRef]
31. Zhao, X.; Guo, P.; Zhang, Z.; Yang, Y.; Zhao, P. Wood density, anatomical characteristics, and chemical components of *Alnus sibirica* used for industrial applications. *Forest Prod. J.* **2020**, *70*, 356–363. [CrossRef]
32. CNS GB/T1933-2009; Method for Determination of the Density of Wood. National Standard of the People's Republic of China: Beijing, China, 2009. (In Chinese)
33. CNS GB/T1935-2009; Method of Testing in Compressive Strength Parallel to Grain of Wood. National Standard of the People's Republic of China: Beijing, China, 2009. (In Chinese)
34. CNS GB/T35816-2018; Standard Method for Analysis of Forestry Biomass—Determination of Extractives Content. National Standard of the People's Republic of China: Beijing, China, 2018. (In Chinese)
35. CNS GB/T35818-2018; Standard Method for Analysis of Forestry Biomass—Determination of Structural Polysaccharides and Lignin. National Standard of the People's Republic of China: Beijing, China, 2018. (In Chinese)
36. Fang, W.; Wu, Y. A study on the rate of hardwood tissue. *J. Fujian Coll. For.* **2006**, *26*, 224–228. [CrossRef]
37. Zheng, J.; Bian, J.; Ge, Q.; Yin, Y. The climate regionalization in China for 1951–1980 and 1981–2010. *Geogr. Res.* **2013**, *32*, 987–997. (In Chinese)
38. Nazari, N.; Bahmani, M.; Kahyani, S.; Humar, M.; Koch, G. Geographic variations of the wood density and fiber dimensions of the Persian oak wood. *Forests* **2020**, *11*, 1003. [CrossRef]
39. Stokke, D.D.; Manwiller, F.G. Proportions of wood elements in stem, branch, and root wood of Black Oak (*Quercus Velutina*). *IAWA J.* **1994**, *15*, 1507–1509. [CrossRef]
40. Horn, R.A. Press-drying of high-yield pulps: The role of parenchyma cells. *J. Tech. Assoc. Pulp Pap. Ind.* **1981**, *64*, 105–108.
41. Kong, B.; Wei, L. Effect of plant fiber morphology on paper. *Southwest Pulp Pap.* **2000**, *28*, 24. (In Chinese)
42. Wheeler, E.; Baas, P.; Gasson, P. IAWA list of microscopic features for hardwood identification. *J. Int. Assoc. Wood Anat.* **1989**, *10*, 219–332.
43. Huang, R.; Chen, C. The research on wood fiber length and width of 42 kinds of hardwood. *J. Minxi Vocat. Tech. Coll.* **2014**, *16*, 97–102. [CrossRef]
44. Cheng, J. *China Wood Records*; China Forestry Press: Beijing, China, 1992. (In Chinese)
45. Zhao, X.; Guo, P.; Zhao, P.; Yang, Y.; Peng, H.; Zhang, Z. Potential of pulp production from whole-tree wood of *Betulaplaty phylla* Roth. based on wood characteristics. *BioResources* **2019**, *14*, 7015–7024. [CrossRef]
46. Kiaei, M.; Tajik, M.; Vaysi, R.; Ciencia y Tecnologia, M. Chemical and biometrical properties of plum wood and its application in pulp and paper production. Maderas. *Cienc. Tecnol.* **2014**, *16*, 313–322. [CrossRef]
47. Aloni, R. Ecophysiological implications of vascular differentiation and plant evolution. *Trees-Struct. Funct.* **2015**, *29*, 1–16. [CrossRef]
48. Charles, A.-P.; Antwi-Boasiako, S. Tissue dimensions and proportions of the stem and branch woods of *Aningeria Robusta* (A. Chev) and *Terminalia Ivorensis* (A. Chev). *Afr. J. Wood Sci. For.* **2016**, *4*, 254–273.
49. Panshin, A.J.; Zeeuw, C.D. *Textbook of Wood Technology*, 4th ed.; McGraw-Hill Book Co.: New York, NY, USA, 1980.
50. West, S.; Finch, J.; Curran, P. Structural Equation Models with Non-Normal Variables: Problems and Remedies. In *Structural Equation Modeling: Concepts, Issues, and Applications*; Hoyle, R.H., Ed.; SAGE Publications, Inc: Newbery Park, CA, USA, 1995.
51. Xu, F.; Zhong, X.C.; Sun, R.C.; Lu, Q. Anatomy, ultrastructure and lignin distribution in cell wall of *Caragana Korshinskii*. *Ind. Crop. Prod.* **2006**, *24*, 186–193. [CrossRef]
52. Amoo, K.; M.A, O.; Omoniyi, T. Fibre characteristics of paper and paper egg tray used In Southwestern Nigeria. *Int. J. Eng. Res. Rev.* **2016**, *4*, 53–60.
53. Stankovská, M.; Fišerová, M.; Gigac, J.; Opálená, E. Blending impact of hardwood pulps with softwood pulp on tissue paper properties. *Wood Res.* **2020**, *65*, 447–458. [CrossRef]
54. Kamijo, Y.; Sugino, M.; Miyanishi, T. Fiber morphologies and sheet properties of hardwood thermomechanical pulp. *Jpn. Tappi J.* **2015**, *69*, 81–89. [CrossRef]
55. Fang, G.-G.; Shen, Z.-B.; Huan, D.-Y. Sustainable development of China's papermaking industry relying on high yield pulping technologies. *J. Chem. Ind. For. Prod.* **2000**, *34*, 7–9. (In Chinese)
56. Pécora Jr, J.E.; Ruiz, A.; Soriano, P. Minimization of the wood density variation in pulp and paper production. *Inf. Syst. Oper. Res.* **2007**, *45*, 187–196. [CrossRef]
57. Colodette, J.L.; Mokfienski, A.; Gomide, J.L.; Oliveira, R.C. Relative importance of wood density and carbohydrate content on pulping yield and product quality. *J. Tianjin Univ. Sci. Technol.* **2004**, *19*, 71–80. (In Chinese)
58. Kennedy, J.F.; Phillips, G.O.; Williams, P.A. *Wood Processing and Utilization*; Ellis Horwood Limited: London, UK, 1989.
59. Wood Science Laboratory of Northeast Forestry University; Research Office of Forest Products Industry of Jilin Forestry Institute. Physical-mechanical properties of seven important hardwoods in Xiaoxing'anling forest. *J. Northeast. For. Univ.* **1978**, *8*, 119–142. (In Chinese)

60. Niklas, K.J.; Spatz, H.C. Worldwide correlations of mechanical properties and green wood density. *Am. J. Bot.* **2010**, *97*, 1587–1594. [CrossRef] [PubMed]
61. Feng, D.J.; Zhao, J.; Chen, W. Properties and fiber morphology of "Qinhei Yang" woods. *J. Northwest For. Univ.* **2021**, *36*, 198–201. [CrossRef]
62. Research Institute of Wood Industry, Chinese Academy of Forestry Sciences. *Wood Physical and Mechanical Properties of Main Tree Species in China*; China Forestry Publishing House: Beijing, China, 1982; p. 154. (In Chinese)
63. Anupam, K.; Sharma, A.K.; Lal, P.S.; Bist, V. Physicochemical, morphological, and anatomical properties of plant fibers used for pulp and papermaking. In *Fiber Plants: Biology, Biotechnology and Applications*; Ramawat, K.G., Ahuja, M.R., Eds.; Springer International Publishing: Cham, Switzerland, 2016; pp. 235–248. [CrossRef]
64. Lu, W.; AN, Y.; Ying, L. A review of anatomy properties and technological characters of woods from four main commercial hardwoods in northeast China. *J. Northeast. For. Univ.* **1991**, *19*, 312–317. (In Chinese)
65. Zhang, H.; Wang, C.; Wang, X. Comparison of concentrations of non-structural carbohydrates between new twigs and old branches for 12 temperate species. *Acta Ecol. Sin.* **2013**, *33*, 5675–5685. [CrossRef]
66. Stalnaker, J.; Harris, E. *Structural Design in Wood*; Springer Science & Business Media: New York, NY, USA, 1997.
67. Iman Suansa, N.; Al-Mefarrej, H. Branch wood properties and potential utilization of this variable resource. *Bioresources* **2020**, *15*, 479–491. [CrossRef]
68. Koljonen, K.; Österberg, M.; Kleen, M.; Fuhrmann, A.; Stenius, P. Precipitation of lignin and extractives on kraft pulp: Effect on surface chemistry, surface morphology and paper strength. *Cellulose* **2004**, *11*, 209–224. [CrossRef]
69. Hoch, G. Cell-wall hemicellulose as mobile carbon stores in non-reproductive plant tissues. *Funct. Ecol.* **2007**, *21*, 823–834. [CrossRef]
70. Schädel, C.; Blöchl, A.; Richter, A.; Hoch, G. Short-term dynamics of nonstructural carbohydrates and hemicellulose in young branches of temperate forest trees during bud break. *Tree Physiol.* **2009**, *29*, 901–911. [CrossRef] [PubMed]

Article

Color and Chemical Composition of Timber Woods (*Daniellia oliveri, Isoberlinia doka, Khaya senegalensis*, and *Pterocarpus erinaceus*) from Different Locations in Southern Mali

Mohamed Traoré [1,2] and Antonio Martínez Cortizas [1,3,*]

[1] CRETUS, EcoPast (GI-1553), Facultade de Bioloxía, Universidade de Santiago de Compostela, 15782 Santiago de Compostela, Spain; mohamed.traore@usc.gal

[2] Department of Geology and Mines, Ecole Nationale d'Ingénieurs Abderhamane Baba Touré (ENI-ABT), 410 Avenue Van Vollenhoven, Bamako P.O. Box 242, Mali

[3] Bolin Centre for Climate Research, Stockholm University, SE-10691 Stockholm, Sweden

*   Correspondence: antonio.martinez.cortizas@usc.es

Abstract: Wood characteristics and properties are related to various factors connected to the biochemical processes that occur in the tree during wood formation, but also, to the interactions with the environmental conditions at the tree growing location. In addition to climatic factors, several investigations drew attention to the significance of the influence of other environmental parameters at the tree growing location. In this perspective, this work aimed to characterize the variation in color and chemical composition of timber wood from different locations in southern Mali, of trees growing under the same climatic conditions. To do so, a total of 68 grounded wood samples, from 4 timber wood species (*Daniellia oliveri, Isoberlinia doka, Khaya senegalensis,* and *Pterocarpus erinaceus*), were analyzed using CIELab color space and FTIR-ATR. Overall, the results indicated that the variation in wood color and chemical properties can be related to the local environmental conditions. *Pterocarpus erinaceus* presented significant differences between samples from the three areas according to the highest number of variables (color parameters, molecular composition determined by FTIR-ATR spectroscopy, and FTIR-ATR ratios). *Daniellia oliveri* and *Khaya senegalensis*, however, showed significant differences between areas of provenance for a lower number of variables. *Isoberlinia doka*, for its part, showed no significant differences and seems to be less sensitive to environmental factors. Furthermore, the results revealed that important differences exist between wood samples from Kati and Kéniéba.

Keywords: tropical timber; color parameters; mid-infrared spectroscopy; FTIR ratios; environmental factors; West African Sahel

Citation: Traoré, M.; Martínez Cortizas, A. Color and Chemical Composition of Timber Woods (*Daniellia oliveri, Isoberlinia doka, Khaya senegalensis*, and *Pterocarpus erinaceus*) from Different Locations in Southern Mali. *Forests* **2023**, *14*, 767. https://doi.org/10.3390/f14040767

Academic Editors: Vicelina Sousa, Helena Pereira, Teresa Quilhó and Isabel Miranda

Received: 4 March 2023
Revised: 4 April 2023
Accepted: 6 April 2023
Published: 8 April 2023

1. Introduction

Wood characteristics and properties (e.g., aesthetical, physical, and chemical) are important parameters that make it valuable for a given end-use [1]. In construction, wood density and other mechanical properties, as well as its natural durability, are among the most considered wood characteristics to face biotic and abiotic threats, whereas, in the art industry, the required characteristics are related to its easy manipulation for carving or a certain specific tune when referring to musical instruments [2–4]. It is well known that wood characteristics and properties are related to various factors connected to the biochemical processes that occur in the tree during wood formation, but also, to the interactions with the environmental conditions at the tree growing location [5]. Over the last decade, numerous studies have been carried out on a large variety of wood species, in temperate regions and in tropical regions, to unravel the biogeochemical processes that are behind the composition of this complex lignocellulosic material [6–8]. Despite the great advances in the field, more work is needed, especially taking advantage of the multidisciplinary aspect of wood research, to obtain a better understanding of the endogenous and exogenous factors controlling wood characteristics and properties.

The effects of the environmental conditions on wood structure and composition come out with various parameters related to the climate, as well as to physical, chemical, and biological processes at the tree growing location [9,10]. Variations of these environmental factors can be related to parameters that reflect such changes in wood structure and composition [11,12]. For instance, tree-ring width is one of the most used proxies in wood research to provide a general overview of tree growth [13,14]. The analysis of anatomical features and the structure of growth rings (earlywood width, latewood width, total ring width) provides insights into the main growth-limiting climatic factors (e.g., temperature, precipitation, drought, etc.), and enables the retrieval of information about the modification of structural features by soil moisture content, depth of groundwater table, slope, etc. [6,15]. Likewise, wood chemical composition (molecular, elemental, and isotopic) constitutes a potential means for the characterization of wood and for the understanding of the effect that the variability in environmental and climatic conditions has on the quality and properties of wood [16,17]. To our understanding, the development of these promising approaches still requires further research work in order to provide both a theoretical basis (i.e., an understanding of the factors behind wood composition and its characteristics) and a well-grounded methodological approach for wood characterization and fingerprinting (i.e., a means to determine the specific features concerning a given factor).

An increasing number of studies are conducted to develop methodological approaches, using accessible and handy analytical techniques. This approach has been important to obtain a better understanding of wood physical and chemical characteristics in relation to environmental influences. To give an example, CIELab color space parameters have been used to study the link between wood color variation and various factors, including genetic, climatic, and edaphic [18–20]. In a like manner, the various applications of infrared spectroscopy revealed the great potential of this technique to investigate the link between wood molecular composition and tree species, as well as the wood provenance [21–24]. Additionally, data from these techniques can be used in parallel with other parameters related to structure and composition to foster the multiproxy potentially associated with wood research disciplines [20,25,26].

The main objective of the present research is to assess the variation in the characteristics (physical and chemical) of timber wood from three different areas of provenance, in a limited geographical area, under the same climatic conditions. To do so, we have applied CIELab color space and FTIR-ATR spectroscopy to wood samples of four timber wood species (*Daniellia oliveri*, *Isoberlinia doka*, *Khaya senegalensis*, and *Pterocarpus erinaceus*) from different locations (Kéniéba, Kita, and Sibi) in southern Mali. Data analysis has been performed using uni- and multivariate statistical methods (one-way ANOVA and principal component analysis).

2. Materials and Methods

2.1. Origin of the Wood Materials and Wood Species

The wood materials used for this research belong to 11 wood cross-sections obtained from individual planks of commercialized timbers, from three different areas in Mali: Kéniéba (12°50′43″ N, 11°14′08″ W), Kita (13°1′41″ N, 9°30′5″ W), and Sibi (12°22′39″ N, 8°19′35″ W) (Figure 1). All three areas are located, under the same climatic conditions, in the Sudano–Guinean climatic zone with average annual rainfall ranging between 800 and 1200 mm, and ferralitic soils predominate [4,27]. The consideration of this geographical delimitation is related to the main purpose of this study, which intends to draw attention to other factors characterizing the local environmental conditions (e.g., location of origin) rather than climatic factors that seem to be the most studied factors driving changes in wood properties.

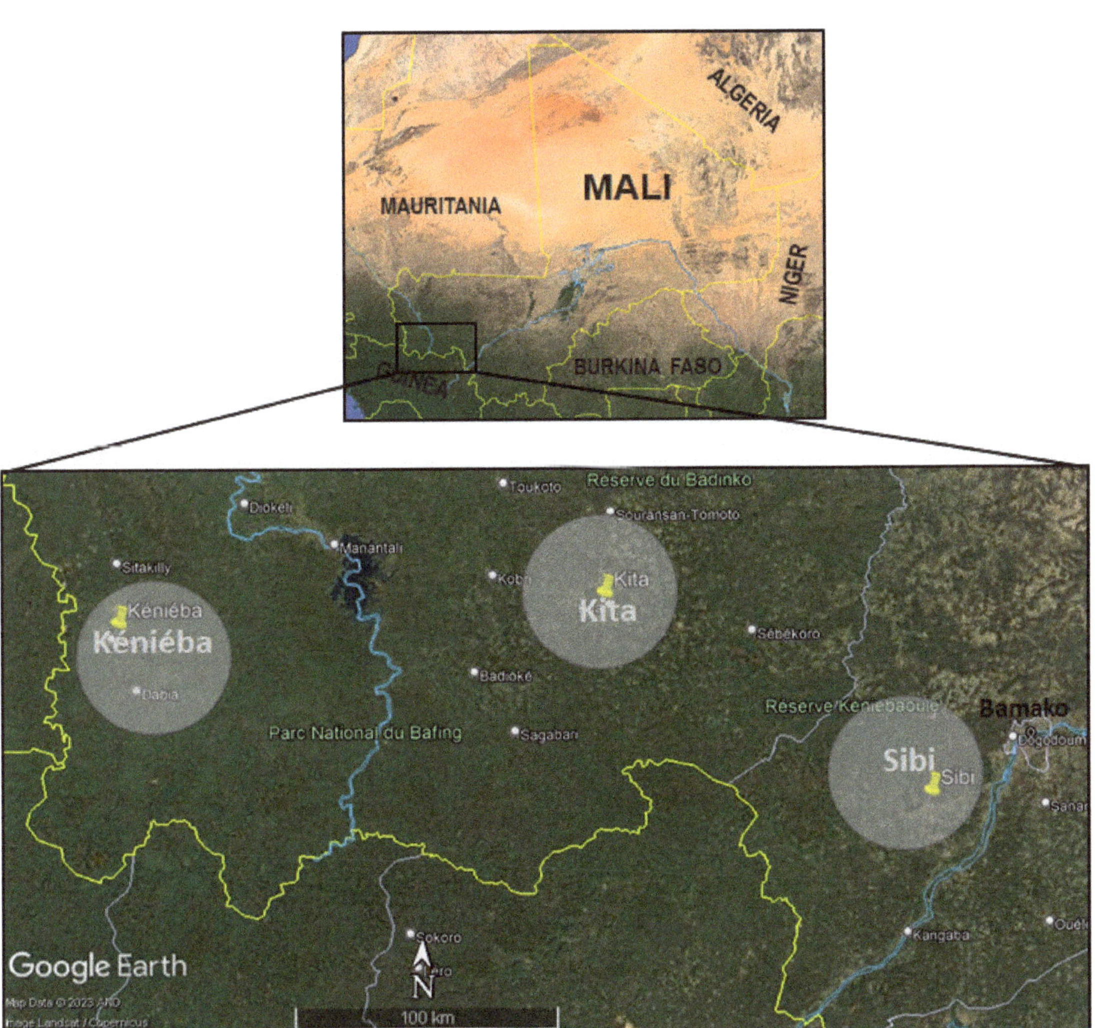

Figure 1. Map showing the area of origin of the studied woods.

The 11 wood cross-sections correspond to four timber wood species: *Daniellia oliveri*, *Isoberlinia doka*, *Khaya senegalensis*, and *Pterocarpus erinaceus*, which are indigenous African tree species with a geographic distribution ranging from West to East Africa [28–30]. These species are associated with high natural resistance to the less favorable climatic conditions of the tropical arid Sahelian and semiarid Sudanian zones in the West African Sahel [31,32]. Further, the selected wood species are among the most used timber wood with vital importance for local economic development [33,34]. According to the IUCN Red List, *D. oliveri* and *I. doka* are registered as the least concerned species, whereas *K. senegalensis* and *P. erinaceus* are, respectively, indicated as vulnerable and threatened species [35–38]. Moreover, these species are being affected by widespread forest degradation because of increasing demand for forest products and excessive exploitation combined with the changing environmental conditions without appropriate regeneration measures [33,39].

2.2. Sample Preparation

Once at the laboratory, for different analytical purposes, each cross-section was divided into two identical parts along the radial direction. The first part was prepared following standard dendrochronological methods to improve the visibility of the growth rings and related macroscopic and microscopic features. The wood surface was manually polished with fine sandpaper (grain sizes from P600 to P1200, Federation of European Producers of Abrasives). This first part enabled us to obtain images of transverse sections, using a Leica Emspira 3 digital microscope (Leica Microsystems GmbH, Wetzlar, Germany), as presented in Figure 2. As for the second part, in the first step, it was divided along the radial direction into small pieces of 3 to 4 cm. Thereafter, a portion (10–15 g) of each wood piece was milled to a very fine powder and homogenized using a Retsch mixer mill MM 301 (Retsch GmbH, Haan, Germany), with final fineness <5 μm, and oven dried at 30 °C for two weeks before the analysis using CIELab color space and infrared spectroscopy techniques. It is worth mentioning that the main reason for milling the wood to very fine powder is related to the later application of the X-ray fluorescence techniques to determine the elemental concentrations, not to the color measurement nor the infrared spectroscopy technique.

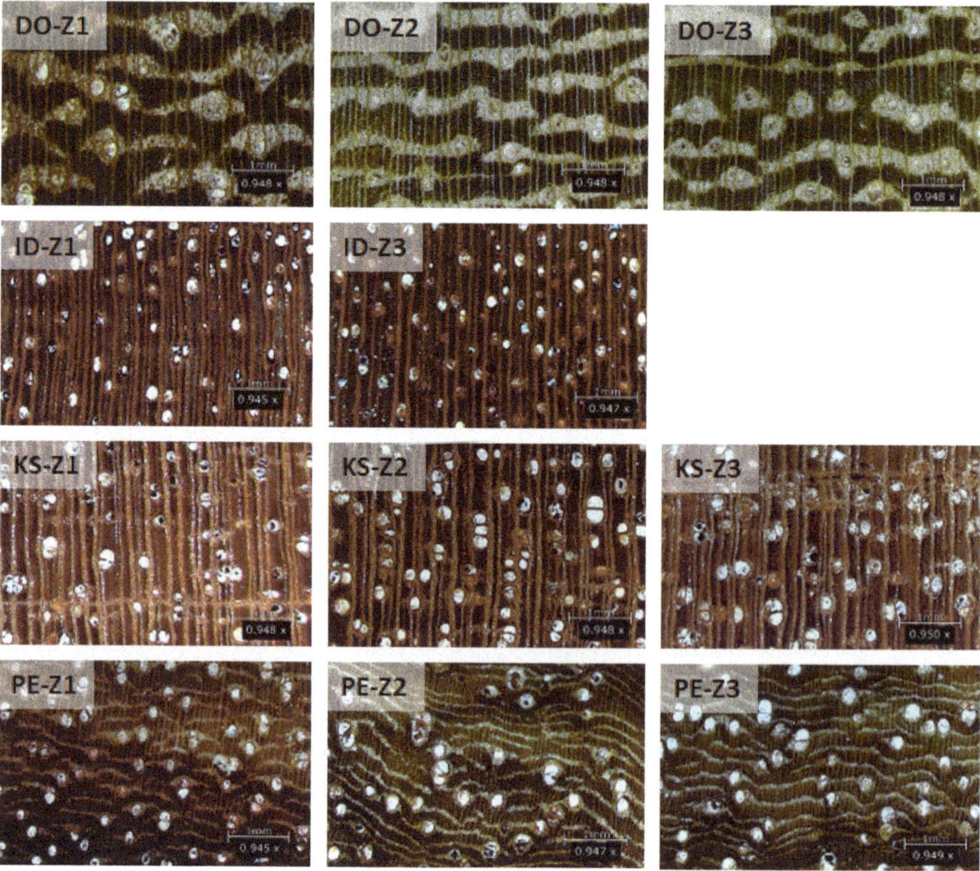

Figure 2. Transverse sections images of the studied wood. DO for *Daniellia oliveri*, ID for *Isoberlinia doka*, KS for *Khaya senegalensis*, and PE for *Pterocarpus erinaceus*. Z1, Z2, and Z3 refer to Kita, Sibi, and Kéniéba, respectively.

2.3. Analytical Techniques

In total, 68 grounded wood samples (see Table 1) were analyzed with two commonly used analytical techniques for wood color and chemical characterization, under room standard temperature and humidity conditions. Color parameters were measured in the CIELab color space using a Konica-Minolta CR-5 Chroma Meter for solids (Konica Minolta Inc., Japan). The CIELab color space provides numerical values for color coordinates: lightness (L*), green–red component (a*), blue-yellow component (b*), chroma (C*), and hue (h°). The infrared spectra were recorded, using an Agilent Cary 630 FTIR Spectrometer (Agilent Technologies Inc., USA), by Fourier-transform infrared spectroscopy in attenuated total reflectance mode (FTIR-ATR). Spectra were obtained in the mid-infrared region (MIR, 4000–400 cm^{-1}) by performing 100 scans at a resolution of 4 cm^{-1}. Further analytical details are provided elsewhere [40,41].

Table 1. Total number of samples (per species and area) analyzed using CIELab and FTIR-ATR.

Wood Species	Kéniéba	Kita	Sibi	Total
Daniellia oliveri	6	5	6	17
Isoberlinia doka	5	5	-	10
Khaya senegalensis	8	8	8	24
Pterocarpus erinaceus	6	6	5	17
	Total			68

2.4. Data Analysis

For the color data, data analysis addressed the color coordinates including lightness (L*), redness (a*), yellowness (b*), chroma (C*), and hue (h°). For the FTIR data, in the first place, we focused on a selected number of infrared bands on the basis of peak identification using the R package *andurinha* [42]. Then, we applied principal component analysis (PCA) using varimax rotation to the selected MIR bands, on the correlation mode. The extracted PCA components were further used, instead of individual FTIR bands. Finally, we calculated the following FTIR ratios: lateral order index (LOI, 1425/895), guaiacyl/syringyl ratios (1265/1230 and 1510/1595), lignin/carbohydrate ratios (1510/1460, 1510/1375, and 1510/895). These FTIR ratios mostly help to determine changes in the molecular structure of polysaccharides and lignin [43,44].

Furthermore, to evaluate whether there were significant differences between wood samples from the three areas (Kéniéba, Kita, and Sibi), for each wood species, we applied one-way analysis of variance (ANOVA) using as variables the CIELab color parameters, the extracted PCA components, and the calculated FTIR ratios.

The statistical tests, one-way ANOVA and principal component analysis, were performed using SPSS 27.

3. Results and Discussion

3.1. Variations in Wood Color Parameters

The color data indicated that the influence of the provenance of the wood sample can depend on the tree species. As presented in Table 2, no significant differences were observed between samples from the three areas where the timbers were collected, for *Daniellia oliveri* and *Isoberlinia doka*. However, for *Khaya senegalensis* and *Pterocarpus erinaceus*, there were significant differences between samples from the three areas. It appears that woods from Kita are highly different from woods of the two other areas (Sibi and Kéniéba). As for *K. senegalensis*, the highest yellowness (b*) values were associated with wood from Kita, the lowest values for Kéniéba, and intermediate values for wood from Sibi (Figure 3). Also, the highest hue (h°) values were related to wood from Kita, and wood from Sibi and Kéniéba similarly indicated the lowest h° values. Regarding *P. erinaceus*, woods from Kita were characterized by the highest values of lightness (L*) and redness (a*), and the lowest values were obtained for woods from Sibi and Kéniéba, which appear to be analogous. The h° values for *P. erinaceus* were found to be the opposite of what has been pointed out by the wood samples of *K. senegalensis*.

Table 2. Summary of the one-way ANOVA using the CIELab color parameters.

Wood Species	Color Parameters	Mean	SD	df	F	Sig.
Daniellia oliveri	Lightness (L*)	60.14	2.10	2	0.78	0.48
	Redness (a*)	8.24	0.29	2	3.01	0.08
	Yellowness (b*)	25.33	1.64	2	0.58	0.57
	Chroma (C*)	26.65	1.57	2	0.54	0.59
	Hue (h°)	71.92	1.23	2	1.39	0.28
Isoberlinia doka	Lightness (L*)	47.82	4.12	1	0.12	0.74
	Redness (a*)	14.88	0.58	1	0.47	0.51
	Yellowness (b*)	18.08	0.93	1	0.33	0.58
	Chroma (C*)	23.42	0.90	1	0.03	0.87
	Hue (h°)	50.52	1.54	1	1.12	0.32
Khaya senegalensis	Lightness (L*)	50.86	3.18	2	2.68	0.09
	Redness (a*)	15.53	0.52	2	0.70	0.51
	Yellowness (b*)	19.09	0.66	2	7.63	0.00 [1]
	Chroma (C*)	24.61	0.66	2	2.16	0.14
	Hue (h°)	50.87	1.23	2	9.11	0.00 [1]
Pterocarpus erinaceus	Lightness (L*)	55.76	3.20	2	4.25	0.04 [2]
	Redness (a*)	11.87	1.09	2	8.29	0.00 [1]
	Yellowness (b*)	23.83	0.85	2	0.37	0.70
	Chroma (C*)	26.65	0.62	2	0.91	0.43
	Hue (h°)	63.52	2.64	2	4.91	0.02 [2]

[1] significant differences at $p < 0.01$; and [2] significant differences at $p < 0.05$.

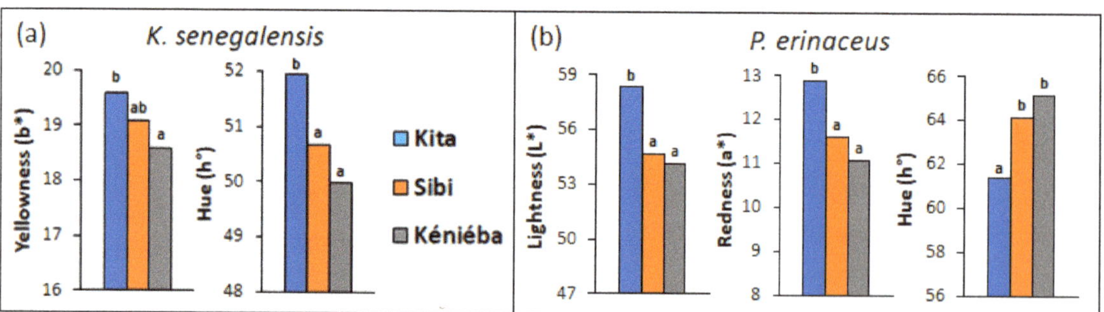

Figure 3. Bar chart of the mean values of color parameters showing significant differences between samples from the three areas; (**a**) for *K. senegalensis* ($n = 8$ for Kita, $n = 8$ for Sibi and $n = 8$ for Kéniéba), (**b**) for *P. erinaceus* ($n = 6$ for Kita, $n = 5$ for Sibi and $n = 6$ for Kéniéba); (the labels a, ab, and b refer to group classification in ascending order of the Student–Newman–Keuls post-hoc test).

These differences observed using the wood color parameters may be related to the variation in wood contents in relation to the tree growing locations. As mentioned in various studies, wood color is connected to its chemical contents, especially the extractive compounds [45,46]. Therefore, the color characteristics of wood can result from processes linked to genetic effects but also the effects of environmental conditions at the tree growing location [20,47]. L* and a* are usually connected to the extractives and phenols, while b* is usually related to lignin and polysaccharides [46,48]. The latter may suggest that, on the one hand, the significant differences observed for the wood samples of *K. senegalensis* may be due to processes that control the chemical content in the wood cell walls (polysaccharides and lignins) and, on the other hand, the significant differences observed between the wood samples of *P. erinaceus* may be due to variation in the wood extractive contents. Similar results have been found for wood from trees growing in different environmental conditions, including different soil properties [19,49,50].

3.2. Variations in Wood FTIR Absorption Bands

The principal component analysis extracted six components (with 26, 25, 14, 13, 10, and 7% of explained variance, respectively) that explained about 95% of the total variance in the dataset. All the selected bands showed a communality value above 0.85. The first component (PC1) is characterized by large positive loadings for bands related to vibrations in aromatic rings and in carboxylic groups in lignin (near 1623, 1606, and 1592 cm^{-1}) and wagging in crystalline cellulose (at 1318 cm^{-1}); and large negative loadings for bands associated with carbonyl vibration in ester and acetyl groups in hemicellulose (near 1735, 1720 and 1703 cm^{-1}) (Table 3). The second component, PC2, shows positive loadings for vibrations related to the molecular structure in guaiacyl and syringyl lignin moieties (1510, 1264, 1245, 1228, and 822 cm^{-1}), C-H deformation in lignin and carbohydrates (1461 cm^{-1}), C-H out of plane in cellulose (895 cm^{-1}), and negative loadings for O-H bond vibration of absorbed water on cellulose (near 3323 and 1655 cm^{-1}) (Table 3). Regarding the third and fourth components (PC3 and PC4), they are characterized by large positive loadings for methyl and methylene stretching of aliphatic structures in cellulose and in extractive compounds (2935, 2922, and 2853 cm^{-1}) and C-O stretching in polysaccharides (1055, 1031, 1019, and 986 cm^{-1}), respectively (Table 3). The fifth component, PC5, is characterized by positive loadings for methoxyl and aromatic skeletal vibrations in lignin and in extractive compounds (near 1560, 1543, and 1422 cm^{-1}). The last extracted component, PC6, presents positive loadings for bands associated with C-O-C stretching in cellulose (1124 and 1107 cm^{-1}) (Table 3). For more insight, the average relative absorption spectra of the wood species with the associated assignment for the absorption bands are presented in Figure 4.

Table 3. List of FTIR bands (with loading values) in relation to the extracted PCA components.

	PC1	PC2	PC3	PC4	PC5	PC6
Bands with positive loadings	1623 (0.94) 1606 (0.87) 1592 (0.86) 1318 (0.74)	822 (0.91) 1245 (0.91) 1228 (0.89) 1264 (0.82) 895 (0.78) 1510 (0.64) 1461 (0.64)	2922 (0.92) 2935 (0.92) 2853 (0.89)	1019 (0.96) 986 (0.91) 1031 (0.91) 1055 (0.71)	1422 (0.79) 1543 (0.76) 1560 (0.67) 1375 (0.55) 1159 (0.52)	1107 (0.77) 1124 (0.61)
Bands with negative loadings	1735 (−0.86) 1720 (−0.86) 1703 (−0.83)	1655 (−0.70) 3323 (−0.62)				

The one-way ANOVA applied to the PCA component scores enabled us to identify the components that show differences between wood samples from the three geographical areas. A summary of this statistical analysis is provided in Table 4. It appears evident that the significance of differences, based on FTIR data, between the sources of the studied samples depends on the wood species. For instance, no significant differences were observed for *I. doka*, whereas significant differences were shown by *D. oliveri* for PC4, by *K. senegalensis* for PC3 and PC4, and by *P. erinaceus* for PC5 and PC6. Overall, samples from Kita and Kéniéba showed high component scores (Figure 5). For *D. oliveri*, the samples from Kéniéba were associated with positive scores of PC4, whereas samples from Kita and Sibi were similarly associated with negative scores of PC4. As for *K. senegalensis*, woods from Kita were associated with negative scores of PC3 and positive scores of PC4, whereas woods from Kéniéba were characterized by positive scores of PC3 and negative scores of PC4. For both PC3 and PC4, intermediate values of component scores were related to wood from Sibi, and regarding *P. erinaceus*, the woods from Kita were associated with negative scores of PC5 and PC6, whereas woods from Sibi and Kéniéba were similarly associated with positive scores of PC5 and PC6.

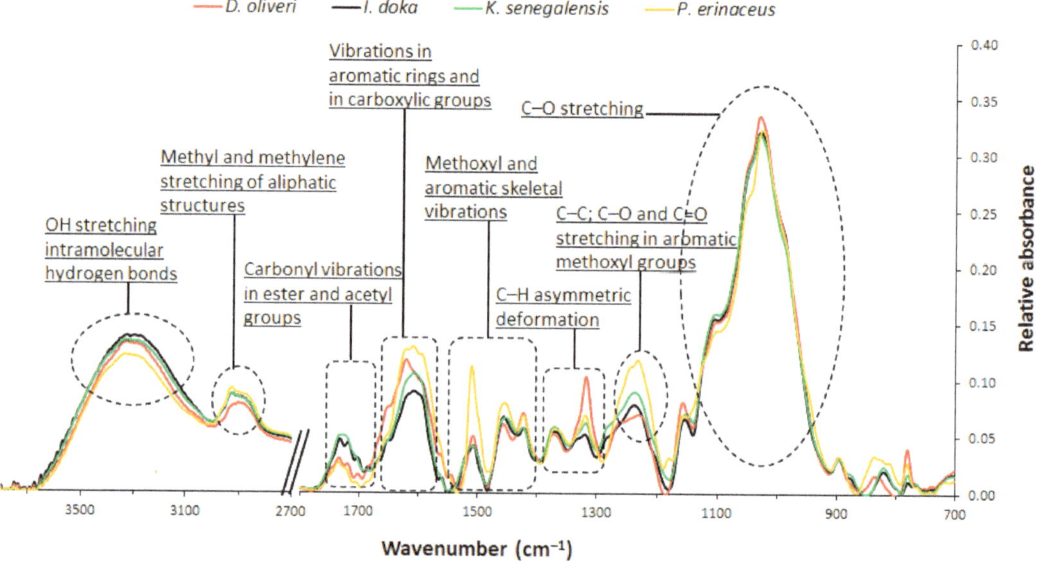

Figure 4. Average FTIR absorption spectra of the studied wood species, indicating the related bond vibrations for FTIR absorption bands.

Table 4. Summary of the one-way ANOVA using component scores of PCA components.

Wood Species	PCA Component	df	F	Sig.
Daniellia oliveri	PC1	2	0.21	0.81
	PC2	2	0.41	0.67
	PC3	2	0.42	0.66
	PC4	2	6.99	0.01 [1]
	PC5	2	0.44	0.65
	PC6	2	3.03	0.08
Isoberlinia doka	PC1	1	1.54	0.25
	PC2	1	0.01	0.92
	PC3	1	0.15	0.71
	PC4	1	0.00	0.99
	PC5	1	2.93	0.13
	PC6	1	0.35	0.57

Table 4. *Cont.*

Wood Species	PCA Component	df	F	Sig.
Khaya senegalensis	PC1	2	1.39	0.27
	PC2	2	1.30	0.29
	PC3	2	5.50	0.01 [2]
	PC4	2	4.51	0.02 [2]
	PC5	2	0.49	0.62
	PC6	2	2.25	0.13
Pterocarpus erinaceus	PC1	2	3.50	0.06
	PC2	2	3.40	0.06
	PC3	2	2.87	0.09
	PC4	2	1.15	0.35
	PC5	2	12.08	0.00 [1]
	PC6	2	4.56	0.03 [2]

[1] significant differences at $p < 0.01$; and [2] significant differences at $p < 0.05$.

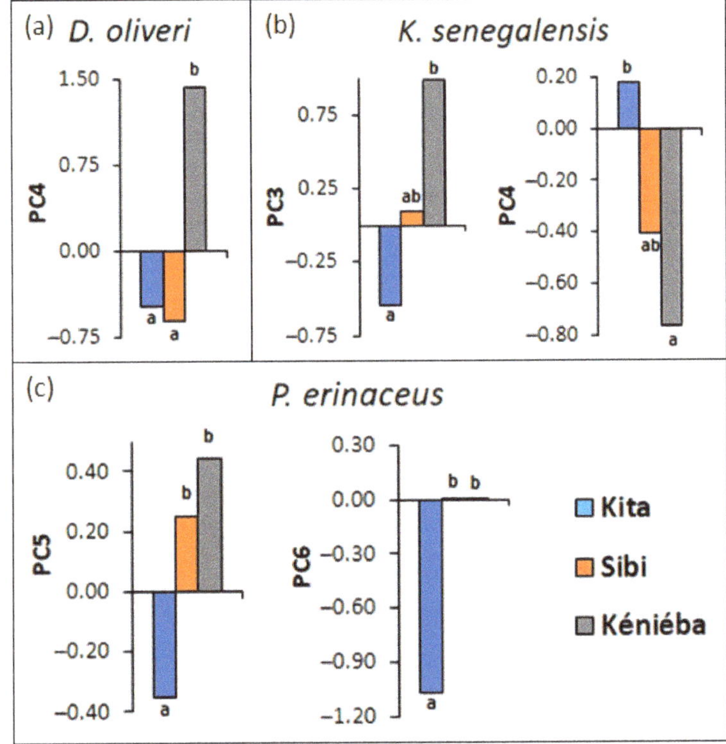

Figure 5. Bar chart of the mean values of component scores of PCA factors showing significant differences between samples from the three areas; (**a**) for *D. oliveri* ($n = 5$ for Kita, $n = 6$ for Sibi and $n = 6$ for Kéniéba), (**b**) for *K. senegalensis* ($n = 8$ for Kita, $n = 8$ for Sibi and $n = 8$ for Kéniéba), and (**c**) for *P. erinaceus* ($n = 6$ for Kita, $n = 5$ for Sibi and $n = 6$ for Kéniéba); (the labels a, ab, and b refer to group classification in ascending order of the Student–Newman–Keuls post-hoc test).

From the results described above, the FTIR signal related to the main macromolecules (cellulose, hemicellulose, and lignin) and extractive contents can be explored to evaluate the relation between the variations in wood characteristics and provenance of the studied timber wood. Again, it seems that the wood of the studied tree species was not equally affected by the different local environmental conditions. These results suggest that glycosidic bonds (in polysaccharides) and aliphatic structures (in polysaccharides and extractives) are the most indicative of the identified chemical variations in relation to the origin of the wood samples for *D. oliveri* and *K. senegalensis*. In fact, the variation in polysaccharide contents in wood can also be linked to the degree of disturbance related to the environmental conditions at the tree growing location [17,51]. Thus, we could conclude that the local environmental conditions at Kéniéba and Kita have a significant influence on the polysaccharide contents of wood of *D. oliveri* and *K. senegalensis*, respectively. This result is also in agreement with what has been described earlier for the color data, for *K. senegalensis*, about the significant differences showed by the yellowness (b*) values, which can be related to the polysaccharide contents [46,48]. Nevertheless, we should also draw attention to the fact that biochemical responses to environmental factors may vary depending on the wood species [51]. Furthermore, for *P. erinaceus*, it seems that extractive compounds with aromatic structures are also indicative of changes that differentiate between wood from different provenance areas. Previous studies on this wood species, in Togo and in Mali (West Africa), revealed that its extractive contents depend on the soil properties at the tree growing location [4,52]. Finally, these findings highlight that wood chemistry and biochemical processes of wood formation are also related to environmental factors other than climate [9,53].

3.3. Variations in Wood FTIR Ratios

The one-way ANOVA applied to the FTIR ratios dataset revealed the possibility to differentiate the wood samples of the three studied areas. The summary of these results is provided in Table 5. No significant differences were found for samples of *I. doka* and *K. senegalensis*, whereas significant differences were found for *D. oliveri* based on the guaiacyl/syringyl (GS2, 1510/1595) and lignin/carbohydrate (LC1, 1510/1460; LC2, 1510/1375; and LC3, 1510/895) ratios, and for *P. erinaceus* for all the considered ratios (Table 5). Again, as illustrated in Figure 6, samples from Kita and Kéniéba stand out by the average values of the FTIR ratios. For *D. oliveri*, samples from the areas of Kita and Sibi appeared to be very much alike with positive values of the FTIR ratios, whereas samples from the area of Kéniéba were associated with negative values of the FTIR ratios. Regarding *P. erinaceus*, all FTIR ratios except for the guaiacyl/syringyl ratio (GS1, 1265/1230) showed significantly lower values for Kita, as compared to Sibi and Kéniéba, which presented similar values. In contrast, the 1265/1230 ratio values were significantly higher for wood from Kita than for the wood from Sibi and Kéniéba.

Table 5. Summary of the one-way ANOVA using FTIR ratios.

Wood Species	FTIR Ratios	df	F	Sig.
	1425/895 (LOI)	2	0.72	0.50
	1265/1230 (GS1)	2	1.18	0.34
Daniellia oliveri	1510/1595 (GS2)	2	4.53	0.03 [2]
	1510/1460 (LC1)	2	5.59	0.02 [2]
	1510/1375 (LC2)	2	4.34	0.03 [2]
	1510/895 (LC3)	2	4.26	0.04 [2]

Table 5. *Cont.*

Wood Species	FTIR Ratios	df	F	Sig.
Isoberlinia doka	1425/895 (LOI)	1	2.38	0.16
	1265/1230 (GS1)	1	0.01	0.94
	1510/1595 (GS2)	1	0.22	0.65
	1510/1460 (LC1)	1	0.00	0.96
	1510/1375 (LC2)	1	0.32	0.59
	1510/895 (LC3)	1	0.32	0.59
Khaya senegalensis	1425/895 (LOI)	2	0.40	0.67
	1265/1230 (GS1)	2	3.42	0.05
	1510/1595 (GS2)	2	2.43	0.11
	1510/1460 (LC1)	2	0.90	0.42
	1510/1375 (LC2)	2	1.80	0.19
	1510/895 (LC3)	2	1.40	0.27
Pterocarpus erinaceus	1425/895 (LOI)	2	18.24	0.00 [1]
	1265/1230 (GS1)	2	18.23	0.00 [1]
	1510/1595 (GS2)	2	6.01	0.01 [2]
	1510/1460 (LC1)	2	5.29	0.02 [2]
	1510/1375 (LC2)	2	3.65	0.05
	1510/895 (LC3)	2	8.19	0.00 [1]

[1] significant differences at $p < 0.01$; and [2] significant differences at $p < 0.05$.

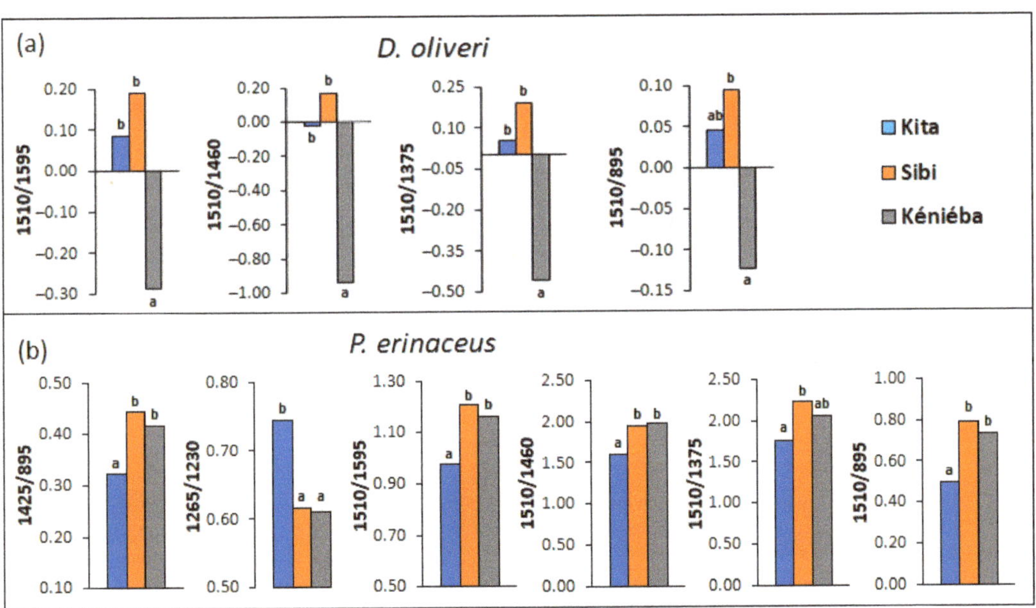

Figure 6. Bar chart of the mean values of FTIR ratios showing significant differences between samples from the three areas; (**a**) for *D. oliveri* (*n* = 5 for Kita, *n* = 6 for Sibi and *n* = 6 for Kéniéba), (**b**) for *P. erinaceus* (*n* = 6 for Kita, *n* = 5 for Sibi and *n* = 6 for Kéniéba); (the labels a, ab, and b refer to group classification in ascending order of the Student–Newman–Keuls post-hoc test).

The results of the FTIR ratios for *D. oliveri* and *P. erinaceus* suggest that they can be used to assess the changes in the molecular structure of the wood cell walls macromolecules related to provenance. Each of the calculated FTIR ratios is somehow connected to specific chemical characteristics. For instance, the ratio 1425/895, also known as the lateral order index (LOI), is an empirical crystallinity index that is associated with the number of crystalline structures of cellulose [43,54]. The guaiacy/syringyl ratios (1265/1230 and 1510/1595) inform about the molecular structure of the lignin contents and the relative number of methoxyl groups in guaiacyl and syringyl nuclei [55,56]. The lignin/carbohydrate ratios (1510/1460, 1510/1375, and 1510/895) are indicative of the relative proportion of lignin and polysaccharide compounds [44,57]. Thus, according to the results of the FTIR ratios, it can be deduced that, for *D. oliveri*, the wood samples from Kéniéba are characterized by a higher proportion of syringyl than guaiacyl groups in the molecular structure of lignin compounds and the highest proportion of polysaccharide compounds. As for *P. erinaceus*, the wood samples from Kita appear to be associated with the less ordered crystalline structure in cellulose, the highest proportion of methoxyl groups in the syringyl nuclei of lignin compounds, and the lowest proportion of polysaccharide compounds. Moreover, the known correlation between the lignin/carbohydrate ratio suggests that *P. erinaceus* wood from Kita may have lower mechanical strength in comparison to *P. erinaceus* wood from Sibi and Kéniéba [58,59].

4. Conclusions

The main conclusion of the study is that the variation in wood color and chemical properties can be related to the local environmental conditions at the tree growing locations. Since the three studied areas are under the Sudano–Guinean climatic zone, the observed variations in wood properties are most likely connected to the local soil properties and/or landforms. Additionally, it was clear that the studied tree species were not equally affected by the different local environmental conditions. *P. erinaceus* appears to be the most sensitive species, as it presented significant differences between samples from the three areas for the largest number of variables (color parameters, PCA components, and FTIR-ATR ratios) used in this research. *D. oliveri* and *K. senegalensis*, however, showed significant differences between sample provenance for a lower number of variables. *I. doka*, for its part, showed no significant differences. Furthermore, the results revealed that important differences exist between wood samples from Kati and Kéniéba; these differences are related to color parameters (redness (a*) and hue (h°)), as well as a number of FTIR signals related to chemical content in the wood cell walls (polysaccharides and lignins).

Finally, from this work, we believe that other factors characterizing the local environmental conditions (e.g., soil properties) rather than climatic factors can drive significant changes in wood properties. Certainly, further research could help to determine specific links between wood properties (physical and chemical) and environmental factors (e.g., soil factors) that are not necessarily climatic.

Author Contributions: Conceptualization, M.T. and A.M.C.; methodology, M.T. and A.M.C.; investigation, M.T. and A.M.C.; resources, M.T. and A.M.C.; data curation, M.T. and A.M.C.; writing—original draft preparation, M.T.; writing—review and editing, M.T. and A.M.C.; supervision, A.M.C. All authors have read and agreed to the published version of the manuscript.

Funding: This research is funded by Grupos de Referencia Competitiva (ED431C 2021/32) by Xunta de Galicia. It was developed within the framework of the visiting fellowship program for African researchers (Programa BECAS ÁFRICA-MED 2021-2022) of the Spanish Agency for International Development Cooperation (AECID). MT is currently funded by Plan Galego I2C Modalidade A (ED481B-2022-017).

Data Availability Statement: The data presented in this study are available on request from the corresponding author.

Acknowledgments: The authors would like to thank Bakary Traoré (Katibougou IPR/IFRA, Mali) who provided the wood samples. Also, the authors would like to thank the editors and reviewers for their time and insightful suggestions that helped to improve our manuscript.

Conflicts of Interest: The authors declare no conflict of interest. And the funders had no role in the design of the study; in the collection, analyses, or interpretation of data; in the writing of the manuscript; or in the decision to publish the results.

References

1. Jozsa, L.A.; Middleton, G.R. *A Discussion of Wood Quality Attributes and Their Practical Implications*; Forintek Canada Corporation Special Publication No. SP-34; Forintek Canada Corporation: Vancouver, BC, Canada, 1944.
2. Rowell, R.M.; Pettersen, R.; Han, J.S.; Rowell, J.S.; Tshabalala, M.A. Cell Wall Chemistry. In *Handbook of Wood Chemistry and Wood Composites*; Rowell, R.M., Ed.; CRC Press: Boca Raton, FL, USA, 2005.
3. Wegst, U.G. Wood for sound. *Am. J. Bot.* **2006**, *93*, 1439–1448. [CrossRef] [PubMed]
4. Traoré, B.; Brancheriau, L.; Perré, P.; Stevanovic, T.; Diouf, P. Acoustic quality of vène wood (*Pterocarpus erinaceus* Poir.) for xylophone instrument manufacture in Mali. *Ann. For. Sci.* **2010**, *67*, 815. [CrossRef]
5. Fengel, D.; Wegener, G. *Wood: Chemistry, Ultrastructure, Reactions*; Walter de Gruyter: Berlin, Germany, 1984.
6. Nabais, C.; Hansen, J.K.; David-Schwartz, R.; Klisz, M.; Lopez, R.; Rozenberg, P. The effect of climate on wood density: What provenance trials tell us? *For. Ecol. Manag.* **2018**, *408*, 148–156. [CrossRef]
7. Hevia, A.; Campelo, F.; Chambel, R.; Vieira, J.; Alía, R.; Majada, J.; Sánchez-Salguero, R. Which matters more for wood traits in *Pinus halepensis* Mill., provenance or climate? *Ann. For. Sci.* **2020**, *77*, 55. [CrossRef]
8. Moya, R.; Gaitan-Alvarez, J.; Berrocal, A.; Caceres, C.B.; Hernandez, R.E. Wood properties of nine acetylated tropical hardwoods from fast-growth plantations in Costa Rica. *Wood Fiber Sci.* **2022**, *54*, 134–148. [CrossRef]
9. Fritts, H.C. *Tree Rings and Climate*; Academic Press: Cambridge, MA, USA, 1976.
10. Côté, B.; Hendershot, W.H.; Fyles, J.W.; Roy, A.G.; Bradley, R.; Biron, P.M.; Courchesne, F. The phenology of fine root growth in a maple-dominated ecosystem: Relationships with some soil properties. *Plant Soil* **1998**, *201*, 59–69. [CrossRef]
11. Fritts, H.C.; Swetnam, T.W. Dendroecology: A tool for evaluating variations in past and present forest environments. *Adv. Ecol. Res.* **1989**, *19*, 111–188.
12. Schweingruber, F.H. *Trees and Wood in Dendrochronology: Morphological, Anatomical, and Tree-Ring Analytical Characteristics of Trees Frequently used in Dendrochronology*; Springer Science & Business Media: Berlin/Heidelberg, Germany, 2012.
13. Hughes, M.K. Dendrochronology in climatology—The state of the art. *Dendrochronologia* **2002**, *20*, 95–116. [CrossRef]
14. Sass-Klaassen, U. Tree physiology: Tracking tree carbon gain. *Nat. Plants* **2015**, *1*, 15175. [CrossRef]
15. Ibanez, T.; Chave, J.; Barrabé, L.; Elodie, B.; Boutreux, T.; Trueba, S.; Vandrot, H.; Birnbaum, P. Community variation in wood density along a bioclimatic gradient on a hyper-diverse tropical island. *J. Veg. Sci.* **2017**, *28*, 19–33. [CrossRef]
16. Jansen, K.; Sohrt, J.; Kohnle, U.; Ensminger, I.; Gessler, A. Tree ring isotopic composition, radial increment and height growth reveal provenance-specific reactions of Douglas-fir towards environmental parameters. *Trees* **2013**, *27*, 37–52. [CrossRef]
17. Kusiak, W.; Majka, J.; Ratajczak, I.; Górska, M.; Zborowska, M. Evaluation of environmental impact on selected properties of lime (*Tilia cordata* Mill.) wood. *Forests* **2020**, *11*, 746. [CrossRef]
18. Sotelo Montes, C.; Hernández, R.E.; Beaulieu, J.; Weber, J.C. Genetic variation in wood color and its correlations with tree growth and wood density of *Calycophyllum spruceanum* at an early age in the Peruvian Amazon. *New For.* **2008**, *35*, 57–73. [CrossRef]
19. Moya, R.; Calvo-Alvarado, J. Variation of wood color parameters of *Tectona grandis* and its relationship with physical environmental factors. *Ann. For. Sci.* **2012**, *69*, 947–959. [CrossRef]
20. Bessa, F.; Sousa, V.; Quilhó, T.; Pereira, H. An Integrated Similarity Analysis of Anatomical and Physical Wood Properties of Tropical Species from India, Mozambique, and East Timor. *Forests* **2022**, *13*, 1675. [CrossRef]
21. Rana, R.; Müller, G.; Naumann, A.; Polle, A. FTIR spectroscopy in combination with principal component analysis or cluster analysis as a tool to distinguish beech (*Fagus sylvatica* L.) trees grown at different sites. *Holzforschung* **2008**, *62*, 530–538. [CrossRef]
22. Santoni, I.; Callone, E.; Sandak, A.; Sandak, J.; Dirè, D. Solid state NMR and IR characterization of wood polymer structure in relation to tree provenance. *Carbohydr. Polym.* **2015**, *17*, 710–721. [CrossRef]
23. Sharma, V.; Yadav, J.; Kumar, R.; Tesarova, D.; Ekielski, A.; Mishra, P.K. On the rapid and non-destructive approach for wood identification using ATR-FTIR spectroscopy and chemometric methods. *Vib. Spectrosc.* **2020**, *110*, 103097. [CrossRef]
24. Toscano, G.; Maceratesi, V.; Leoni, E.; Stipa, P.; Laudadio, E.; Sabbatini, S. FTIR spectroscopy for determination of the raw materials used in wood pellet production. *Fuel* **2022**, *313*, 123017. [CrossRef]
25. Schollaen, K.; Baschek, H.; Heinrich, I.; Slotta, F.; Pauly, M.; Helle, G. A guideline for sample preparation in modern tree-ring stable isotope research. *Dendrochronologia* **2017**, *44*, 133–145. [CrossRef]
26. Domínguez-Delmás, M.; Rich, S.; Traoré, M.; Hajj, F.; Poszwa, A.; Akhmetzyanov, L.; García-González, I.; Groenendijk, P. Tree-ring chronologies, stable strontium isotopes and biochemical compounds: Towards reference datasets to provenance Iberian shipwreck timbers. *J. Archaeol. Sci. Rep.* **2020**, *34*, 102640. [CrossRef]

27. Konaré, D. Équations allométriques pour l'évaluation de la biomasse foliaire de trois espèces ligneuses fourragères. Cas de *Afzelia africana*, *Ficus gnaphalocarpa* et *Pterocarpus erinaceus*, des parcours naturels du cercle de Kéniéba au Sud-Ouest du Mali. *Rev. Afr. Des Sci. Soc. Et De La Sante Publique* **2019**, *1*, 29–52.

28. Orwa, C.; Mutua, A.; Kindt, R.; Jamnadass, R.; Simons, A. *Agroforestree Database: A Tree Reference and Selection Guide Version 4.0.*; World Agroforestry Centre: Nairobi, Kenya, 2009.

29. Segla, K.N.; Adjonou, K.; Rabiou, H.; André Bationo, B.; Mahamane, A.; Guibal, D.; Kokou, K.; Chaix, G.; Kokutse, A.D.; Langbour, P. Relations between the ecological conditions and the properties of *Pterocarpus erinaceus* Poir. wood from the Guinean-Sudanian and Sahelian zones of West Africa. *Holzforschung* **2020**, *74*, 999–1009. [CrossRef]

30. Sanogo, K.; Gebrekirstos, A.; Bayala, J.; van Noordwijk, M. Climate-growth relationships of *Daniellia oliveri* (Rolfe) Hutch. & Dalziel in the Sudanian zone of Mali, West Africa. *Trees For. People* **2022**, *10*, 100333.

31. Hänke, H.; Börjeson, L.; Hylander, K.; Enfors-Kautsky, E. Drought tolerant species dominate as rainfall and tree cover returns in the West African Sahel. *Land Use Policy* **2016**, *59*, 111–120. [CrossRef]

32. Hérault, B.; N'guessan, A.K.; Ahoba, A.; Bénédet, F.; Coulibaly, B.; Doua-Bi, Y.; Koffi, T.; Koffi-Konan, J.C.; Konaté, I.; Tiéoulé, F.; et al. The long-term performance of 35 tree species of sudanian West Africa in pure and mixed plantings. *For. Ecol. Manag.* **2020**, *468*, 118171. [CrossRef]

33. Brandt, M.; Romankiewicz, C.; Spiekermann, R.; Samimi, C. Environmental change in time series–An interdisciplinary study in the Sahel of Mali and Senegal. *J. Arid Environ.* **2014**, *105*, 52–63. [CrossRef]

34. Faye, M.D.; Weber, J.C.; Abasse, T.A.; Boureima, M.; Larwanou, M.; Bationo, A.B.; Diallo, B.O.; Sigué, H.; Dakouo, J.M.; Samaké, O.; et al. Farmers' preferences for tree functions and species in the West African Sahel. *For. Trees Livelihoods* **2011**, *20*, 113–136. [CrossRef]

35. World Conservation Monitoring Centre. Khaya senegalensis. In *The IUCN Red List of Threatened Species*; e.T32171A9684583; World Conservation Monitoring Centre: Cambridge, UK, 1998.

36. Contu, S. Isoberlinia doka. In *The IUCN Red List of Threatened Species*; e.T19892774A20090324; World Conservation Monitoring Centre: Cambridge, UK, 2012.

37. Barstow, M. Pterocarpus erinaceus. In *The IUCN Red List of Threatened Species*; e.T62027797A62027800; World Conservation Monitoring Centre: Cambridge, UK, 2018.

38. Botanic Gardens Conservation International (BGCI); IUCN SSC Global Tree Specialist Group. Daniellia oliveri. In *The IUCN Red List of Threatened Species*; e.T144315321A149026856; World Conservation Monitoring Centre: Cambridge, UK, 2019.

39. Ky-Dembele, C.; Bayala, J.; Kalinganire, A.; Traoré, F.T.; Koné, B.; Olivier, A. Vegetative propagation of twelve fodder tree species indigenous to the Sahel, West Africa. *South. For. A J. For. Sci.* **2016**, *78*, 185–192. [CrossRef]

40. Martínez Cortizas, A.; López-Costas, O.; Orme, L.; Mighall, T.; Kylander, M.E.; Bindler, R.; Gallego Sala, Á. Holocene atmospheric dust deposition in NW Spain. *Holocene* **2020**, *30*, 507–518. [CrossRef]

41. Traoré, M.; Kaal, J.; Martínez Cortizas, A. Variation of wood color and chemical composition in the stem cross-section of oak (*Quercus spp.*) trees, with special attention to the sapwood-heartwood transition zone. *Spectrochim. Acta Part A Mol. Biomol. Spectrosc.* **2023**, *285*, 121893. [CrossRef] [PubMed]

42. Álvarez Fernández, N.; Martínez Cortizas, A. *Andurinha: Make Spectroscopic Data Processing Easier*; R Package Version 0.0.2.; R Foundation for Statistical Computing: Vienna, Austria, 2020.

43. Auxenfans, T.; Crônier, D.; Chabbert, B.; Paës, G. Understanding the structural and chemical changes of plant biomass following steam explosion pretreatment. *Biotechnol. Biofuels* **2017**, *10*, 36. [CrossRef] [PubMed]

44. Popescu, M.C.; Popescu, C.M.; Lisa, G.; Sakata, Y. Evaluation of morphological and chemical aspects of different wood species by spectroscopy and thermal methods. *J. Mol. Struct.* **2011**, *988*, 65–72. [CrossRef]

45. Klumpers, J.; Janin, G.; Becker, M.; Lévy, G. The influences of age, extractive content and soil water on wood color in oak: The possible genetic determination of wood color. In *Annales des Sciences Forestières*; EDP Sciences: Les Ulis, France, 1993; Volume 50, pp. 403s–409s.

46. Gierlinger, N.; Jacques, D.; Grabner, M.; Wimmer, R.; Schwanninger, M.; Rozenberg, P.; Pâques, L.E. Colour of larch heartwood and relationships to extractives and brown-rot decay resistance. *Trees* **2004**, *18*, 102–108. [CrossRef]

47. Moya, R.; Berrocal, A. Wood colour variation in sapwood and heartwood of young trees of *Tectona grandis* and its relationship with plantation characteristics, site, and decay resistance. *Ann. For. Sci.* **2010**, *67*, 109. [CrossRef]

48. Moya, R.; Fallas, R.S.; Bonilla, P.J.; Tenorio, C. Relationship between wood color parameters measured by the CIELab system and extractive and phenol content in *Acacia mangium* and *Vochysia guatemalensis* from fast-growth plantations. *Molecules* **2012**, *17*, 3639–3652. [CrossRef]

49. Nelson, N.D.; Maeglin, R.R.; Wahlgren, H.E. Relationship of black walnut wood color to soil properties and site. *Wood Fiber Sci.* **1969**, *1*, 29–37.

50. Sotelo Montes, C.; Weber, J.C.; Silva, D.A.; Andrade, C.; Muñiz, G.I.; Garcia, R.A.; Kalinganire, A. Effects of region, soil, land use, and terrain type on fuelwood properties of five tree/shrub species in the Sahelian and Sudanian ecozones of Mali. *Ann. For. Sci.* **2012**, *69*, 747–756. [CrossRef]

51. Zhang, P.; Wu, F.; Kang, X. Chemical properties of wood are under stronger genetic control than growth traits in *Populus tomentosa* Carr. *Ann. For. Sci.* **2015**, *72*, 89–97. [CrossRef]

52. Segla, K.N.; Kokutse, A.D.; Adjonou, K.; Langbour, P.; Chaix, G.; Guibal, D.; Kokou, K. Caractéristiques biophysiques du bois de *Pterocarpus erinaceus* (Poir.) en zones guinéenne et soudanienne au Togo. *Bois Forêts Trop.* **2015**, *324*, 51–64. [CrossRef]
53. Creber, G.T.; Chaloner, W.G. Influence of environmental factors on the wood structure of living and fossil trees. *Bot. Rev.* **1984**, *50*, 357–448. [CrossRef]
54. Nelson, M.L.; O'Connor, R.T. Relation of certain infrared bands to cellulose crystallinity and crystal lattice type. Part II. A new infrared ratio for estimation of crystallinity in celluloses I and II. *J. Appl. Polym. Sci.* **1964**, *8*, 1325–1341. [CrossRef]
55. Colom, X.; Carrillo, F. Comparative study of wood samples of the northern area of Catalonia by FTIR. *J. Wood Chem. Technol.* **2005**, *25*, 1–11. [CrossRef]
56. Popescu, C.M.; Singurel, G.; Popescu, M.C.; Vasile, C.; Argyropoulos, D.S.; Willför, S. Vibrational spectroscopy and X-ray diffraction methods to establish the differences between hardwood and softwood. *Carbohydr. Polym.* **2009**, *77*, 851–857. [CrossRef]
57. Pandey, K.K. A note on the influence of extractives on the photo-discoloration and photo-degradation of wood. *Polym. Degrad. Stab.* **2005**, *87*, 375–379. [CrossRef]
58. Whetten, R.W.; MacKay, J.J.; Sederoff, R.R. Recent advances in understanding lignin biosynthesis. *Annu. Rev. Plant Biol.* **1998**, *49*, 585–609. [CrossRef]
59. Novaes, E.; Kirst, M.; Chiang, V.; Winter-Sederoff, H.; Sederoff, R. Lignin and Biomass: A Negative Correlation for Wood Formation and Lignin Content in Trees. *Plant Physiol.* **2010**, *154*, 555–561. [CrossRef]

Article

Influence of Wood Knots of Chinese Weeping Cypress on Selected Physical Properties

Jianhua Lyu [1,2,†], Hongyue Qu [1,2,†] and Ming Chen [1,2,*]

1   Department of Product Design, College of Forestry, Sichuan Agricultural University, Chengdu 611130, China
2   Wood Industry and Furniture Engineering Key Laboratory of Sichuan Provincial Department of Education, Chengdu 611130, China
*   Correspondence: chenming@sicau.edu.cn
†   These authors contributed equally to this work.

Abstract: The effects of wood knots of Chinese weeping cypress (*Cupressus funebris* Endl.) wood on selected physical and color properties were investigated. Thirty samples of live knots, dead knots, and clear wood groups were selected for experiments to determine the physical properties of wood density, wood shrinkage, wood swelling, and wood color. The experimental analysis results showed that the wood density values are in the order: dead knots > live knots > clear wood, with a significant difference in wood density between different groups ($p < 0.01$). In addition, the values of the air-dry volumetric wood shrinkage, air-dry volumetric wood swelling, oven-dry volumetric wood shrinkage, and oven-dry volumetric wood swelling ratios are in the order: dead knots > live knots > clear wood, being consistent with a variation in wood density. Three groups of wood colors were provided: the color of clear wood is light, the color of live knots is reddish, and the color of live knots is blackish, in relative terms. The chromatic aberration between the three groups can be identified, and the wood color difference resulted from the discrepancy in the lightness index.

Keywords: *Cupressus funebris* Endl. wood; wood knots; physical properties; wood color properties

Citation: Lyu, J.; Qu, H.; Chen, M. Influence of Wood Knots of Chinese Weeping Cypress on Selected Physical Properties. *Forests* **2023**, *14*, 1148 . https://doi.org/10.3390/f14061148

Academic Editors: Vicelina Sousa, Helena Pereira, Teresa Quilhó and Isabel Miranda

Received: 27 February 2023
Revised: 29 May 2023
Accepted: 31 May 2023
Published: 1 June 2023

1. Introduction

Cupressus L. (common name Cypress) is a genus of evergreen trees belonging to the family Cupressaceae, endemic in the Mediterranean region and includes 20 species distributed in temperate and subtropical regions, such as the southern part of North America, East Asia, the Himalayan region, and the Mediterranean region. Members of the genus *Cupressus* are mostly tall evergreen conifers with short scale-like leaves when the trees are adult, whereas they are bigger and needle-like when the trees are young, and they have good wood properties, which can be used for the construction of buildings, ships, furniture, and coffins. The branches and leaves of *Cupressus* trees can provide essential oil, and the genus *Cupressus* is generally regarded as one of the main tree species that is a commercial Chinese cedarwood oil extraction source. In addition to forest product applications, the use of *Cupressus* trees as ornamental plants, wind breaks, and hedges has increased all over the world, which mainly includes *Cupressus glabra* Sudw., *Cupressus macrocarpa* A.Cunn., *Cupressus lusitanica* Mill., *Cupressus sempervirens* Linn., etc. It is possible that *Cupressus* trees were planted in formal gardens built in proximity to funerary temples and cemeteries in many cultures, including ancient Greece, Rome, and China. This might link the tree to the concept of immortality and solemnity.

China, the main area of natural distribution for *Cupressus funebris* Endl., known also as Chinese weeping cypress, has four endemic species in the genus *Cupressus*, one of which is *C. funebris* [1]. *C. funebris* is one of the representative evergreen coniferous species in China's subtropics; it is found in major vegetation restoration forests and is a timber forest species in the upper reaches of the Yangtze River of China, such as in the Sichuan, western Hubei, and Guizhou provinces [2–5]. Some botanists accept Franco's reclassification of

C. funebris as *Chamaecyparis funebris* (Endl.) primarily because of its flattened foliage sprays and atypically small ovulate cones [6]. *C. funebris* wood is moderately hard, with a delicate structure, low wood shrinkage, and high resistance to decay, making it an excellent material for producing durable furniture, handicrafts, and other wood products. Woodwork made of *C. funebris* is popular with consumers because of its strength, resistance to decay, unique texture, and fragrant smell. Overall, *C. funebris* wood has been used for shipbuilding, construction, packaging, and woodworking since ancient times [7,8].

Trees spontaneously produce some natural defects during growth, and knots are one of the most common defects [9]; they are also one of the important factors that directly change the timber yield, grade, and price [10–12]. Knots play an important role in the physiological growth process of trees, and external branches are the main source of a tree's nutritional needs as it grows. The leaves on living branches convert external CO_2 into organic matter and O_2 through photosynthesis, and the organic matter is transported to the tree roots through the sieve tubes of the phloem and then absorbed, contributing to tree growth [13]. Knots arise mainly from decaying parts caused by fungal infections and greatly affect the wood's appearance, quality, and mechanical properties [14]. Therefore, searching for ways to minimize the occurrence of knots and expand the use of knot-related wood has become an important research topic.

Knots are one of the most common wood defects and are specific imperfections in the wood that reduce its strength, which can also be exploited for artistic effects. Knots can be divided into live and dead knots, where live knots, formed from the living part of a branch, are intergrown with the wood, and dead knots, formed from the dead part of a branch, have tracheids that are structurally disjunct from tracheids in the stem [15–17].

Knots are the main influencing factor of wood quality, regarding chemical composition and physical and mechanical properties. Knots have an important influence on the grade of structural lumber via reducing stiffness and strength through local distortion and the deviation of wood fibers from the longitudinal axis of the stem, resulting in weak mechanical areas [18]. Cherry et al. found that knots were significantly different from clear wood for all test types of mechanical properties, within a range of 48% to 196% [19].

Wood density, wood shrinkage, and wood swelling are the most important physical properties that affect processing performance. Yong et al. found that the average width of the annual rings of branch wood was 69% smaller than that of tree trunk wood, and the air-dry wood density of branch wood was 70% greater than that of tree trunk wood [20]. The wood density of knots varies depending on where it is located. Wang et al. found that the closer the knot was to the branch parts, the higher the wood density; the wood density was higher in the upper part of the whole knot and lower in the lower part [21]. Knots reduce the strength of wood, and the effect of knots on strength depends on the proportion of the cross-section of the given piece occupied by the knots and its relative location; as the proportion of knots on the cross-section increases, wood density increases accordingly [22]. In short, knots will increase the wood density of corresponding wood parts. Some studies have found that changes in the orientation of ducts, fibers, and other tissues distorted the grain around the knots, causing the knot grain to be different from normal wood. In contrast, the shrinkage and swelling of wood are anisotropic, resulting in different dry wood shrinkage coefficients between the knots and the surrounding wood in all directions, which can easily cause wood cracking during the drying process [23]. Knots do not shrink as much laterally or in the diameter of the surrounding wood. Studies have shown that knots can influence the wood swelling coefficients in a longitudinal direction, while the influence on wood swelling coefficients in tangential and radial directions is negligible [24]. Knots fall out because the knots and the surrounding clear wood have different wood shrinkage, which can occur during thermal modification processing [25]. The characteristics of the knots, such as volume, affect the wood shrinkage value; studies have shown that wood shrinkage is not different, especially in specimens containing small knots [21].

Knots affect not only the physical properties of wood but also the appearance quality of wood products. It is refereed that checks in knots are the worst defect for a wood's appearance grade [26]; furthermore, wood color is an important appearance quality characteristic for the appearance grade of wood. Some studies have shown that the color of a wood's knots is darker than the wood [27,28]; the main causes are that dead knots are separated from the surrounding tissues, and the detached edges varied in color, showing hard black scars with clear boundaries in color from the surrounding area. Further, the color of the live knots underwent gradual changes and deepened [29]. The determination of wood colors after treatment and wood color metrics are increasingly used in wood technology [30–32].

Most studies have focused on forest resource management, wood anatomy, physical and mechanical wood properties, forest genetics and breeding, essential oil properties and extraction, and mechanical properties of wood components of *C. funebris* [7,33–37]. Few studies have been conducted on the effects of knots on physical properties [21,38], and no research specifically on *C. funebris* wood has been reported.

This study had two objectives: the first was to investigate the relationship between different types of knots and the main physical properties of *C. funebris* wood, and the second was to provide certain instructions to realize the commercial application of *C. funebris* wood with knots.

2. Materials and Methods

2.1. Site Sampling

Thirty-three-year-old *Cupressus funebris* Endl. trees from pure forest plantations located in Yongxin Town, Jingyang District, Deyang City, Sichuan Province, China ($31°1'$–$31°19'$ N and $104°15'$–$104°35'$ E) were used in this study. This area is a subtropical humid and semi-humid climate zone with little sunshine, abundant rainfall, and four distinct seasons, with an average annual temperature of 16–17 °C. The forest stands were planted in the mid-1980s, during which time no forestry management and cultivation management measures were conducted. A sample site with the same growth conditions on the mountain's western slope of about 20–30° was selected for harvesting, with a sample area of 0.04 hm^2 (20 m × 20 m).

2.2. Physical Tests

Five *C. funebris* trees with normal growth, complete trunks, straightness, and no obvious defects were randomly selected from the sample area, marked with north–south direction, and measured for age. The diameter at breast height (DBH, i.e., diameter at 1.30 m of stem height) was 14.24 ± 0.84 cm and the tree height was 13.11 ± 0.58 m. One log was cut from 1.3 m to 3.3 m (with a length of 2 m), a second from 3.3 to 5.3 m of the trunk height (with a length of 2 m), and a third from 5.3 to 7.3 m of the trunk height (with a length of 2 m). The sets of 30 wood samples with dimensions of 20 mm × 20 mm × 30 mm for each group, including dead knots, live knots, and clear wood (knot-free), were selected from the 10 cut logs and prepared for the different studies (Figure 1). The physical properties of all the samples were tested according to Chinese National Standards GB/T 1928–2009 "General requirements for physical and mechanical tests of wood" [39], GB/T 1929–2009 "Method of sample logs sawing and test specimens selection for physical and mechanical tests of wood" [40], GB/T 1932–2009 "Method for determination of the shrinkage of wood" [41], GB/T 1933–2009 "Method for determination of the density of wood" [42], and GB/T 1934.2–2009 "Method for determination of the swelling of wood" [43]. The only difference between testing the physical properties of knot samples and clear wood samples was that the knot samples were cut from the inside of the tree, and their volume was measured using the drainage method.

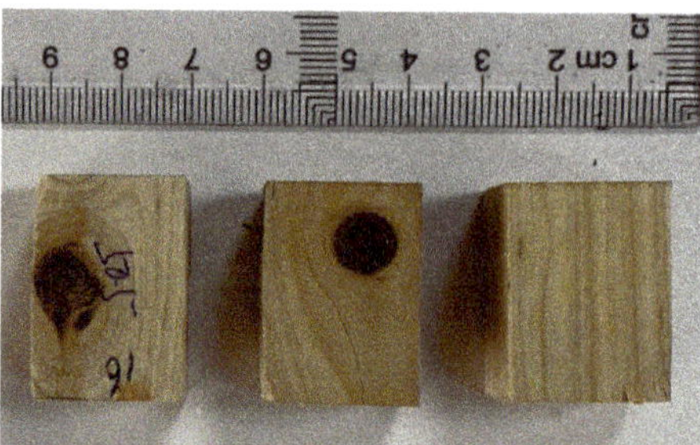

Figure 1. Test specimens from left to right: dead knots, live knots, and clear wood.

First, the specimen was exposed to the conditions of 20 °C temperature and 65% relative humidity to reach equilibrium moisture content, and its weight and radial, tangential, and longitudinal dimensions were measured. Then, the samples were placed at 60 °C for 4 h, and the temperature was raised to 103 °C for 8 h. Then, the samples were measured every 2 h, and when the difference was less than 0.5% of the sample weight, radial, tangential, and longitudinal dimensions were measured. The air-dry wood density was calculated according to Equation (1) and the oven-dry wood density was calculated according to Equation (2). Accuracy index is defined as the overall distance between estimated or observed values and the true value; accuracy index was calculated according to Equation (3).

$$\rho_w = M_w / V_w \tag{1}$$

where ρ_w (g/cm^3) is air-dry wood density when the moisture content is W; M_w (g) is the air-dry mass of each sample when the moisture content is W; V_w (cm^3) is air-dry volume of samples when the moisture content is W.

$$\rho_0 = M_0 / V_0 \tag{2}$$

where ρ_0 (g/cm^3) is oven-dry wood density; M_0 (g) is oven-dry mass of each sample; V_0 (cm^3) is oven-dry volume of samples.

$$P = 2(S_T / X) \tag{3}$$

where P is accuracy index; S_T is the standard error of the mean; X is mean value.

2.3. Surface Color Test

Wood color is closely related to the quality assessment of wood products, which is related to the visual aesthetics of wood products and their decorative properties, thus affecting their economic, collecting, and artistic value. Wood color is a fundamental property of wood, and its chromaticity index is distributed among wood species. There is a great difference in the distribution characteristics of chromatic index among wood species, and even in the same species. Therefore, it is recommended to measure the color variation of wood with knots and its influencing factors to guide the selection and processing of high-value wood products.

A Minolta spectrophotometer model CM-700D (Konica Minolta, Osaka, Japan) was used for color measurement (10° standard observer, D65 standard illumination, color difference format ΔE^*ab). We randomly selected an area on the samples for color testing of clear wood; knot areas were tested on dead and live knots samples for color testing. Tangential sections were used for color testing; if there was no knot area in the tangential section of dead and live knotted wood samples, the color test was performed in the radial or transverse section with the knot. The color of live knots, dead knots, and clear wood of *C. funebris* wood was measured using the spectrophotometer and described in the CIE L*a*b* color system. In this system, color was defined as three numerical values known as the trichromatic coordinates (L*, a*, and b*). The coordinate L* refers to the lightness of a given sample (scored from 0, which represents the black color, to 100, which represents the white one), a* is the coordinate that defines the degree of approximation to the red color when a* takes positive values and green when it takes negative values, and the coordinate b* indicates yellow when it takes positive values and blue when it takes negative values. Each color test was repeated 30 times, and the average was calculated.

2.4. Statistical Analyses

Data were analyzed using Microsoft Excel (version 2013; Microsoft Corp., Redmond, WA, USA) and SPSS (version 19.0; IBM Corp., Armonk, NY, USA), and graphs were generated using Origin (version 9.1; OriginLab Corp., Northampton, MA, USA).

3. Results and Discussion

3.1. Wood Density

The moisture content of the samples was tested. The equilibrium moisture content (EMC) of the dead knot, live knot, and clear wood groups are 14.44%, 14.73%, and 15.08%, respectively. Oven-dry moisture content of the dead knot, live knot, and clear wood groups are 12.25%, 12.41%, and 12.69%, respectively.

Air-dry wood density affects the hardness and strength of wood, and at the same moisture content, it is positively correlated with wood hardness and strength, i.e., the higher the air-dry wood density, the harder and stronger the wood, and vice versa. Furthermore, air-dry wood density is an important factor affecting the final product quality of wood [44]. Whiskers are lines that extend from the top and bottom of the box to the adjacent values.

It can be seen that the box area of clear wood is the largest and the box area of dead knots is the smallest, indicating that the sample data values of clear wood have the largest range of variation and the sample data of dead knots have the smallest range of variation. The mean values of live knots, dead knots, and clear wood are below the median line, indicating that the sample data of all three are clustered at smaller values (Figure 2). Related studies have shown that the wood density is greatest in the branch part of the tree, followed by the knots, and lowest in the trunk. There is an obvious variation in the wood density of trunk [21], for example, in *Pinus taeda*. However, the wood density variation of different parts of the knots has not been reported yet.

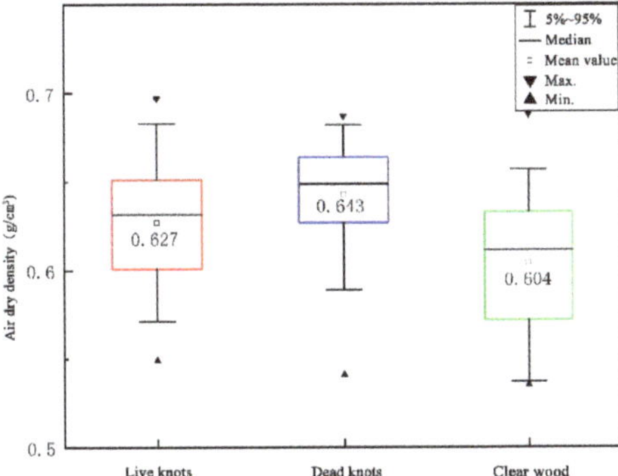

Figure 2. Boxplots of air-dry wood density (g/cm^3) for the live knots, dead knots, and clear wood. The 95% confidence interval, median, mean values, maximum (Max), and minimum (Min) positioning are shown.

In Figure 3, it can be seen that the box area of clear wood is the largest, and the box area of dead knots is the smallest, indicating that the range of variation of sample data for clear wood is the largest among the three groups, and the range of variation of sample data for dead knots is the smallest among the three groups. The mean values of live and dead knots are above the median line, and the mean values of clear wood are below the median line, indicating that the sample data values of live and dead knots are clustered at larger values, contrary to the air-dry wood density of both, while the sample data values of clear wood are clustered at smaller values, consistent with their air-dry wood density.

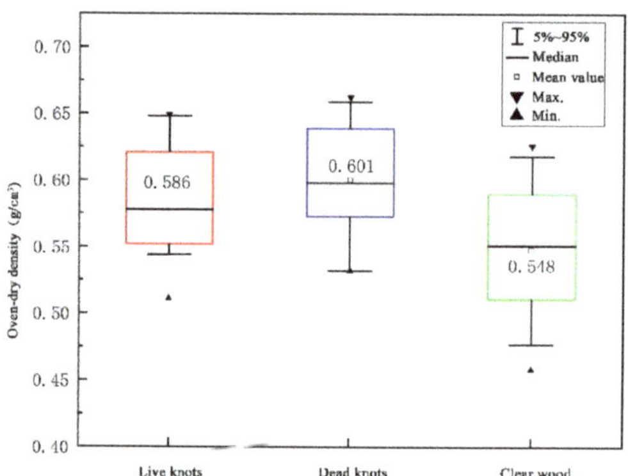

Figure 3. Boxplots of oven-dry wood density (g/cm^3) for the live knots, dead knots, and clear wood. The 95% confidence interval, median, mean values, maximum (Max), and minimum (Min) positioning are shown.

The mean values of live and dead knots are above the median line, and the mean values of clear wood are below the median line, indicating that the sample data values of live and dead knots are clustered in larger values and the sample data values of clear wood are clustered in smaller values, which is consistent with the distribution of the oven-dry wood density data of the three groups. The box area of clear wood is the largest among the three groups, and the box area of dead knots is the smallest among the three groups, indicating that the range of variation of sample data for clear wood is the largest among the three groups and the range of variation of sample data for dead knots is the smallest among the three groups (Figure 4).

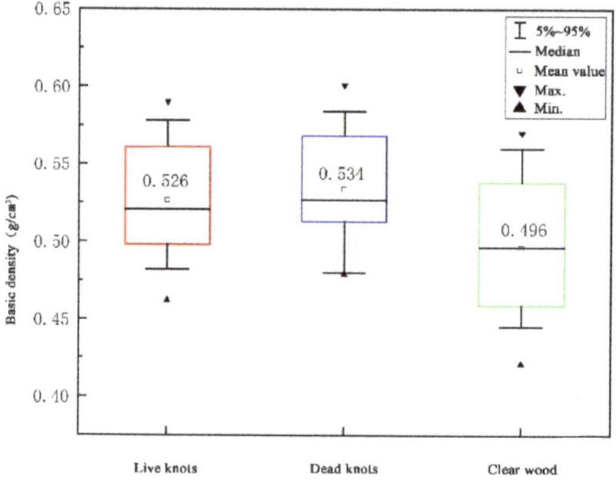

Figure 4. Boxplots of basic wood density (g/cm^3) for the live knots, dead knots, and clear wood. The 95% confidence interval, median, mean values, maximum (Max), and minimum (Min) positioning are shown.

Air-dry, oven-dry, and basic wood density are all medium densities, i.e., they are easy to process and can meet the mechanical requirements of the material used. The three-wood density-changes law for different knots is dead knots > live knots > clear wood. Since knots are the best choice for trees in resisting external changes, and wood density is positively correlated with the resistance of trees, the wood density of knots should be greater than the wood density of clear wood, consistent with the above test results.

From Table 1, it can be seen that all correlations among air-dry wood density, oven-dry wood density, and basic wood density of live knots, dead knots, and clear wood were highly significant ($p < 0.01$), except for the correlation between the basic wood density and the air-dry wood density of dead knots, which was significant ($p < 0.05$). The coefficient of determination of the regression equation between the basic wood density and the oven-dry wood density of dead knots was 0.759 with average regression results. Moreover, the coefficient of the determination of regression equation between the basic wood density and the air-dry wood density of dead knots was 0.184, and the oven-dry wood density and the air-dry wood density of dead knots was 0.273, both with poor regression results.

Table 1. Linear regression equations among wood density at the different studied conditions for dead knots, live knots, and clear wood.

Knots Type	Regression Equation of Basic Wood Density (X) and Air-Dry Wood Density (Y)	Regression Equation of Basic Wood Density (X) and Oven-Dry Wood Density (Y)	Regression Equation of Air-Dry Wood Density (X) and Oven-Dry Wood Density (Y)
Live knots	Y = 0.206 + 0.799X $R^2 = 0.557$ ** ($p < 0.01$)	Y = 0.046 + 1.025X $R^2 = 0.884$ ** ($p < 0.01$)	Y = 0.057 + 0.844X $R^2 = 0.675$ ** ($p < 0.01$)
Dead knots	Y = 0.437 + 0.385X $R^2 = 0.184$ * ($p < 0.05$)	Y = 0.084 + 0.967X $R^2 = 0.759$ ** ($p < 0.01$)	Y = 0.185 + 0.646X $R^2 = 0.273$ ** ($p < 0.01$)
Clear wood	Y = 0.196 + 0.834X $R^2 = 0.672$ ** ($p < 0.01$)	Y = 0.003 + 1.098X $R^2 = 0.942$ ** ($p < 0.01$)	Y = 0.181 + 0.621X $R^2 = 0.685$ ** ($p < 0.01$)

** indicates that the significance test is significant at the $p = 0.01$ level, * indicates significance test is significant at $p = 0.05$ level.

3.2. Wood Shrinkage and Wood Swelling

Since the growth direction of wood fibers in live and dead knots differs from that of clear wood, the radial, tangential, and longitudinal directions of wood cannot be accurately distinguished, so only the volumetric wood shrinkage and wood swelling ratio of different types of knots were investigated. Air-dry wood shrinkage ratio is the percentage change in volume from the free-drying shrinkage of green or wet wood under no external force to the air-drying state, and oven-dry wood shrinkage ratio is the percentage change in volume from dry-shrinkage of green or wet wood to the oven-dry state. Air-dry wood swelling ratio is the percentage change in volume from the oven-dry state to the air-dry state, and oven-dry wood swelling ratio is the percentage change in volume from the oven-dry state to when water is absorbed to the dimensional stability state.

The data given in Table 2 show the volumetric wood shrinkage and wood swelling ratios of the live knot, dead knot, and clear wood groups. Combined with Figure 5, it can be seen that the volumetric shrinkage and swelling ratio of dead knots are the highest and those of clear wood are the lowest. The reason for the high air-dry and oven-dry wood shrinkage and wood swelling ratio of dead knots may be due to the following:

(1) Wood density is shown to affect the various wood shrinkages and wood swellings that occur during wood drying. Wood density is positively correlated with both radial and tangential wood shrinkage, but is also negatively correlated with longitudinal wood shrinkage [45]. The wood density is determined by the proportion of cell wall material. The proportion of tissue is a key anatomical factor in the volume shrinkage of wood [46]. The greater the wood density of the wood, which generally represents a greater number of cell walls, the greater the wood shrinkage and wood swelling ratio, and vice versa. Wood density decreases within the three groups, from the dead

knots to the live knots to clear wood, so the wood shrinkage and wood swelling ratio are consistent with the wood density ranking.

(2) Dead knots contain more extracts, and the chemical composition of the extracts has a greater influence on water absorption and loss, similar to other observations and studies [47]. The coefficients of variation of air-dry and oven-dry wood shrinkage and wood swelling ratios of all three groups were high and belonged to the medium variation.

Table 2. Volumetric wood shrinkage and wood swelling ratio of different knot types.

Properties Type	Knots Type	Number of Samples	Mean Value	Standard Deviation	Standard Error	Coefficient of Variation	Accuracy Index
Air-dry volumetric wood shrinkage ratio	Live knots	30	4.33	2.08	0.38	22.14%	18.49%
	Dead knots	30	5.89	1.38	0.25	15.38%	8.36%
	Clear wood	30	4.03	1.25	0.23	16.65%	10.32%
Oven-dry volumetric wood shrinkage ratio	Live knots	30	9.50	2.90	0.53	18.33%	11.42%
	Dead knots	30	11.22	2.84	0.52	14.24%	9.05%
	Clear wood	30	9.05	1.75	0.32	13.54%	6.64%
Air-dry volumetric wood swelling ratio	Live knots	30	5.74	2.98	0.54	21.90%	18.58%
	Dead knots	30	6.20	3.46	0.63	19.05%	19.95%
	Clear wood	30	5.68	2.00	0.37	19.97%	12.61%
Oven-dry volumetric wood swelling ratio	Live knots	30	10.45	3.52	0.64	18.72%	12.42%
	Dead knots	30	12.79	3.70	0.68	14.89%	10.36%
	Clear wood	30	10.18	2.16	0.39	14.53%	7.38%

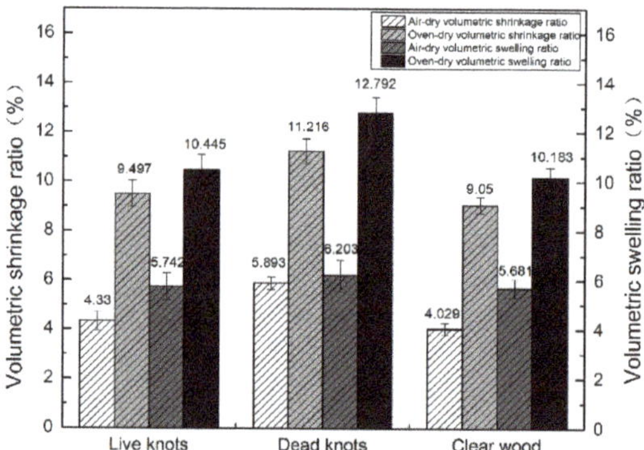

Figure 5. Histogram of the volumetric wood shrinkage and wood swelling ratios. Mean value (n = 30) and standard deviation of dead knots, live knots, and clear wood.

As shown in Table 3, the differences in the air-dry and oven-dry volumetric wood swelling ratio between live knots and clear wood are not significant. The differences in the air-dry and oven-dry volumetric wood shrinkage ratio between live knots and dead knots are highly significant, the differences in the oven-dry volumetric wood swelling ratio between live knots and dead knots are significant, and the differences in the air-dry volumetric wood swelling ratio between live knots and dead knots are not significant. In addition, the differences between the air-dry volumetric wood shrinkage ratio and the oven-dry volumetric wood swelling ratio between dead knots and clear wood are highly significant, and the air-dry volumetric wood swelling ratio between them is not significant. In general, different knots of *C. funebris* wood have different effects on the air-dry and oven-dry volumetric wood shrinkage and wood swelling ratio; specifically, dead knots have a greater effect on the air-dry and oven-dry volumetric wood shrinkage and wood swelling ratio than live knots.

Table 3. Summary of the results for the significance test of differences between volumetric wood shrinkage and wood swelling ratios of live knots, dead knots, and clear wood.

Properties Type	Live Knots and Clear Wood	Live Knots and Dead Knots	Dead Knots and Clear Wood
Air-dry volumetric wood shrinkage ratio	−0.695	−4.102 **	4.602 **
Oven-dry volumetric wood shrinkage ratio	−0.581	−2.702 **	2.698 **
Air-dry volumetric wood swelling ratio	−0.059	−0.362	0.36
Oven-dry volumetric wood swelling ratio	−0.487	−2.660 *	2.701 **

** indicates that the significance test is significant at the $p = 0.01$ level. * indicates significance test is significant at $p = 0.05$ level.

From Figure 5, it is clear that the change rules of the air-dry and oven-dry volumetric wood shrinkage ratio and air-dry and oven-dry volumetric wood swelling ratio are consistent among the live knots, dead knots, and clear wood: air-dry volumetric wood shrinkage < air-dry volumetric wood swelling < oven-dry volumetric wood shrinkage < oven-dry volumetric wood swelling.

3.3. Wood Color Parameters

The test results of the color parameters of different knot types of *C. funebris* wood are shown in Table 4.

Table 4. Wood color parameters of different knot types.

Type of Wood	Number of Samples	Wood Color Parameters	Mean Value	Standard Deviation	Standard Error	Coefficient of Variation	Accuracy Index
Live knots	30	L*	55.29	7.76	1.74	14.03%	6.15%
		a*	10.30	1.24	0.28	12.02%	5.27%
		b*	21.58	1.69	0.38	7.83%	3.43%
Dead knots	30	L*	49.17	6.70	1.34	13.63%	5.34%
		a*	9.60	1.19	0.24	12.40%	4.86%
		b*	18.10	2.90	0.58	16.04%	6.29%
Clear wood	30	L*	73.83	5.52	1.23	7.47%	3.27%
		a*	7.70	0.75	0.17	9.74%	4.27%
		b*	24.99	3.20	0.72	12.80%	5.61%

As can be seen from Figures 1 and 6, the lightness of dead knots is the smallest and the lightness of clear wood is the largest, indicating that the visible light reflection of dead knots is the smallest among the three groups and its color is more biased toward black, and vice versa for clear wood. The a* values of live knots, dead knots, and clear wood are all positive. The a* value of clear wood is the smallest and the a* value of live knots is the largest, indicating that they are all biased towards red. Clear wood has the weakest reflection of red light, but live knots have the strongest reflection of red light.

The b* values of live knots, dead knots, and clear wood are all positive. The b* value of clear wood is the largest, and the b* value of dead knots is the smallest, indicating that the wood color of all three groups is more biased toward yellow; clear wood has the strongest reflection of yellow light, but dead knots have the weakest reflection of yellow light. The chromatic aberration between the three groups was 19.341, 29.642, and 10.301, respectively, which is identifiable in [48]. The magnitude of variation of the wood color parameter L* among different groups is significantly larger than that of b* and a*, indicating that the chromatic aberration among the three groups is mainly due to lightness (L*).

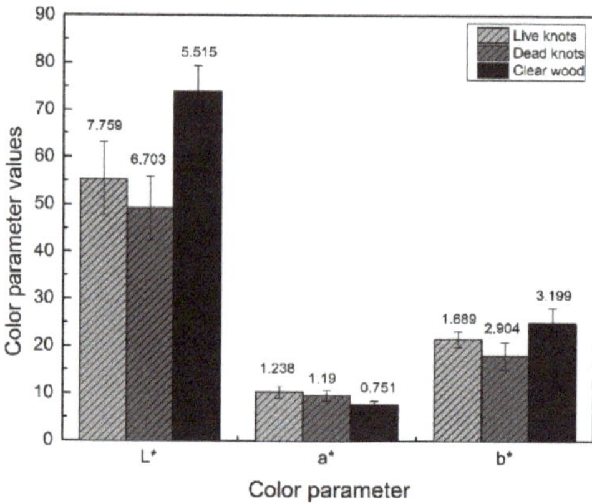

Figure 6. Wood color parameter histogram of mean value and standard deviation of different knot types (L* value refers to the lightness of the sample, a* value refers to the redness of the sample, and the b* value gives the yellowness of the sample).

Pearson's correlation analysis was performed to define the relationship between the color parameters of different knot types in *C. funebris* wood (see Figure 7). This analysis is classified into correlation categories: very weak ($r < 0.2$), weak ($0.2 \leq r < 0.4$), moderate ($0.4 \leq r < 0.6$), strong ($0.6 \leq r < 0.8$), and very strong ($r \geq 0.8$) [49]. As can be seen from Figure 7, the correlation coefficients between L* and a* and L* and b* are $r = -0.622$ and $r = 0.616$, respectively, indicating that L* is strongly correlated with both a* and b*, and combined with the results of the paired *t*-test, L* is significantly negatively correlated with a* and significantly positively correlated with b* at the 0.01 level. The correlation coefficient between a* and b* is $r = -0.354$, which indicates that a* is weakly correlated with b*, and combined with the results of the paired t-test results, a* and b* are significantly negatively correlated at the 0.01 level.

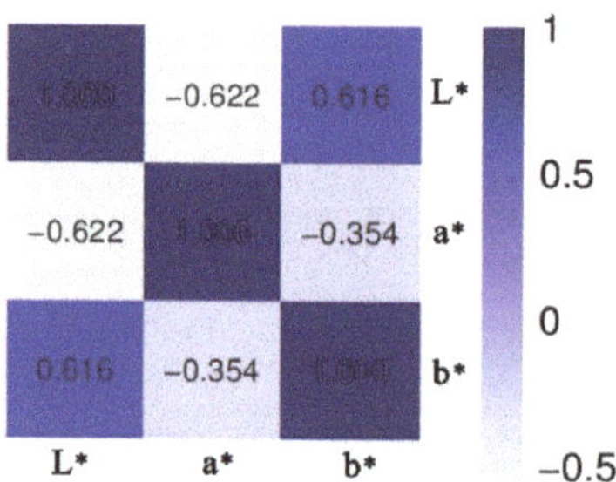

Figure 7. Correlation heat map of wood color parameters of dead knots, live knots, and clear wood (L* value refers to the lightness of the sample, a* value refers to the redness of the sample and the b* value gives the yellowness of the sample).

4. Conclusions

In this study, the main differences in physical properties, including wood density, wood shrinkage, wood swelling, and wood color of *C. funebris* wood were investigated, and the main conclusions obtained are as follows.

The air-dry, oven-dry, and basic wood density differed among dead knots, live knots, and clear wood. Overall, the descending order of wood density is dead knots, live knots, then clear wood. The changing order of air-dry and oven-dry volumetric wood shrinkage and wood swelling among dead knots, live knots, and clear woods group is consistent with the changing order of wood density among them: dead knots > live knots > clear wood.

The decreasing order of lightness and yellow–blue color chromaticity is from clear wood to live knots to dead knots, and the decreasing order of red–green color chromaticity is from live knots to dead knots to clear wood. The color of clear wood is bright, the color of live knots is reddish, and the color of dead knots is blackish, in relative terms. The chromatic aberration between the three groups is recognizable, and the difference in color results from the difference in L* values.

Therefore, when wood with knots is used in products, the wood density of knots has less influence on the wood from the physical aspect of wood; wood with knots and clear wood may be applied without distinction, in terms of wood density. However, the wood shrinkage and wood swelling of dead knots and live knots have more influence on wood compared to clear wood. If wood is used as a joint material for buckets, boats, etc., wood with knots should be selected to increase the tightness between the materials when wet. In short, in terms of wood shrinkage and wood swelling, the effect of knots should be fully considered.

Author Contributions: Conceptualization, J.L. and M.C.; methodology, J.L. and M.C.; writing—original draft preparation, H.Q. and J.L.; writing—review and editing, J.L. and M.C. All authors have read and agreed to the published version of the manuscript.

Funding: This research was supported by the Sichuan Science and Technology Program (Grant No. 2023YFS0462), the Ministry of Education Humanities and Social Sciences Research Project of China (Grant No. 19YJC760009), and the Double Support Plan of Sichuan Agricultural University (Grant No. 2022SYZD06).

Data Availability Statement: Not applicable.

Conflicts of Interest: The authors declare no conflict of interest.

References

1. Fu, L.; Yu, Y.; Farjon, A. *Flora of China 4*; Science Press: Beijing, China, 1999; pp. 62–77.
2. Yang, Y.; Yang, H.; Wang, Q.; Dong, Q.; Yang, J.; Wu, L.; You, C.; Hu, J.; Wu, Q. Effects of Two Management Practices on Monthly Litterfall in a Cypress Plantation. *Forests* **2022**, *13*, 1581. [CrossRef]
3. Wang, Y.; Chen, S.; He, W.; Ren, J.; Wen, X.; Wang, Y.; Li, X.; Chen, G.; Feng, M.; Fan, C. Shrub Diversity and Niche Characteristics in the Initial Stage of Reconstruction of Low-Efficiency *Cupressus funebris* Stands. *Forests* **2021**, *12*, 1492. [CrossRef]
4. Wu, B.; Qi, S. Effects of Underlay on Hill-Slope Surface Runoff Process of *Cupressus funebris* Endl. Plantations in Southwestern China. *Forests* **2021**, *12*, 644. [CrossRef]
5. Fu, L.; Yu, Y.; Mill, R.R. Taxodiaceae. In *Flora of China*; Wu, Z.Y., Raven, P.H., Eds.; Science Press: Beijing, China, 1999; Volume 4, pp. 54–61.
6. Cool, L.G.; Hu, Z.L.; Zavarin, E. Foliage terpenoids of Chinese Cupressus species. *Biochem. Syst. Ecol.* **1998**, *26*, 899–913. [CrossRef]
7. Lyu, J.; Wang, J.; Chen, Z.; Chen, M. Influence of Dowel Center Spacing on Chamfered-Joint Components Made by *Cupressus funebris* Wood. *J. Renew. Mater.* **2023**, *11*, 309–319. [CrossRef]
8. Chen, M.; Li, S.; Lyu, J. Effects of selected joint parameters on tensile strength of steel bolt-nut connections in *Cupressus funebris* wood. *BioResources* **2019**, *14*, 5188–5211. [CrossRef]
9. Karaszewski, Z.; Bembenek, M.; Mederski, P.S.; Szczepanska-Alvarez, A.; Byczkowski, R.; Kozlowska, A.; Michnowicz, K.; Przytula, W.; Giefing, D.F. Identifying beech round wood quality-distribution of beech timber qualities and influencing defects. *Drew. Pr. Nauk. Doniesienia Komun.* **2013**, *56*, 39–54.
10. Vongkhamho, S.; Imaya, A.; Takenaka, C.; Yamamoto, K.; Yamamoto, H. Correlations among Tree Quality, Stand Characteristics, and Site Characteristics in Plantation Teak in Mountainous Areas of Lao PDR. *Forests* **2020**, *11*, 777. [CrossRef]
11. Midgley, S.; Mounlamai, K.; Flanagan, A.; Phengsopha, K. *Global Markets for Plantation Teak; Implications for Growers in Lao PDR*; Australian Centre for International Agricultural Research: Canberra, Australia, 2015.
12. Baillères, P.H.; Durand, P.Y. Non-destructive techiques for wood quality assessment of plantation-grown teak. *Bois Forêts Trop.* **2000**, *263*, 17–29.
13. Longuetaud, F.; Mothe, F.; Kerautret, B.; Krähenbühl, A.; Hory, L.; Leban, J.M.; Debled-Rennesson, I. Automatic knot detection and measurements from X-ray CT images of wood: A review and validation of an improved algorithm on softwood samples. *Comput. Electron. Agric.* **2012**, *85*, 77–89. [CrossRef]
14. Macdonald, E.; Hubert, J. A review of the effects of silviculture on timber quality of Sitka spruce. *For. Int. J. For. Res.* **2002**, *75*, 107–138. [CrossRef]

15. Qin, G.; Hao, J.; Yang, J.; Li, R.; Yin, G. Branch Occlusion and Discoloration under the Natural Pruning of *Mytilaria laosensis*. *Forests* **2019**, *10*, 892. [CrossRef]
16. *GB/T 155–2017*; Defects in Logs. National Standard of the People's Republic of China: Beijing, China, 2009. (In Chinese)
17. Lowell, E.C.; Maguire, D.A.; Briggs, D.G.; Turnblom, E.C.; Jayawickrama, K.J.S.; Bryce, J. Effects of Silviculture and Genetics on Branch/Knot Attributes of Coastal Pacific Northwest Douglas-Fir and Implications for Wood Quality—A Synthesis. *Forests* **2014**, *5*, 1717–1736. [CrossRef]
18. Tong, Q.J.; Duchesne, I.; Belley, D.; Beaudoin, M.; Swift, E. Characterization of knots in plantation white spruce. *Wood Fiber Sci.* **2013**, *45*, 84–97.
19. Cherry, R.; Karunasena, W.; Manalo, A. Mechanical Properties of Low-Stiffness Out-of-Grade Hybrid Pine—Effects of Knots, Resin and Pith. *Forests* **2022**, *13*, 927. [CrossRef]
20. Yong, C.; Li, Y. Physical and mechanical properties and utilization of Chinese fir branch wood. *Guizhou For. Sci. Technol.* **1991**, *4*, 69–70. (In Chinese)
21. Wang, Y.R.; Zhao, R.J.; Jiang, Z.H.; Hse, C.Y. Density analysis of stem, knot and branch of *Pinus taeda*. *China For. Prod. Ind.* **2016**, *43*, 14–17. (In Chinese)
22. As, N.; Goker, Y.; Dundar, T. Effects of knots on the physical and mechanical properties of Scots pine (*Pinus sylvestris* L.). *Wood Res.* **2006**, *51*, 51–58.
23. Zhou, Y.; Sun, F.; Lü, J.; Li, X. Veneer drying quality comparison of six Eucalyptus species. *Sci. Silvae Sin.* **2014**, *50*, 104–108. (In Chinese) [CrossRef]
24. Bengtsson, C. Variation of moisture induced movements in Norway spruce (*Picea abies*). *Ann. For. Sci.* **2001**, *58*, 568–581. [CrossRef]
25. Forest Products Laboratory. *Wood Handbook-Wood as an Engineering Material*; Forest Products Laboratory: Madison, WI, USA, 2010; pp. 1–46.
26. Shelbourne, C.J.A. Cupressus and Other Conifers. In *Tree Breeding and Genetics in New Zealand*; Shelbourne, C.J.A., Carson, M., Eds.; Springer: Cham, Switzerland, 2019; pp. 93–96. [CrossRef]
27. Gao, M.; Qi, D.; Mu, H.; Chen, J. A Transfer Residual Neural Network Based on ResNet-34 for Detection of Wood Knot Defects. *Forests* **2021**, *12*, 212. [CrossRef]
28. Hălălișan, A.-F.; Dinulică, F.; Gurean, D.M.; Codrean, C.; Neykov, N.; Antov, P.; Bardarov, N. Wood Colour Variations of *Quercus* Species in Romania. *Forests* **2023**, *14*, 230. [CrossRef]
29. Li, J. Ecological Properties of Wood: Wood as the Contributor to the Green Environment and Human Health. *J. Northeast For. Univ.* **2010**, *38*, 1–8. (In Chinese) [CrossRef]
30. Budakci, M.; Ozcifci, A.; Cinar, H.; Sonmez, A. Effects of application methods and species of wood on color changes of varnishes. *Afr. J. Biotechnol.* **2009**, *8*, 5964–5970. [CrossRef]
31. Rüther, P.; Jelle, B.P. Color changes of wood and wood-based materials due to natural and artificial weathering. *Wood Mater. Sci. Eng.* **2013**, *8*, 13–25. [CrossRef]
32. Kržišnik, D.; Lesar, B.; Thaler, N.; Humar, M. Influence of Natural and Artificial Weathering on the Colour Change of Different Wood and Wood-Based Materials. *Forests* **2018**, *9*, 488. [CrossRef]
33. Li, P.; Li, F.T.; Fan, C.; Li, X.W.; Zhang, J.; Huang, M.L. Effects of plant diversity on soil organic carbon under different reconstructing patterns in low efficiency stands of *Cupressus funebris* in the hilly region of central Sichuan. *Acta Ecol. Sin.* **2015**, *35*, 2667–2675. [CrossRef]
34. Zhang, Z.; Jin, G.; Chen, T.; Zhou, Z. Effects of CaO on the Clonal Growth and Root Adaptability of Cypress in Acidic Soils. *Forests* **2021**, *12*, 922. [CrossRef]
35. Lyu, J.; Zhao, J.; Xie, J.; Li, X.; Chen, M. Distribution and composition analysis of essential oils extracted from different parts of *Cupressus funebris* and *Juniperus chinensis*. *BioResources* **2018**, *13*, 5778–5792. [CrossRef]
36. Lyu, J.; Huang, W.; Chen, M.; Li, X.; Zhong, S.; Chen, S.; Xie, J. Analysis of tracheid morphological characteristics, annual rings width and latewood rate of *Cupressus funebris* in relation to climate factors. *Wood Res.* **2020**, *65*, 565–578. [CrossRef]
37. Tang, J.; Yang, D. The macro-ecological variation of wood quality factors with *Cupressus funebris* Endl. in Sichuan. *J. Sichuan Agric. Univ.* **1993**, *11*, 138–144. (In Chinese) [CrossRef]
38. Li, M.Y.; Chen, D.S.; Wang, X.Y.; Ren, H.Q. Nature of the Knot and Its Impact on Wood Properties. *J. Northwest For. Univ.* **2020**, *35*, 197–204. (In Chinese) [CrossRef]
39. *GB/T 1928–2009*; General Requirements for Physical and Mechanical Tests of Wood. National Standard of the People's Republic of China: Beijing, China, 2009. (In Chinese)
40. *GB/T 1929–2009*; Method of Sample Logs Sawing and Test Specimens Selection for Physical and Mechanical Tests of Wood. National Standard of the People's Republic of China: Beijing, China, 2009. (In Chinese)
41. *GB/T 1932–2009*; Method for Determination of the Shrinkage of Wood. National Standard of the People's Republic of China: Beijing, China, 2009. (In Chinese)
42. *GB/T 1933–2009*; Method for Determination of the Density of Wood. National Standard of the People's Republic of China: Beijing, China, 2009. (In Chinese)
43. *GB/T 1934.2–2009*; Method for determination of the swelling of wood. National Standard of the People's Republic of China: Beijing, China, 2009. (In Chinese)

MDPI

St. Alban-Anlage 66

4052 Basel

Switzerland

www.mdpi.com

Forests Editorial Office

E-mail: forests@mdpi.com

www.mdpi.com/journal/forests

44. Zhao, L.; Qiu, X. Study on the variation of physical and mechanical properties of Chinese fir seedling forest in different stand ages. *J. Anhui Agric. Univ.* **2021**, *48*, 726–732. (In Chinese) [CrossRef]
45. Pliura, A.; Yu, Q.; Zhang, S.Y.; MacKay, J.; Perinet, P.; Bousquet, J. Variation in Wood Density and Shrinkage and Their Relationship to Growth of Selected Young Poplar Hybrid Crosses. *For. Sci.* **2005**, *51*, 472–482. [CrossRef]
46. Zhang, S.Y.; Zhong, Y. Structure-property relationship of wood in East-Liaoning oak. *Wood Sci. Technol.* **1992**, *51*, 139–149. [CrossRef]
47. Zhou, F.; Fu, Z.; Gao, X.; Zhou, Y.; Jiang, J. Within-tree variation of physical properties of *Acacia melanoxylon* wood. *Chin. J. Wood Sci. Technol.* **2021**, *35*, 30–36. (In Chinese) [CrossRef]
48. Duan, X. *Wood Color Control Technology*; China Building Materials Industry Press: Beijing, China, 2002; pp. 45–47. (In Chinese)
49. Evans, J.D. *Straightforward Statistics for the Behavioral Sciences*; Thomson Brooks/Cole Publishing Co.: Three Lakes, WI, USA, 1996.